AF695853

Structural Analysis 1

In memory of my father, Tahar
In memory of my wife, Mrs Nadjet Khalfallah-Boudaa
To my family
20 Ramadan 1437 h

Structural Analysis 1

In memory of my father, Tahar
In memory of my wife, Mrs Nadjet Khalfallah-Boudaa
To my family
20 Ramadan 1437 h

Series Editor
Gilles Pijaudier-Cabot

Structural Analysis 1

Statically Determinate Structures

Salah Khalfallah

iSTE

WILEY

First published 2018 in Great Britain and the United States by ISTE Ltd and John Wiley & Sons, Inc.

Apart from any fair dealing for the purposes of research or private study, or criticism or review, as permitted under the Copyright, Designs and Patents Act 1988, this publication may only be reproduced, stored or transmitted, in any form or by any means, with the prior permission in writing of the publishers, or in the case of reprographic reproduction in accordance with the terms and licenses issued by the CLA. Enquiries concerning reproduction outside these terms should be sent to the publishers at the undermentioned address:

ISTE Ltd
27-37 St George's Road
London SW19 4EU
UK

www.iste.co.uk

John Wiley & Sons, Inc.
111 River Street
Hoboken, NJ 07030
USA

www.wiley.com

© ISTE Ltd 2018
The rights of Salah Khalfallah to be identified as the author of this work have been asserted by him in accordance with the Copyright, Designs and Patents Act 1988.

Library of Congress Control Number: 2018943087

British Library Cataloguing-in-Publication Data
A CIP record for this book is available from the British Library
ISBN 978-1-78630-338-7

MIX
Paper from responsible sources
FSC® C013604

Contents

Preface

The main objective of this volume is to provide students of civil, mechanical, aeronautical and marine engineering as well as those interested in structural analysis with the essentials for analyzing statically determinate structures. This book allows them to acquire sufficient knowledge to study and analyze statically determinate structures. The reader will find a series of exercises at the end of each chapter that can be used to deepen their knowledge and improve their ability to master statically determinate structure analysis methods.

This volume covers aspects of structural analysis such as trusses, beams, frames, cable structures and arch structures. The last chapter looks at influence lines for live loads, which play a role in the design phase of statically determinate structures.

This book is the first of two volumes. It consists of two parts. Each chapter of the book is constructed in a specific way: an illustration of the objectives and the parts covered, a general introduction, a theory of the proposed approach, a numerical study of some examples and a summary at the end. Each chapter ends with a series of problems and exercises.

The first part includes a general introduction to structural analysis (Chapter 1) and illustrates in detail the different types of loads that can be applied to a structure (Chapter 2). The load magnitude is evaluated according to European standards, including those adopted by France. The second part describes statically determinate structure analysis methods.

This part is divided into eight separate chapters. Chapter 3 offers an overall analysis of determinate structures and looks into their external equilibrium. Chapter 4 looks at how to analyze trusses and several methods are given, such as the method of joint equilibrium, the method of sections, the graphical method and the matrix method. In this context, complex and spatial trusses are analyzed. In the same way, the analysis of beams and frames is developed in Chapter 5. Chapter 6 discusses the calculation of beam deflections using geometric methods. Here, elastic line deflection, direct integration and fictitious beam methods are reviewed. For the same purpose, Chapter 7 also discusses the calculation of deflections using energy methods, emphasizing the virtual work method. Chapter 8 describes the methodology for analyzing cables acted upon by concentrated and distributed loads. Also, arc analysis is described in Chapter 9, in which two types of arches are selected: semicircular and parabolic. The loads proposed here are concentrated and uniformly distributed. The last chapter illustrates the influence lines of the internal or external actions that are built under the effect of a moving unitary load along a beam, a frame or a truss.

Finally, we hope that our approach in this book's publication will meet the needs of students interested in this scientific and technical subject. Nevertheless, we are very aware that the work presented is not exempt from mistakes. For this reason, we warmly welcome any corrections and comments, which will improve future editions of this book. Comments or suggestions can be sent to the email addresses found on the website www.freewebs.com/khalfallah/index.htm.

Salah KHALFALLAH
May 2018

1

Introduction to Structural Analysis

The teaching objectives for this chapter are as follows:

– the role of structural analysis teaching;

– the concept of a structure;

– the development of structural analysis methods;

– the distinction between categories of structures;

– the calculation of a statically indeterminate structure.

This chapter is descriptive and gives a general presentation of the preliminary aspects of statically determinate structure analysis. In the first part, we present the concept of a structure, the objectives to be achieved during structural analysis teaching and the history of its development. In the second part, we look at structural classification based on the structural dimension. Finally, we give the calculation of the degree of static indeterminacy of the structures.

1.1. Introduction

The primary role of structural study and analysis is to determine the internal actions and the support reactions of a structure subjected to mechanical loads, imposed deformations and settlements of supports. An action can mean either a force and/or a moment. In the same way, a deformation can mean a displacement and/or a rotation.

Structures are classified into two broad categories: (1) statically determinate structures and (2) statically indeterminate structures. The three static equilibrium equations are used to analyze statically determinate structures. In this case, the support reactions can be determined using only static equations. As a result, internal actions, such as a bending moment, a torsion moment, a shear force, and a normal force, can be deduced using the internal equilibrium equations.

On the contrary, for statically indeterminate structures equilibrium equations are not sufficient to calculate the unknowns of the problem. This means that the number of unknowns (the support reactions) is strictly greater than that of the equations. The difference between the number of unknowns of the problem and the equations is called the degree of static indeterminacy of the system or structure.

1.2. Concept of a structure

The word "structure" in the field of mechanics indicates any solid body that satisfies the following conditions:

1) the presence of a material characterized by mechanical properties;

2) the structure studied has a geometry or a form;

3) the structure is linked to the external environment through the supports;

4) the body is subjected to external loading.

The definition of a structure is clearly projected in Figure 1.1.

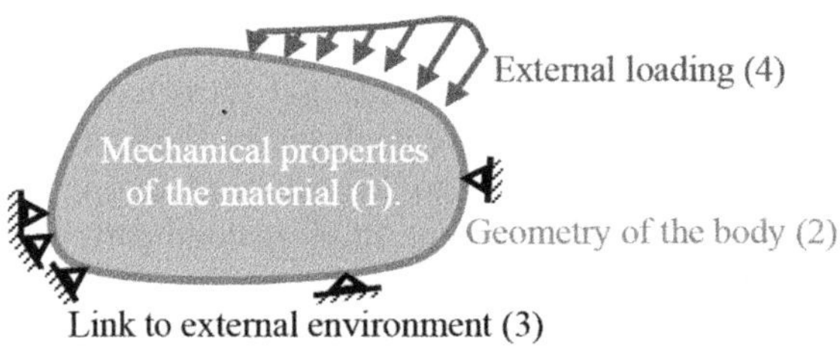

Figure 1.1. *The conditions for defining a structure*[1]

1.3. Structural analysis

Structural analysis is the calculation of responses when a structure is subjected to a set of external loads. In general, the structural response is characterized by the determination of internal actions or deflections at any point in the structure. To achieve this goal, it is necessary to use a mathematical method, an experimental test, or an analytical or numerical model.

1 All of the figures in this chapter are available to view in full color at www.iste.co.uk/khalfallah/analysis1.zip.

In addition, the main role of structural analysis is to study structural performance when it is subjected to the effects of the external environment, such as a set of external actions, a movement of the support or a temperature change. The common features of structural performance, which are important during the design phase, are as follows:

– internal forces: axial force, shear force, bending moment and torsion moment;

– support reactions;

– the deflections that occur after external loads are applied.

The classical methods of structural analysis have calculation limitations that depend in general on the geometry of the structure or the applied loading (standard problems). Recently, the evolution of computing machines has led to an enormous development in methods of analysis, especially matrix methods.

The purpose of the two categories of calculation methods is to arrive at a structural design that responds to criteria of resistance and economy, simultaneously.

1.4. History of structural analysis

Structural design and analysis is an ancient art and many civilizations have contributed to its development. For example, the Egyptian pyramids were built around 2000 BC with a civilization of the time.

The first constructions were based on empirical experiments and rules constituting the fundamental basis of structural analysis. In this subject, the first principles of statics appeared toward 300–400 BC. Thereafter, the Romans contributed to the evolution of structural analysis by using stone in constructions up to 500 AD and by developing new forms of construction in arches and vaults.

During the Middle Ages (500–1500), the Greeks and Romans collaborated in the development of structural analysis through the construction of cathedrals. During the Renaissance period, Leonardo Da Vinci (1452–1519) described the first theories of structures. However, Galileo (1564–1642) initiated the theory of mechanics of materials by studying the breaking of cantilever beams. Then Hooke (1635–1703) established the law of material elasticity for the first time. Johann Bernoulli (1667–1748) formulated the principle of virtual displacements. Daniel Bernoulli (1700–1782) studied elastic curves and strain energy in bending. Navier (1785–1836) initiated statically indeterminate structural analysis, and then Coulomb (1736–1806) published his work on the resistance of materials, in which he analyzed flexed beams.

The noble period of structural analysis is the 19th Century. During this period, the theoretical foundations of the mechanics of materials and structures were widely developed. We can distinguish the following developments:

1) the principles of the analysis of statically determinate truss systems, Whipple (1804–1888);

2) the theorem of three moments, Clapeyron (1799–1864);

3) the theorem of reciprocal displacements, Maxwell (1831–1879);

4) the study of influence lines, Mohr (1835–1918);

5) Castigliano's theorem (1847–1884);

6) the slope deflection method, Maney (1888–1947);

7) the moment distribution method, Cross (1885–1959);

8) the relaxation method, Southwell (1888–1970).

Several researchers in structural mechanics subsequently participated in the development of matrix and modern analysis methods. In addition, the 19th Century saw the development of new materials, new techniques and structures with complex shapes.

Over the past few decades, there have been considerable developments in the field of digital mechanics after the revolution in computer science and digital methods.

1.5. Classification of structures

Choosing which type of structure to use depends on several factors, such as the operating characteristics, the condition of resistance, the economy criterion, the aesthetics and the availability of the construction material. It is worth emphasizing that the loading system plays a major role in selecting the type of structure.

The structures used in civil engineering can be divided into six categories. This classification is based on the structural dimension and the law of stress distribution in the structure. For these reasons, they can be classified as follows.

1.5.1. *Plane trusses*

The plane trusses are systems made up of several interconnected bars at the articled joints. The bars of a truss are subjected only to traction or compression.

When the elements of a truss and the load are applied to a single plane, the truss is called a plane (Figure 1.2).

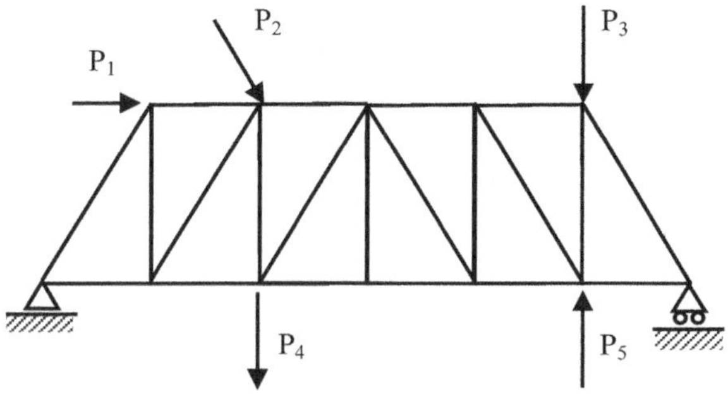

Figure 1.2. *Truss*

In this type of structure, it is only the nodal loads that can be considered in the calculation and distributed loads can't be considered.

In the same context, we can distinguish cable structures that are subject only to pure traction. This category is used in the construction of bridges or in the floors of special structures (Figure 1.3).

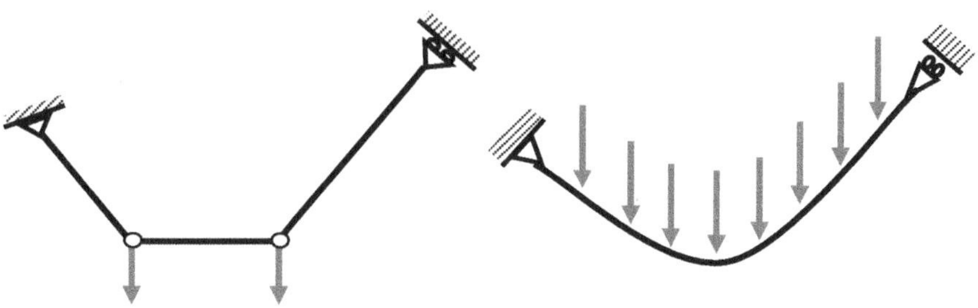

Figure 1.3. *Membranous structures*

In general, analyzing a truss cannot be established by a consideration of the plane due to the size of the structure and the applied loading system. In this case, it can be studied as a spatial structure and the load is distributed according to the three geometrical dimensions. Similar to plane systems, external forces generate only normal stresses in the bars of the truss (Figure 1.4).

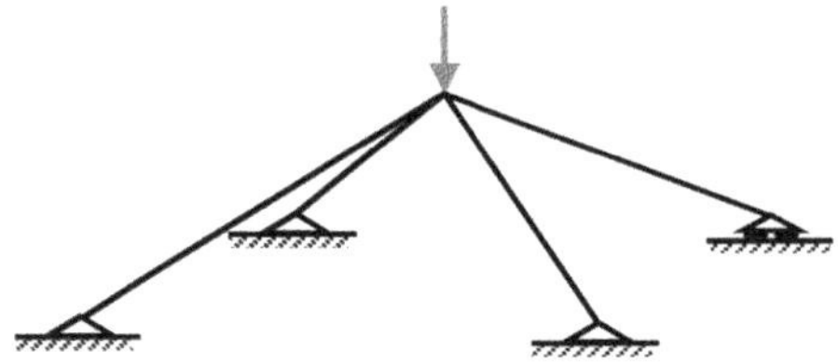

Figure 1.4. *Spatial trusses*

1.5.2. *Beams*

A beam is a rectilinear structural element. In general, beams are solicited by loads applied on the same geometric plane and perpendicular to the middle axis. In this type of structure, the applied loads cause bending and shearing forces. The normal force is neglected when the applied loads are perpendicular to the axis of the beam (Figure 1.5).

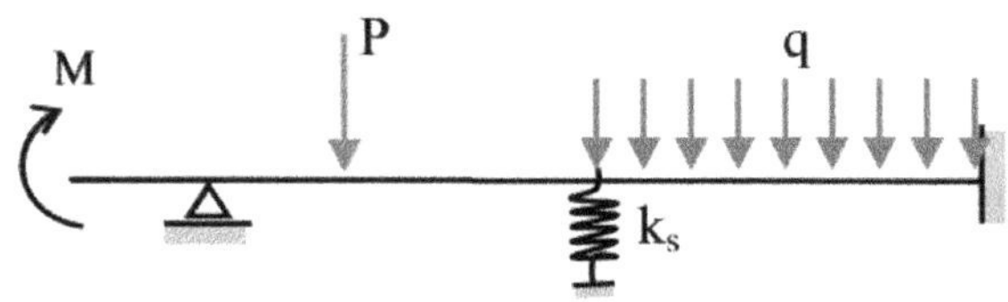

Figure 1.5. *Beams*

The internal actions at any section are generally a bending moment, a tangential force and a normal force (Figure 1.6).

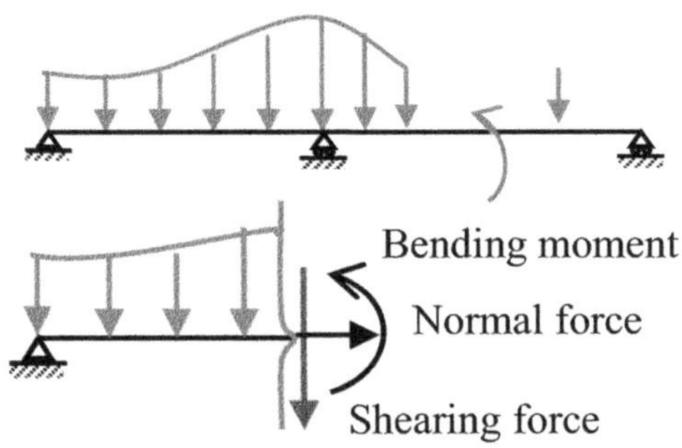

Figure 1.6. *Internal actions*

1.5.3. *Frames*

Frames are systems composed of several rectilinear and/or oblique elements interconnected by rigid joints. In this case, external loads can be applied to the joints and on structural elements. In plane frames, the elements and the loads are linked to a single plane. The structural elements are acted on by bending, shear force and normal force due to the external loading system (Figure 1.7).

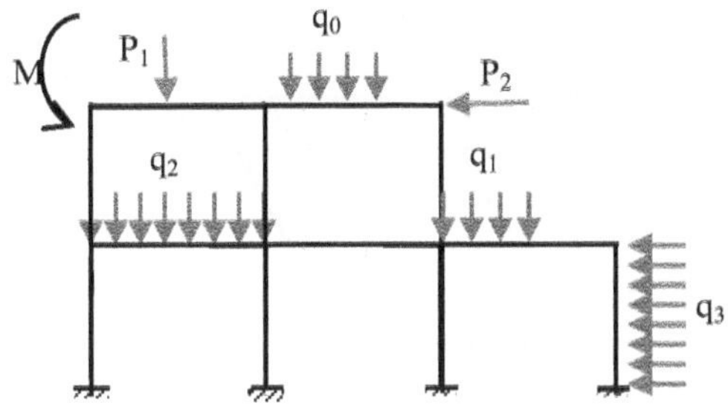

Figure 1.7. *Frames*

Space frames are the most appropriate in civil engineering buildings. The loading system is applied along the structural elements and at the joints. Generally, the loads generate bending, shear, normal force and torsion as internal actions. (Figure 1.8).

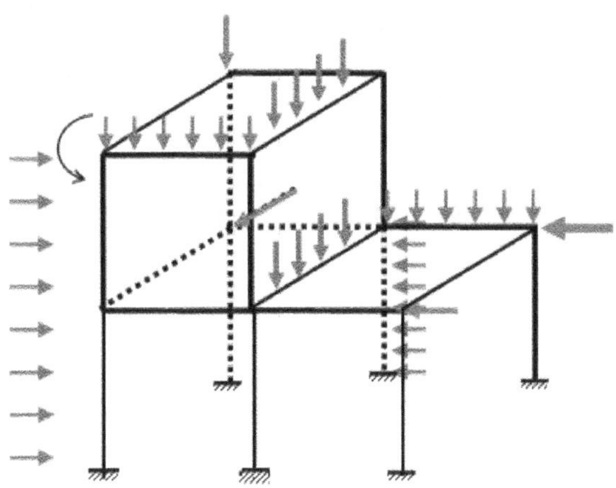

Figure 1.8. *Three-dimensional frames*

1.5.4. *Crossbeams*

The crossbeams are composed of a layer of rectilinear elements arranged on the same plane constituting a series of beams. The loading system is applied perpendicular to the structural plane. The internal loads that can be envisaged are bending, shear, normal force and torsion. The crossbeams are designed to bear heavy loads of slabs, sports halls, auditoriums, etc. (Figure 1.9).

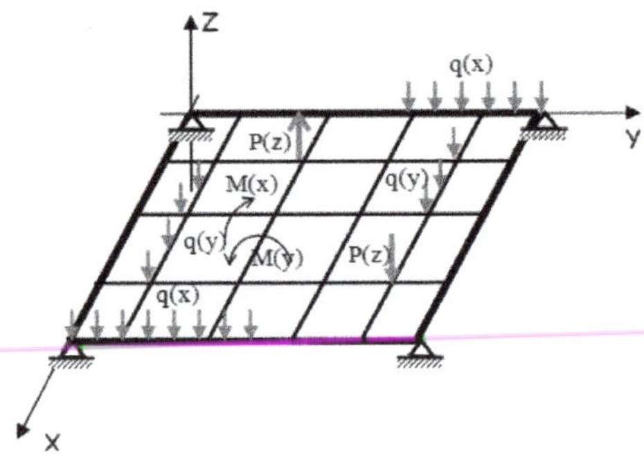

Figure 1.9. *Crossbeam structures*

1.5.5. *Arches*

Arches are structures with an inverted curvature in relation to cable structures (Figure 1.10). Arches are designed to bear loads with a long span and they must be rigid enough to maintain their shapes. The applied load generates a normal force, a shear force and a bending moment that are considered during the design phase. Arches are used in the construction of bridges, domes, sports halls, department stores, etc.

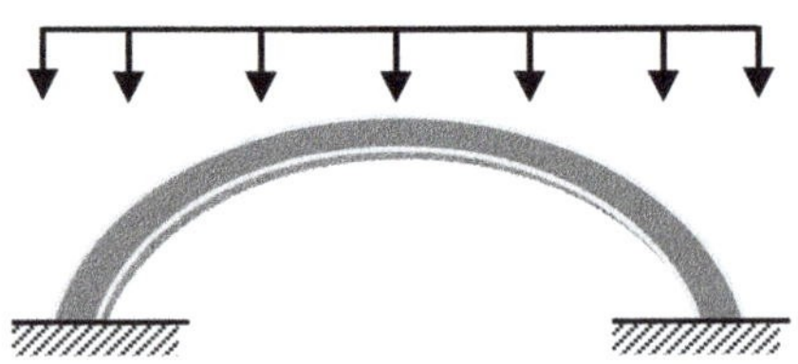

Figure 1.10. *A given arch*

1.6. Static indeterminacy of structures

Structures are grouped into two categories: (1) statically determinate structures and (2) statically indeterminate structures. For statically determinate structures, static equations are sufficient to determine the support reactions and consequently the internal actions: the bending moment, the shear force and the normal force.

In the opposite case, equilibrium equations are insufficient to determine the support reactions. This means that the number of unknowns of the given problem is much greater than that of the independent equations.

Every structure is in equilibrium under the effect of the applied loads when the result of forces and the sum of the moments on a point is zero. For plane structures, the three static equations must be fulfilled, namely:

$$\sum F_x = 0 \qquad [1.1]$$

$$\sum F_y = 0 \qquad [1.2]$$

$$\sum M_i = 0 \qquad [1.3]$$

Therefore, the plane structure is called statically determinate if equations [1.1]–[1.3] are sufficient to determine the support reactions; otherwise it is called statically indeterminate. To analyze a statically indeterminate structure, it is necessary to add other independent equations so that the number of unknowns is equal to the degree of static indeterminacy of the problem. One thus defines the degree of static indeterminacy of a structure by the difference between the number of unknowns and that of the static equations.

1.6.1. *Trusses*

In the case of truss systems, the nature of the support designates the number of corresponding support reactions. However, in structural mechanics, a hinge is used with a vertical and horizontal reaction and a roller with a single vertical reaction (Figure 1.11).

Figure 1.11. *Types of support and associated reactions*

In general, for a truss, which has b bars, n joints and r support reactions, the nature of the structure can be identified as following:

The structure is statically determinate externally if

$$b-(2n-3)=0 \qquad [1.4]$$

It is statically indeterminate, "f", if

$$f=b-(2n-3)\succ 0 \qquad [1.5]$$

It is unstable when

$$f=b-(2n-3)\prec 0 \qquad [1.6]$$

f is the degree of static indeterminacy of the system or structure.

EXAMPLE 1.1.–

Analyze the static indeterminacy of trusses (Figure 1.12).

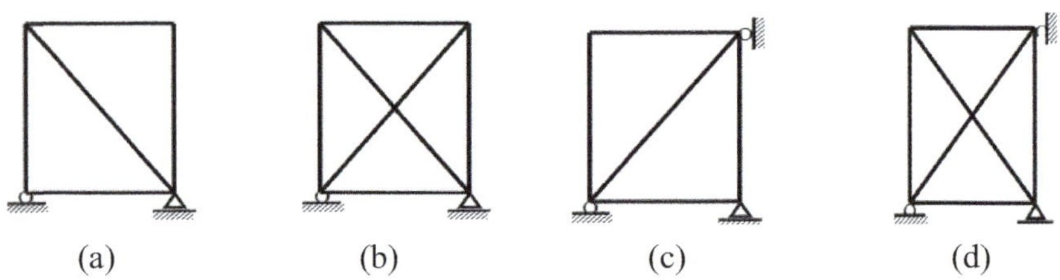

Figure 1.12. *Different structures of strusses*

System	(a)	(b)	(c)	(d)
b	5	6	5	6
$2n-3$	5	5	5	5
r	3	3	4	4
f	0	1	0	1

System	Internal analysis	External analysis
(a)	Statically determinate	Statically determinate
(b)	Once statically indeterminate	Statically determinate
(c)	Statically determinate	Once statically indeterminate
(d)	Once statically indeterminate	Once statically indeterminate

EXAMPLE 1.2.–

Analyze the static indeterminacy of trusses (Figure 1.13).

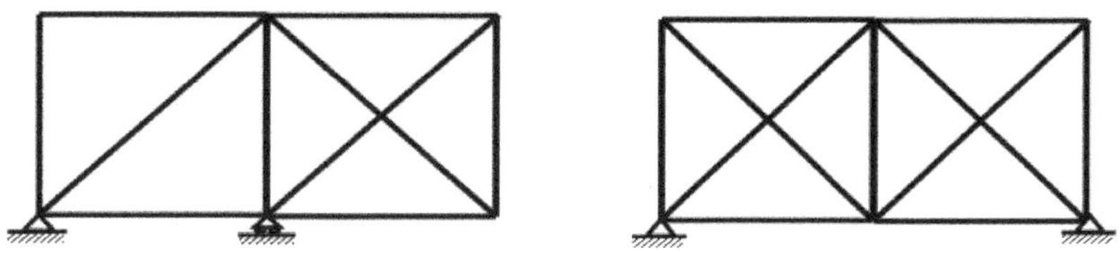
(I) (II)

Figure 1.13. *Given trusses*

– System (I)

$b = 10,\ 2n - 3 = 9$ and $r = 3$

The system is statically determinate externally and once statically indeterminate internally.

– System (II)

$b = 11, 2n - 3 = 9$ and $r = 4$

So, the system is once statically indeterminate externally and twice statically indeterminate internally.

1.6.2. *Beam and frames*

Aside from hinged and roller supports, beams and frames may contain recesses that are characterized by three independent actions (a horizontal reaction, a vertical reaction and a bending moment) (Figure 1.14).

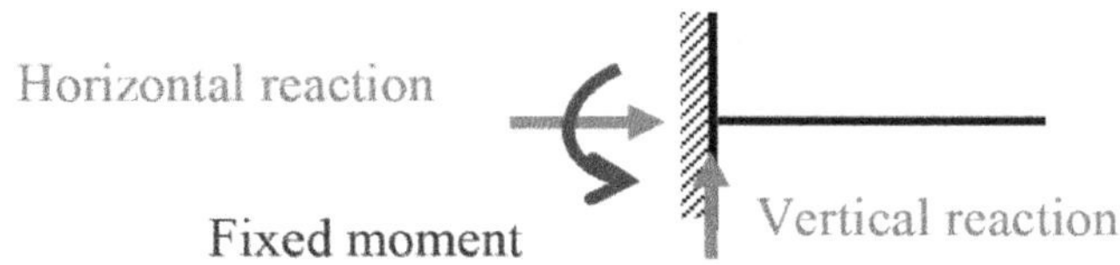

Figure 1.14. *Reaction at fixed node*

In the same way, for a frame or a beam with n joints, b bars, r support reactions and k additional conditions (hinges), the degree of static indeterminacy of the system is given by the following formula:

$$f = (3b + r) - (3n + k) \quad [1.7]$$

Determine the degree of static indeterminacy of the frames (Figure 1.15).

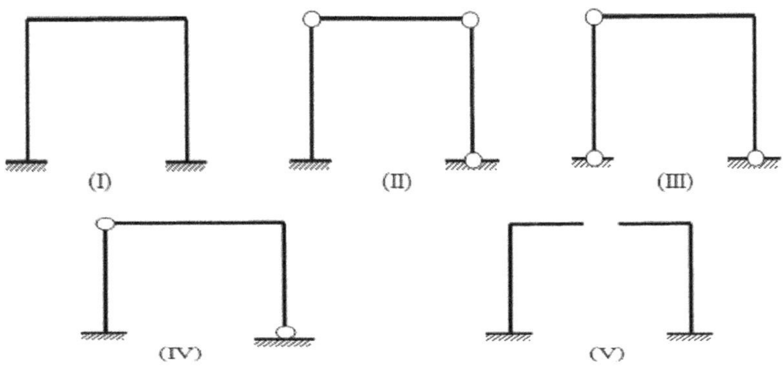

Figure 1.15. *Single span frames*

System	(I)	(II)	(III)	(IV)	(V)
b	3	3	3	3	3
r	6	5	4	4	6
n	4	4	4	4	4
k	0	2	1	1	3
f	3	0	0	0	0

EXAMPLE 1.3.–

Determine the degree of static indeterminacy of the frames (Figure 1.16).

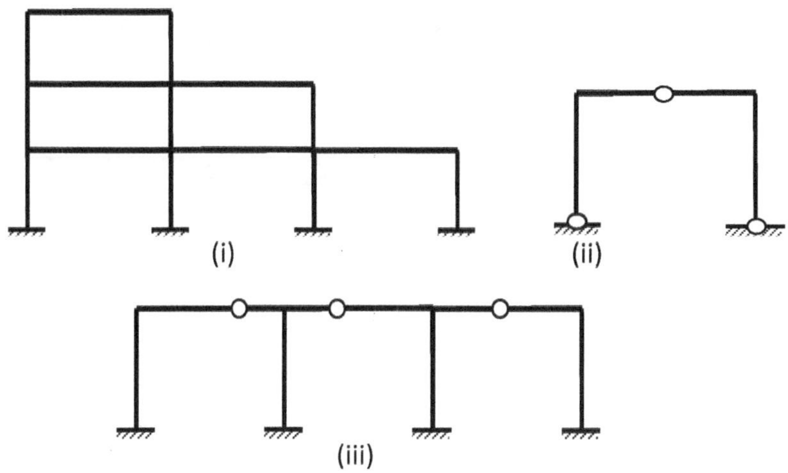

Figure 1.16. *Single and multiple span frames*

Applying equation [1.7], the degree of static indeterminacy of each system is:

– Structure (i)

$f = (3b + r) - (3n + k) = (3 \times 15 + 12) - (3 \times 13 + 0) = 18$, once statically indeterminate.

– Structure (ii)

$f = (3 \times 3 + 4) - (3 \times 4 + 1) = 0$, the system is statically determinate.

– Structure (iii)

$f = (3 \times 7 + 12) - (3 \times 8 + 3) = 6$, once statically indeterminate.

In the case of beams and when the vertical loads are perpendicular to the bar axes, the degree of static indeterminacy is written in the form

$$f = (2b + r) - (2n + k) \quad [1.8]$$

EXAMPLE 1.4.–

Evaluate the degree of static indeterminacy of the beams given by Figure 1.17.

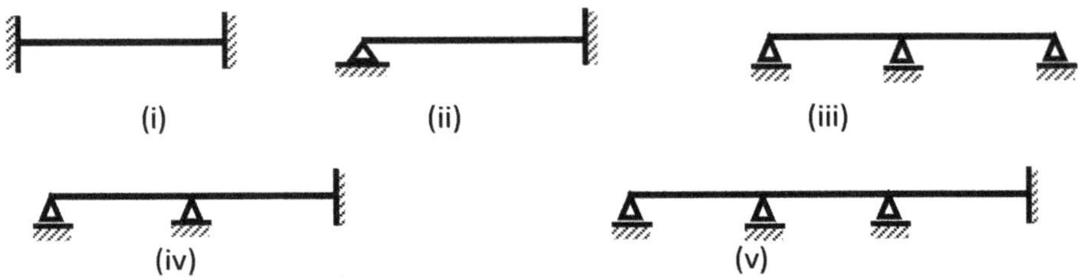

Figure 1.17. *Given beams*

– Beam (i): $f = (2b + r) - (2n + k) = (2 \times 1 + 4) - (2 \times 2 + 0) = 2$, once statically indeterminate.

– Beam (ii): $f = (2 \times 1 + 3) - (2 \times 2 + 0) = 1$, once statically indeterminate.

– Beam (iii): $f = (2 \times 2 + 3) - (2 \times 3 + 0) = 1$, once statically indeterminate.

– Beam (iv): $f = (2 \times 2 + 4) - (2 \times 3 + 0) = 2$, once statically indeterminate.

– Beam (v): $f = (2 \times 3 + 5) - (2 \times 4 + 0) = 3$, once statically indeterminate.

1.7. Sign convention

To establish the structural analysis methods presented in this work, we adopt *a priori* the following sign convention:

– normal force is positive when it causes traction. On the contrary, it is negative if it generates a compression;

– the bending moment is considered positive if its rotation is counterclockwise;

– the shear force is assumed to be positive if it is oriented downwards to the left side.

This sign convention is represented in Figure 1.18. The actions shown are considered positive.

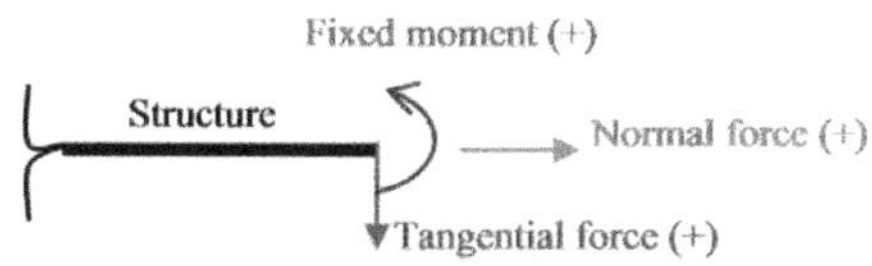

Figure 1.18. *Sign conventions of internal actions*

In the same context, the kinematic sign convention can be deduced based on the sign convention of the actions cited above (Figure 1.19).

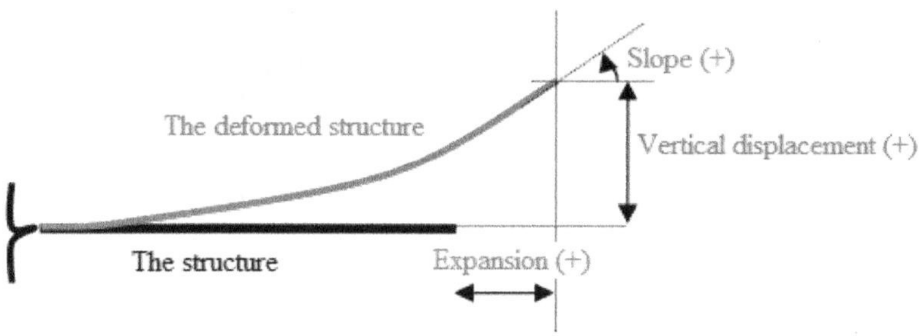

Figure 1.19. *Sign conventions of deflections*

1.8. Conclusion

In this chapter, we presented the preliminary conceptual components of the structural analysis material. The teaching aims first of all to determine the internal actions at every point of a structure. This can be used for the design and dimensioning of the different elements of a structure.

The distinction of the static indeterminacy of a structure plays a very important role in leading to the appropriate method of analysis. This is the distinction between the methods of analysis for statically determinate and statically indeterminate structures. The second part of the book is devoted to presenting methods of analysis for statically determinate structures. The teaching in this chapter makes it possible to differentiate between the nature of the structure and its application in the industrial sector. Finally, this chapter proposes a sign convention for internal actions and their corresponding kinematic consequences.

1.9. Problems

Exercise 1

Specify the analysis conditions for each structure.

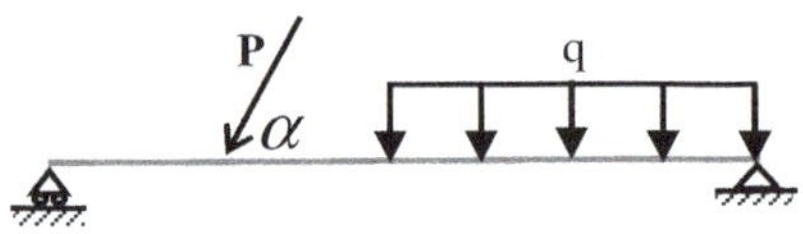

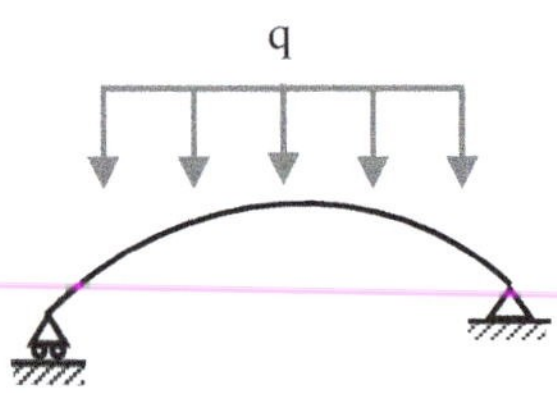

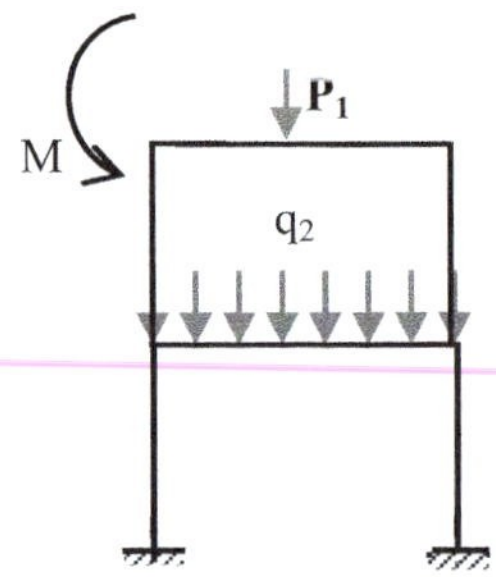

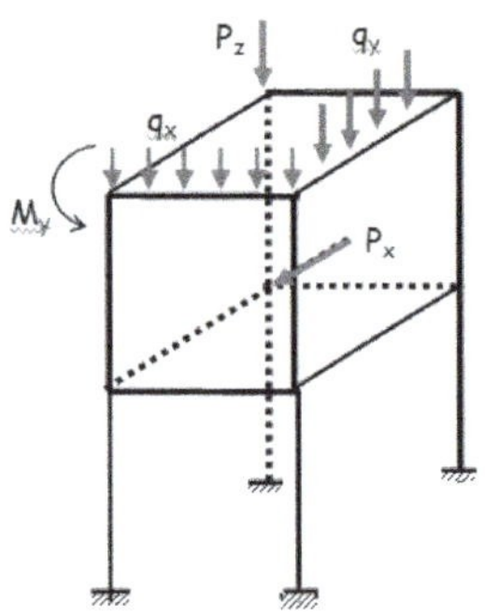

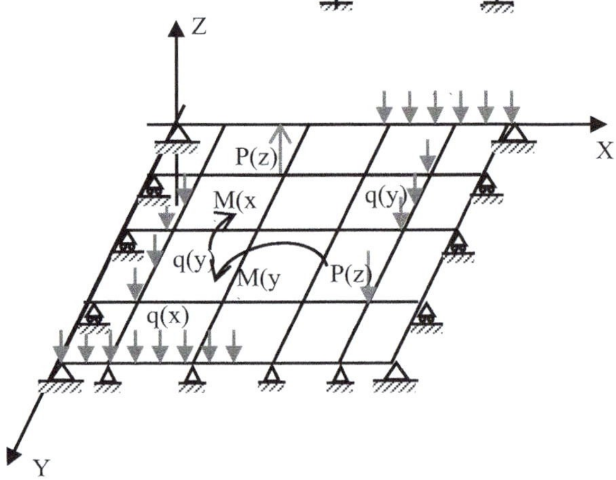

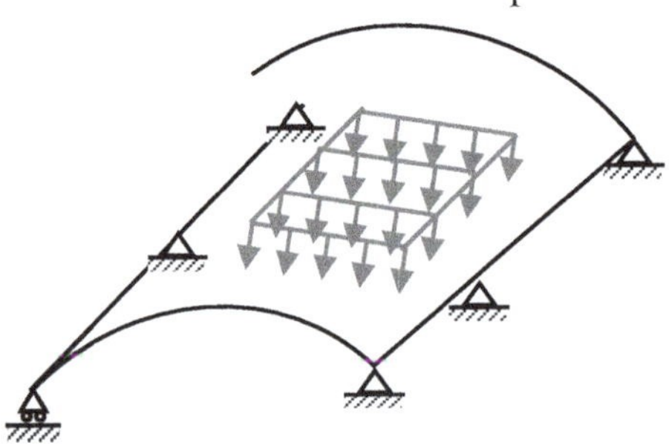

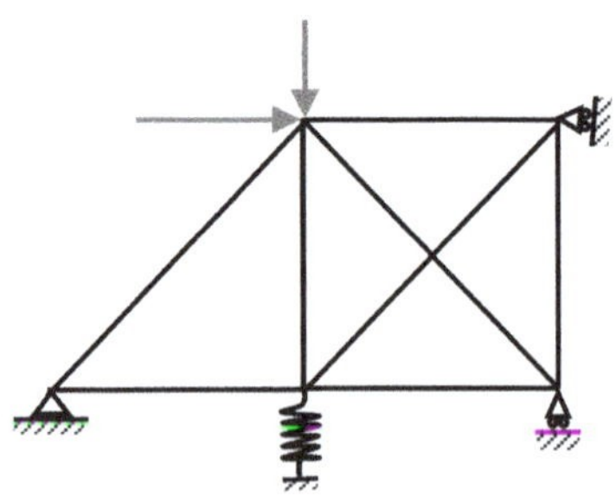

Exercise 2

Distinguish between statically determinate, statically indeterminate and unstable trusses.

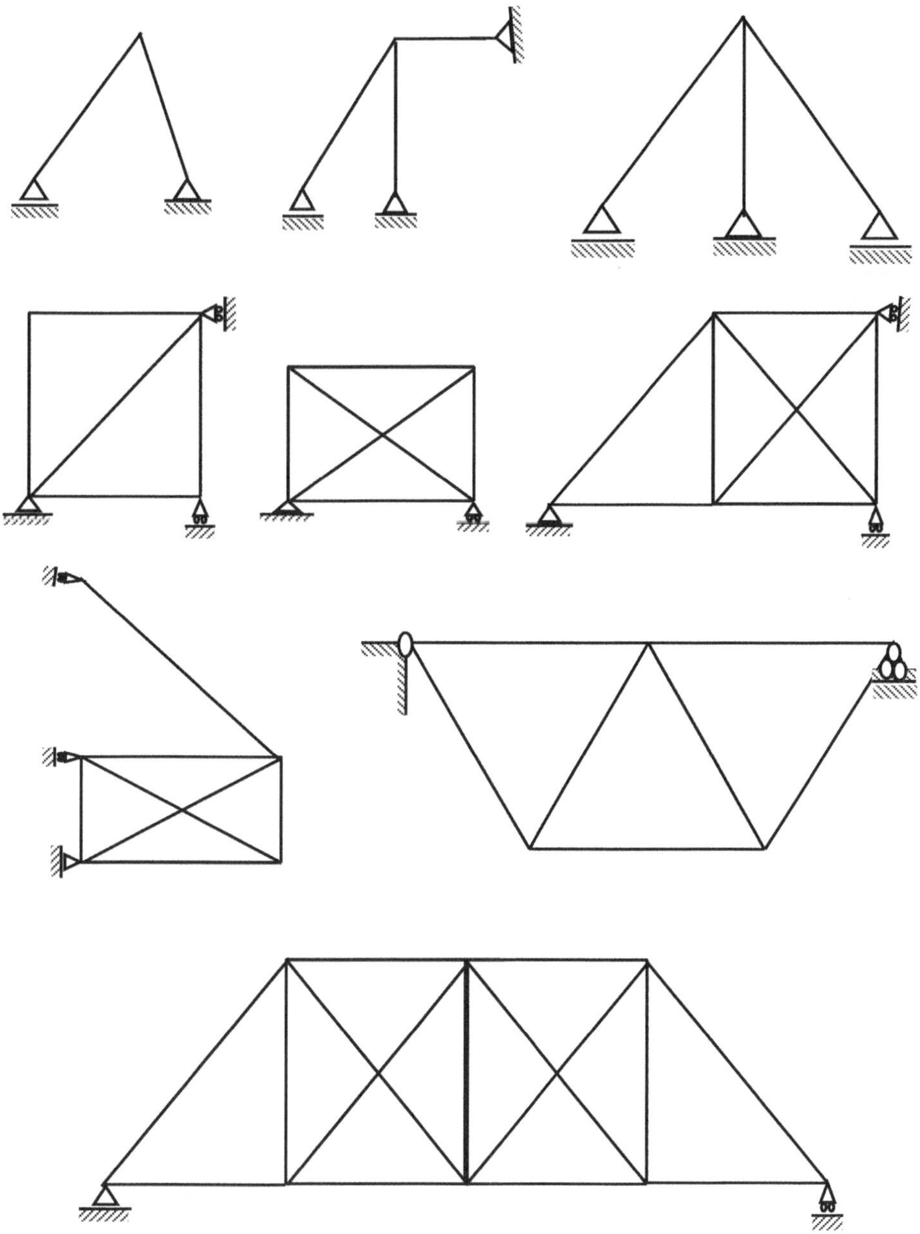

Exercise 3

Differentiate the static indeterminacy of each beam.

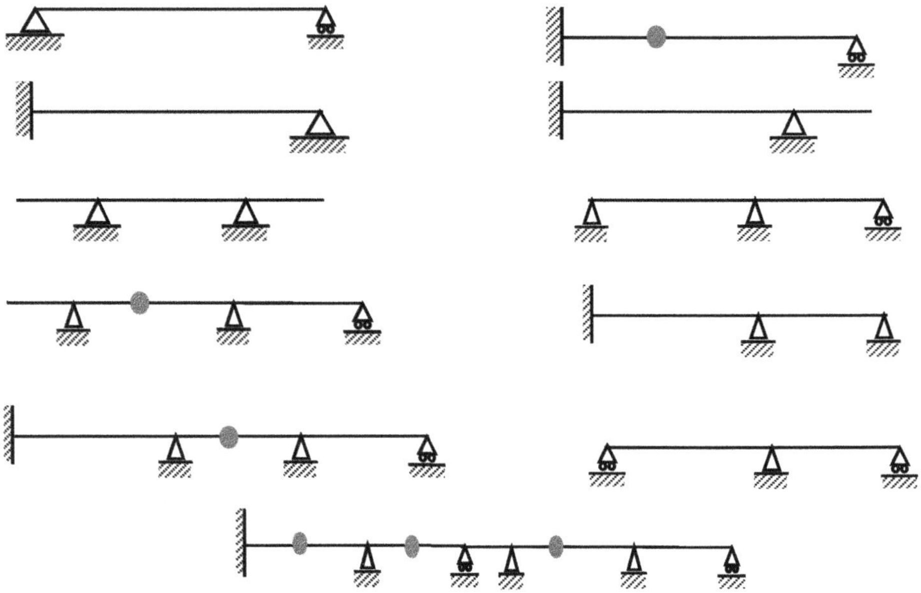

(designates a hinge)

Exercise 4

Determine the degree of static indeterminacy of the following frames:

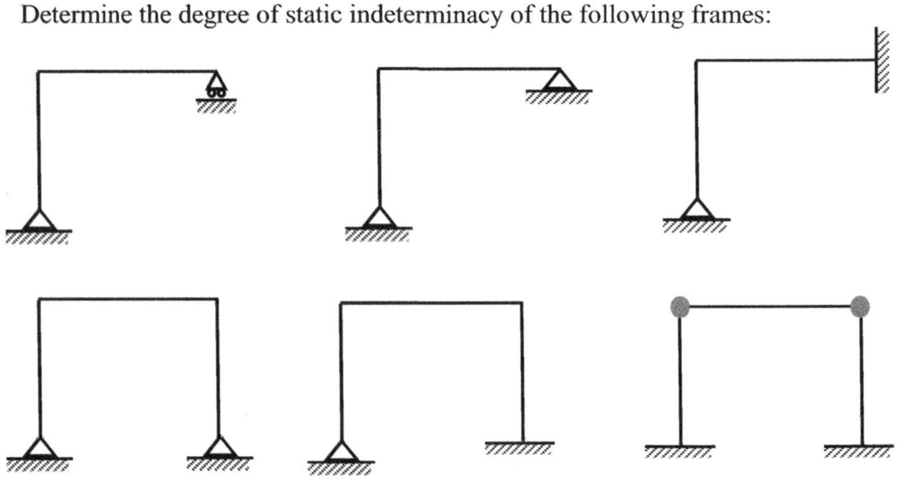

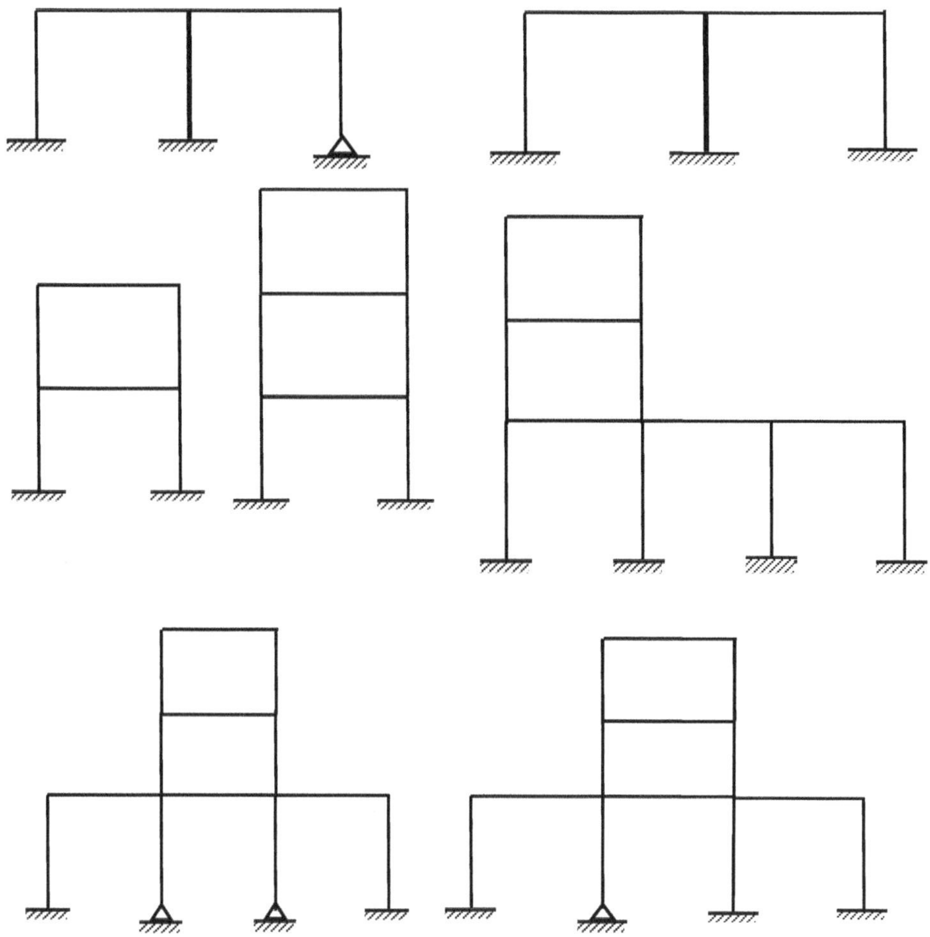

2

Actions on Structures

The teaching objectives for this chapter are as follows:

– evaluation of design loads;

– differentiation between different types of load;

– distinction between permanent, operating and environmental loads;

– distinction between static and dynamic loads;

– calculation of hydrostatic pressure and land thrust pressure.

This chapter describes the load assessment methodology, which is a key element of structural design. Forces acting on a structural element of a civil construction can be (1) permanent (dead) loads, (2) variable (live) loads and (3) environmental loads, such as snow, wind, seismic, and thermal loads. The magnitudes of the loads are quantified according to the European rules, which are used for evaluating the load calculation.

2.1. Loads

All structures and constructions must be designed to withstand the different actions applied to them. The loads applied to a structure or a construction are multiple and can be classified into several families: (1) mechanical loads, (2) thermal loads, (3) environmental loads, (4) seismic or dynamic loads, (5) loads with differential settlements, etc.

To build a structure that meets the criteria of resistance, shape stability and economy, the civil engineer must carefully evaluate all of the actions that potentially impact the studied structure. During the design phase, the load calculations are generally increased to respond to the recommendations and requirements of the calculation codes. For these reasons, the accurate evaluation of load calculations plays a very important role in the design and analysis phase.

2.2. Permanent loads

Permanent loads are all actions that do not depend on time and position. They can be point forces, surface forces or volume forces. Permanent loads include the weight of the structural system and the weight of the equipment permanently attached to it. The weight of the structural system contains the weight of the frame, supporting walls, partitions, slabs, heating and air conditioning systems, plumbing, electric systems, etc.

After the design phase, the dimensions of the different elements can be established. During this phase, it is possible to calculate the total weight of the structure by multiplying the unit weights of the different building materials used by the cross-sectional areas of the elements. Table 2.1 groups the weights of some building materials.

Material	Unit weight (kN/m^3)	Material	Unit weight (kN/m^3)
Lightweight	9–20	Steel	77
Concrete	20–22	Aluminum	27
Reinforced concrete	25	Glass	25
Cement mortar	19–23	Fresh water	10
Lime mortar	12–18	Milk	10
Wood	3.50–10.8	Propane	5
Aggregates	20–30	Mercury	26.6

Table 2.1. *Unit weights of some materials*

EXAMPLE 2.1.–

Evaluate the total weight of the construction in Figure 2.1. The slab is made up of wood that is 8 cm thick, the beams are made of reinforced concrete and the posts are made of metal frames. The dimensions of the sections are shown on the figure itself. Deduct the load on each post.

The dead weight of the wooden slab: $W_1 = 6.3 \times 6 \times 4 \times 0.08 = 12.10$ kN.

The dead weight of the metallic post: $W_2 = 4 \times 77 \times 0.10 \times 0.15 \times 3 = 13.86$ kN.

The dead weight of the reinforced concrete beams:

$$W_3 = 2 * 25 * (0.30 \times 0.30 \times 6 + 0.30 \times 0.40 \times 4) = 51.00 \text{ kN.}$$

The total load is:

$$W = W_1 + W_2 + W_3 = 76.96 \text{ kN.}$$

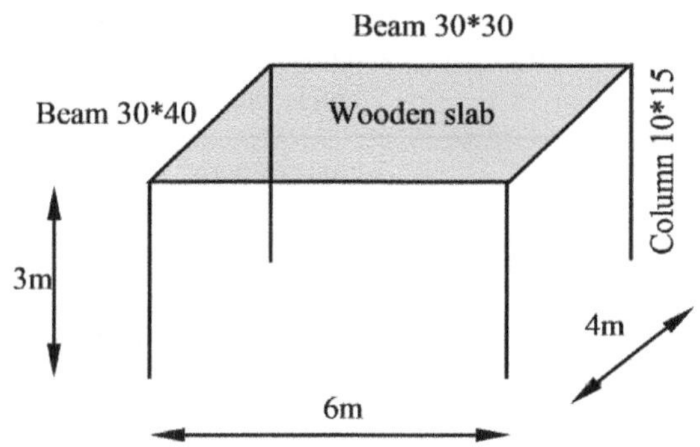

Figure 2.1. *Given construction*

The load on each post, assuming an even distribution between the posts, is:

$$W_P = \frac{W}{4} = 19.24 \text{ kN}$$

Finally, each post must be dimensioned so that it can support the W_P load.

2.3. Live loads

Live loads are actions that vary in time and position. To put it another way, they are loads that are not listed in the category of permanent loads, including environmental loads such as snow, earthquake and wind. The rules and calculation codes designate the live loads or variable loads based on the use of the construction.

The live load values are specified in the calculation codes or in the documents prepared by the project manager. This assessment is based on the size and the type of utility in construction. Variable loads can change position, and each structural element must be studied to see how it withstands the most adverse internal stresses.

Live loads are assumed to be uniformly distributed over the structural elements and their values used in the calculation depend on the nature and use of the structure. Table 2.2 shows some live load standards, according to the use of the structure.

Use group	Names	Constructions		Operating load value (kN/m^2)
A	Domestic activities	Residential buildings, houses, hospitals, hotels, etc.		1.50
B	Offices	Administrations		2.50
C	Gathering places	Exception of use groups A, B, D and E		
		C1	Schools, cafes, restaurants, banks, reading rooms	2.50
		C2	Theaters, cinemas, conference rooms, waiting rooms	4.00
		C3	Museum rooms, exhibition halls, access rooms to buildings, administrations and hotels	4.00
		C4	Dancing halls, gyms, venues	5.00
		C5	Concert halls, sports halls, terraces and accessible areas	5.00
D	Shops	Department stores, retail premises		5.00
E	Storage areas	Storage areas, industrial premises, warehouses, libraries		7.50
F		Traffic or parking areas for light vehicles with a weight of $\leq$ 30 kN		2.30
G		Traffic or parking areas for vehicles with a weight of between 30 kN and 160 kN		5.00

H		Roofs inaccessible except for normal maintenance	1.00
I		Inaccessible roofs of A to D category buildings	A to D
J		Accessible roofs with special function (helicopter landing, etc.)	Case by case

Table 2.2. *Live load values*

For use groups A, B, C3, D and F, reducing the operating load is recommended. The reduction coefficient depends on the reference area $A_0 = 3.50$ m² and that of construction A. The reduction coefficient is defined by

$$\alpha_A = 0.77 + \frac{A_0}{A} \leq 1. \qquad [2.1]$$

Live loads are also due to the weight of objects that are temporarily installed on the structure, such as loads on the slabs, the weight of a vehicle on a bridge and equipment and tools.

EXAMPLE 2.2.–

Evaluate the total load of a construction for school use due to the variable load. The dimensions of the slab are shown in Figure 2.2. The terrace floor is inaccessible except for normal maintenance.

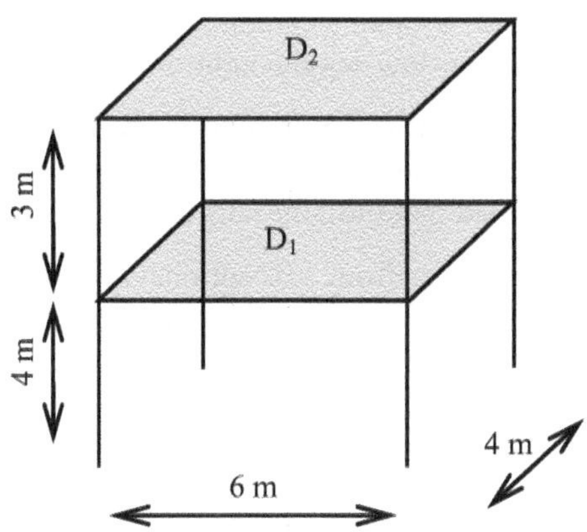

Figure 2.2. *Construction – live loads*

Due to the live load, we identify:

The load relative to slab 1: $Q_1 = 6 \times 4 \times 2.50 = 60$ kN.

The load relative to slab 2: $Q_2 = 6 \times 4 \times 1.00 = 24$ kN.

The total load is: $Q = Q_1 + Q_2 = 84$ kN.

The load on each post due to the operating load is:

$$Q_P = \frac{Q}{4} = 21 \text{ kN}$$

2.4. Snow loads

Snow loads are considered to be variable, fixed and static actions. They are assumed to be evenly distributed over the roof elements in a vertical direction.

In many parts of the world, snow loads must be taken into account when designing structures. The snow load depends on several parameters, such as the region identifying its snow cover, which can be evaluated by the calculation codes or by the meteorological data of each region as well as the region's altitude. In this case, normal and extreme snow loads must be increased according to the construction's elevation.

European standards identify different categories of snow loads. These are:

1) The normal snow load without accumulation that takes into account the amount of snow falling in calm or low-wind conditions.

2) The normal snow load with accumulation where there is a redistribution of snow in the particularly windy areas of the roof. Locally, you can get a snow load that is much higher than the load that initially falls on the ground.

3) The snow load resulting from exceptional snowfall with or without accumulation. Such snowfall is an exceptionally rare occurrence that corresponds to a much higher value than the typical snow load.

To determine the snow load, it is necessary to consider the uniform layer of snow which has accumulated during a snowfall in calm weather, the shape of the roof and the distribution of the snow under the effect of the wind with normal accumulation.

In France, for example, the snow load on the roof is given by

$$S = \mu_i .C_e .C_t .S_k + S_1 \quad \text{(long-term situation)} \qquad [2.2a]$$

$$S = \mu_i .S_A + S_1 \quad \text{(accidental situation)} \qquad [2.2b]$$

1) μ_i is the shape coefficient of a pitched roof α and the wind redistribution of snow (Table 2.3).

2) S_k is the characteristic value of the snow load on the ground that depends on the geographical location and an altitude of 200 m (Table 2.4).

Slope angle	$0 < \alpha \leq 30°$	$30° < \alpha < 60°$	$\alpha \geq 60°$
μ_i	0.80	$0.80.\dfrac{60-\alpha}{30}$	0.0

Table 2.3. *Shape coefficients*

Region	A1	A2	B1	B2	C1	C2	D	E
S_{k200} (kN/m²)	0.45	1.00	0.55	1.00	0.65	0.65	0.90	1.40

Table 2.4. *Characteristic values of snow on the ground*

The snow load standards are given in accordance with the altitude by the following formulas:

$$S_k = S_{k200} + \frac{h}{1,000} - 0.20 \qquad 200 < h \leq 500 \qquad [2.3a]$$

$$S_k = S_{k200} + 1.5\frac{h}{1,000} - 0.45 \qquad 500 < h \leq 1,000 \qquad [2.3b]$$

$$S_k = S_{k200} + 3.5\frac{h}{1,000} - 2.45 \qquad 1,000 < h \leq 2,000 \qquad [2.3c]$$

where h is the elevation of the place of construction.

3) S_A is the accidental value of the snow load on the ground corresponding to an exceptional snowfall. The snow load standards are given depending on the region of the construction (Table 2.5).

4) $C_e \geq 1.0$ is the exposure coefficient, which is generally 1, except when there are neighboring buildings that prevent wind movement of the snow. Therefore, we use $C_e \geq 1.0$.

Region	A2	B1	B2	C2	D
S_A (kN/m²)	1.00	1.00	1.35	1.35	1.80

Table 2.5. *Exceptional values of the snow load*

5) $C_t \leq 1.0$ is the thermal coefficient that equals 1 since the heated buildings are systematically isolated. A lower value may be used if it is justified by a specific study.

6) S_1 is an increase for low slopes where rainwater evacuation is made difficult by the presence of snow.

$$S_1 = \begin{cases} 0.20 \text{ kN/m}^2 \text{ if the slope is less than or equal to 3\%.} & [2.4a] \\ 0.10 \text{ kN/m}^2 \text{ if the slope is between 3 and 5\%.} & [2.4b] \end{cases}$$

EXAMPLE 2.3.–

Calculate the snow load per m² of an industrial building whose roof is inclined at 20° to the horizontal. The building site is located 300 m above sea level.

The intensity of the snow load is given by [2.2].

$$S = \mu_i \,.C_e \,.C_t \,.S_k + S_1$$

The roof is inclined at a 20° angle, $\mu_i = 0.8$.

We use $C_e = C_t = 1.0$ and $S_1 = 0.0$.

The effect of the altitude is calculated by :

$$S_k = S_{k200} + \frac{h}{1,000} - 0.20.$$

With $S_{k200} = 0.55\,\mathrm{kN/m^2}$,

$$S_k = 0.45\,\mathrm{kN/m^2}.$$

The snow load is therefore $S_k = 0.36\,\mathrm{kN/m^2}$.

We assume that the roof is subject to a snow load of intensity $S_k = 0.36\,\mathrm{kN/m^2}$.

2.5. Wind loads

European rules describe the method of determining wind actions for calculating the structures of buildings and constructions. They include the whole structure and some of its constituent elements. The rules in force apply to buildings and civil engineering constructions whose height is less than 200 m and bridges where no span exceeds 200 m provided that they satisfy the criteria for dynamic response. Special constructions such as lattice towers, masts and guyed chimneys are not covered by these rules.

Wind loads act directly on the walls of a construction and their intensity depends on several parameters: the geographical location; the geometry of the construction; obstruction in its surroundings, such as nearby buildings; and structural vibration characteristics.

Similarly, wind loads vary with time and are considered to be variable and fixed. The action of the wind acts as pressure producing perpendicular forces on the walls of the construction. The pressures generated act on the external surfaces of a closed construction and also act on the interior walls where there are openings.

European standards also deal with chimneys, lattice towers and bridges. The external aerodynamic pressure is given by the following formula:

$$q_e = q_p(z_e).C_{pe} \qquad [2.5]$$

$q_p(z_e)$ is the peak dynamic pressure at height z_e which depends on the local climate of the region, the roughness, the orography of the ground and the height in relation to the ground. The dynamic pressure is written as:

$$q_p(z_e) = C_e(z_e).q_b \qquad [2.6]$$

$C_e(z_e)$ is the operating coefficient that takes into account local forces that influence wind action and q_b is the dynamic reference pressure. It is given according to the geographical location of the construction.

$$q_b = \frac{1}{2}\rho(C_{dir}.C_{season}.V_{b0})^2, \qquad [2.7]$$

where V_{b0} is the reference wind speed at the base, ρ is the density of the air, C_{dir} is the coefficient of privileged directions ($C_{dir} \leq 1.0$) and C_{saison} is the seasonal coefficient that takes into account wind modulations according to each season. Table 2.6 groups the values of the reference speed at the base and the dynamic pressure of each region.

Region	1	2	3	4
$V_{b,0}$ (m/s)	22	24	26	28
q_b (kN/m²)	0.30	0.35	0.41	0.48

Table 2.6. *Reference velocities and pressures*

The external pressure coefficient C_{pe} is one of the aerodynamic coefficients, which depend on the size of the loaded surface. We identify $C_{pe,1}$ and $C_{pe,10}$, which are the coefficients of external pressure for a surface of 1 m² and 10 m², respectively. Values for other surfaces are obtained by logarithmic interpolation.

$$C_{pe} = C_{pe,1} \qquad A \leq 1.m^2 \qquad [2.8a]$$

$$C_{pe} = C_{pe,1} + (C_{pe,10} - C_{pe,1})LogA \qquad 1.m^2 < A < 10.m^2 \qquad [2.8b]$$

$$C_{pe} = C_{pe,10} \qquad A \geq 10.m^2 \qquad [2.8c]$$

Finally, the exposure coefficient $C_e(z)$ depends on the roughness of the land and the topography of the building's vicinity. The roughness is generated by trees, hedges and other constructions, etc. The topography of the building's neighborhood, through the orography coefficient, quantifies the variations in altitude of the land surface within a radius of 1 km around the construction. The formula giving the coefficient of exposure is

$$C_{pe} = C_r^2(z).C_0^2(z)\left[1+7\,\mathrm{I}_V(z)\right], \qquad [2.9]$$

where $C_r(z)$ is the roughness coefficient of the land following the direction of the wind action.

$$C_r(z) = K_r \ \ln\frac{z}{z_0} \qquad z_{\min} \le z \le 200\,\mathrm{m} \qquad [2.10a]$$

$$C_r(z) = K_r \ \ln\frac{z_{\min}}{z_0} \qquad z \le z_{\min} \qquad [2.10b]$$

The values of K_r, z_0, $z_{\min}$ and K_l are grouped in Table 2.7.

Class	Land categories	z_0(m)	$z_{\min}$(m)	K_r	K_l
0	Sea, lake or body of water covered by wind over a distance of at least 5 km	0.005	1.0	0.16	1.0
II	Open countryside, with or without some isolated obstacles (trees, buildings), airports	0.05	2.0	0.19	0.995
IIIa	Countryside with hedges, orchards, small woods, hedged fields, dispersed habitat	0.20	5.0	0.21	0.97
IIIb	Industrial, urban or forest zones	0.50	9.0	0.22	0.923
IV	Urban areas in which buildings occupy at least 15% of the surface and have an average height greater than 15 m.	1.00	15.0	0.23	0.854

Table 2.7. *Values of K_r, z_0, $z_{\min}$ and K_l*

The orography coefficient $C_0(z)$ $(1 \leq C_0(z) \leq 1.15)$ shows the impact of relief on wind speed variation. It is given by

$$C_0(z) = 1 + \frac{4}{1{,}000} \Delta A_c e^{-0.014.(z-10)} \qquad z \geq 10\text{m} \qquad [2.11a]$$

$$C_0(z) = 1 + \frac{4}{1{,}000} \Delta A_c e^{-0.014.(z-10)} \qquad z < 10\text{m} \qquad [2.11b]$$

where ΔA_c is the altitude relative to the construction.

$I_v(z)$ is the turbulence intensity coefficient, which measures the change in wind speed. The standard deviation of wind turbulence σ_v is defined by the average value of the wind speed $V_m(z)$.

$$I_v(z) = \frac{\sigma_V}{V_m(z)} \qquad [2.12a]$$

In another form, we can write:

$$I_v(z) = \frac{K_I}{C_0(z).\ln \frac{z}{z_0}} \qquad z_{\min} \leq z \leq 200\text{m} \qquad [2.12b]$$

$$I_v(z) = \frac{K_I}{C_0(z_{\min}).\ln \frac{z_{\min}}{z_0}} \qquad z < z_{\min} \qquad [2.12c]$$

The internal hydrodynamic pressure acting on the interior surfaces of a construction should be determined by:

$$q_i = q_p(z_e).C_{pi}, \qquad [2.13]$$

where $q_p(z_e)$ is the peak dynamic pressure and C_{pi} is the internal pressure coefficient.

The coefficient of internal pressure C_{pi} depends on the surface and the distribution of the openings in the building's walls. We must take into account the fact that permanent openings, windows and doors are considered closed and are not included in the surface of openings. It is possible to consider them as open but the wind must then be considered as an accidental load in an accidental combination.

For buildings that do not have a dominant face where all of the openings are well-distributed on all facades, the coefficient C_{pi} relates to permeability μ (Figure 2.3). The permeability is calculated by relation [2.14].

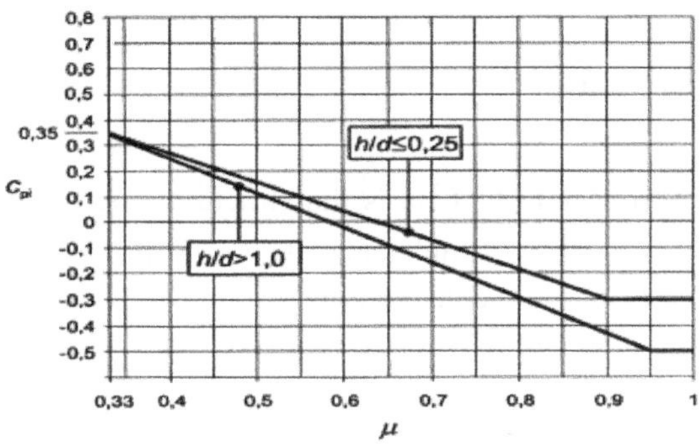

Figure 2.3. *Coefficient of internal pressure according to permeability*

$$\mu = \frac{\sum \text{area of openings where } C_{pe} \leq 0}{\sum \text{area of all openings}} \qquad [2.14]$$

For $0.25 \leq \dfrac{h}{d} \leq 1$, a linear interpolation can be used (h and d are, respectively, the height and the length of the building).

For buildings with a dominant face, we identify:

1) $C_{pi} = 0.75\ C_{pe}$, if the area of the openings on this face is greater than twice that of all of the other faces.

2) $C_{pi} = 0.90\ C_{pe}$, if the area of the openings on this face is at least equal to three times the value of all of the other faces.

3) When the area of the openings on the dominant face is between two and three times the area of the openings on the other faces, the coefficient of internal pressure is calculated by a linear interpolation.

To evaluate the total force exerted by the wind on a structure or on a structural element, we can calculate the forces F_{we} and F_{wi} resulting from the pressures acting on all of the external and internal walls that make up the structure, without forgetting the forces of friction F_{fr}, given that:

$$F_{we} = C_s C_d \sum_{surfaces} q_e . A_{ref} ; \quad [2.15]$$

$$F_{wi} = \sum_{surfaces} q_i . A_{ref} ; \quad [2.16]$$

$$F_{fr} = C_{fr} . q_p(z_e) . A_{fr} . \quad [2.17]$$

The coefficient produced ($C_s C_d$) is the structural coefficient intended to take into account the absence of simultaneity between the pressure peaks on the construction surface C_s and the construction vibrations generated by turbulence C_d. The value of the structural coefficient is given by:

$$C_s C_d = \frac{1 + 2.k_p . I_v(z_s)\sqrt{B^2 + R^2}}{1 + 7.I_v(z_s)}, \quad [2.18]$$

where z_s is the reference height, which can be equal to h (construction height). K_p is the peak factor defined by the ratio of the maximum value of the variable part of the response to its standard deviation. I_v is the intensity of the turbulence.

B^2 is the quasi-static response coefficient reflecting the correlation of the pressure on the surface of the construction and R^2 is the resonant response coefficient reflecting the resonance of the vibration mode generated by the turbulence.

C_{fr} and A_{fr} are, respectively, the friction coefficient and surface area friction.

The coefficient of friction is given as a function of the roughness of the building surface. Its value is given by

– $C_{fr} = 0.01$ for smooth surfaces.

– $C_{fr} = 0.02$ for rough surfaces.

– $C_{fr} = 0.04$ for very rough surfaces.

Finally, the dynamic pressure $q_p(z_e)$ at the reference height z_e is given by:

$$q_p(z_e) = C_e(z_e).q_b \quad (\text{kN/m}^2) \qquad [2.19]$$

Bearing in mind that the exposure coefficient is $C_e(z_e) = C_r^2.C_0^2(z).(1+7.I_v(z))$ and C_r is the coefficient of roughness of the land according to the direction of the wind, $C_0(z)$ is the orography coefficient of the relief and $I_v(z)$ is the coefficient of the turbulence intensity.

EXAMPLE 2.4.–

A building with the dimensions 20 × 14 × 45 m is built in zone 4. The wind acts along the height of the building (Figure 2.4(b)).

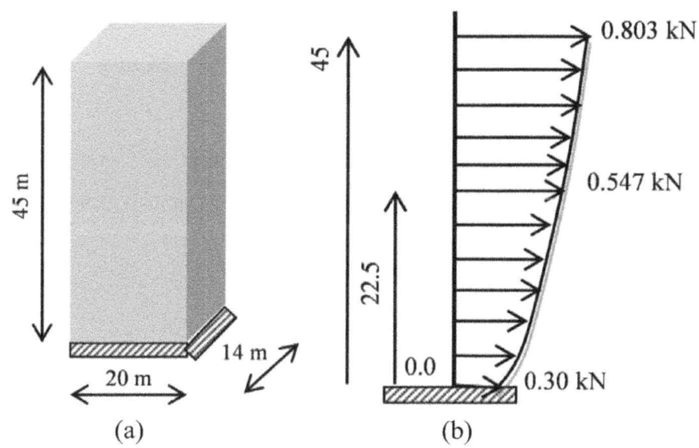

Figure 2.4. *Given building and wind diagram*

The geometric description of the building (Figure 2.4(a)) and zone type are given as follows:

– Wind width, $b = 20$ m

– Building depth, $d = 14$ m

– Total height, $h = 45$ m

– Calculation height, $z = 45$ m

– Fill rate, $\varphi = 1$

– Wind zone = 1

– Land type = 4

– Slope gradient of the wind = 0

– Equivalent mass per unit length, m_e = 2,100 kN/mL

The data set is :

– $Z_{0,II}$ = 0.05 m

– Reference height, Z_t = 200 m

– Reference scale, L_t = 300 m

– Air density, $\rho = 0.001$ kg/m^3

– Reference period, T = 600 s

Calculation of parameters:

Reference wind speed at the base of the building, V_{b0} = 24 m/s

Roughness length, z_0 = 1 m

Minimum height, z_{min} = 10 m

$$\alpha = 0.67 + 0.05\ln(z_0) = 0.67$$

$$L(z) = L_t.(\frac{z_{min}}{h})^{\alpha} = 40.31m$$

Reference height, z_e = 0.6*h = 27 m

Turbulence scale, $L(z_e) = L_t.(\frac{z}{z_t})^{\alpha} = 78.42$ m

Quasi-static response coefficient, $B^2 = \dfrac{1}{1+0.9(\dfrac{b+h}{L(z_s)})^{0.63}} = 0.556$

Basic vibration frequency, $n_{x,1} = (\dfrac{46}{h}) = 1.022\,\text{Hz}$

Rectangular section strength coefficient, $C_{f,0} = f(\dfrac{d}{b}) = 2.40$

Reduction factor (non-rounded edges), $\psi_r = 1$

Effective slenderness, $\lambda = 0.013$

End effect factor, $\psi_\lambda = 0.589$

Strength coefficient, $C_f = C_{f,0}.\psi_r.\psi_\lambda = 1.414$

Land factor, $k_r = 0.19(\frac{z_0}{z_{0,II}})^{0.07} = 0.234$

Roughness coefficient, $C_r(z_{min}) = k_r.\ln(\frac{z_{min}}{z_0}) = 0.54$

$C_r(z\ ) = 0.892$

$C_r(z_e) = 0.772$

Orography coefficient (flat land), $C_0(z) = 1$

Direction coefficient, $C_{dir} = 1$

Seasonal coefficient, $C_{season} = 1$

Wind speed reference, $V_b = C_{dir}.C_{season}.V_{b0} = 24\,\text{m/s}$

Wind speed average, $V_m(z_e) = C_r(z).C_0(z).V_b = 21.41\,\text{m/s}$

Basic frequency, $n_1 = \frac{46}{h} = 1.022\,\text{Hz}$

Logarithmic aerodynamic decrement, $\delta_a = \frac{C_t.\rho.b.V_m(z_s)}{2n_1.m_e} = 2.10^{-4}$

Logarithmic structural decrement, $\delta_s = 0.10$

Logarithmic damping decrement due to dampers, $\delta_d = 1$

Frequency, $n = n_{1,x} = 1.022\,\text{Hz}$

Turbulence scale, $L(z) = L_t.(\frac{z}{z_t})^{\alpha} = 110.40$

Average wind speed on side z, $V_m(z) = C_r(z).C_0(z).V_b = 18.54\,\text{m/s}$

Dimensionless frequency, $F_L(z,n) = \frac{n.L(z)}{V_m(z)} = 5.273$

Spectral density of dimensionless power, $S_L(z_e, n_{1,x}) = \frac{6.8 F_L(z,n)}{(1+10.2F_L(z,n))} = 0.045$

$$n_h = \frac{4.6h}{L(z_s)} F_L(z_s, n_{1,x}) = 13.92$$

$$R_h = \frac{1}{n_h} - \frac{1}{2n_h^2}(1 - e^{-2n_h}) = 0.069$$

$$n_b = \frac{4.6b}{L(z_s)} F_L(z_s, n_{1,x}) = 6.186$$

$$R_b(n_b) = \frac{1}{n_b} - \frac{1}{2n_b^2}(1 - e^{-2n_b}) = 0.149$$

$$R^2 = \frac{\pi^2}{4\delta^2} S_L^2(z_s, n_{1,x}).R_h^2(n_h).R_b^2(n_b) = 0.002$$

The frequency is $\nu = n_{1,x}\sqrt{\frac{R^2}{B^2 + R^2}} = 0.063\,\text{Hz}$

$$k_1 = 1$$

$$I_v(z_{\min}) = \frac{k_1}{C_0(z_{\min})\ln(\frac{z_{\min}}{z_0})} = 0.434$$

$$I_v(z\ ) = \frac{k_1}{C_0(z\ )\ln(\frac{z}{z_0})} = 0.263$$

$$I_v(z_e) = \frac{k_1}{C_0(z_e)\ln(\frac{z_e}{z_0})} = 0.303$$

$$q_p(z) = [1+7I_v(z)]\frac{1}{2}\rho.V_m^2(z) = 0.813$$

$$q_p(z_e) = [1+7I_v(z_e)]\frac{1}{2}\rho.V_m^2(z_e) = 0.895$$

$$A_{ref} = 1$$

$$k_p = \sqrt{2\ln(\nu T)} + \frac{0.6}{\sqrt{2\ln(\nu T)}} = 2.916$$

$$C_s.C_d = \frac{1+2k_p.I_v(z_s).\sqrt{B^2+R^2}}{1+7I_v(z_s)} = 0.635$$

The wind force of 1 m² on dimension 45 m is $F_w = C_s.C_d.C_f.q_P(z_e).A_{ref} = 0.803$ kN

We can conclude that:

– the wind force of 1 m² on the 45 m side is 0.803 kN;

– the wind force of 1 m² on the 22.5 m side is 0.547 kN;

– the wind force of 1 m² on the 0 side is 0.30 kN.

The horizontal force diagram is shown in Figure 2.4(b).

This procedure can be applied in the same way to evaluate the wind force in the second direction.

2.6. Seismic loads

The seismic force at the base of a construction is mainly due to the transmission of seismic waves from the ground to the construction. These waves induce accelerations of the construction mass, giving rise to internal actions in the various elements that make up this construction. Internal actions due to an earthquake depend on:

1) the seismic intensity;

2) the behavior of the ground acting as a shock absorber or amplifier of the seismic wave;

3) the construction class;

4) structural design; flexible or rigid behavior.

In addition, the seismic force at the base of the construction generates a horizontal acceleration in each direction of the building. This calculation of seismic acceleration is given by the expression

$$S_d = a_{gR} \cdot \gamma_I \cdot \eta . S_d(T), \qquad [2.20]$$

where a_{gR} is the zonal or reference acceleration at ground level of class A (m/s^2), γ_I is the coefficient of importance, η is the viscous damping correction coefficient and $S_d(T)$ is the response of the spectrum calculation for an elastic analysis at the period T. It should be emphasized that the expression of the acceleration giving the seismic force takes into account the basic acceleration.

1) Maximum reference acceleration at ground level of class A, a_{gR}.

The maximum acceleration at ground level depends on the mechanical characteristic of the foundations. For example, French land is divided into five zones. Table 2.8 gives the value of the coefficient of zonal acceleration according to the nature of the foundation.

Seismicity zone	Very weak	Weak	Moderate	Medium	Strong
a_{gR}	0.40	0.70	1.10	1.60	3.0

Table 2.8. *The maximum reference acceleration at ground level of class A*

2) Coefficient of importance, γ_1.

The buildings are classified into four categories of importance that depend on the impact on human life if the building collapses, the importance of the building for public security and civil protection immediately after an earthquake and the economic and social consequences.

This classification is grouped in Table 2.9. In addition, the value of the coefficient of importance according to the purpose of the building is given in the same table.

Category of importance	Buildings	Coefficient of importance γ_1
I	Works whose failure represents a minimal risk to persons or economic activity	0.80
II	Risk is "current" for people: houses, offices, commercial premises, workshops, factories, shared garages, etc.	1.00
III	High risk for people because of their attendance or socioeconomic importance: schools, stadiums, halls, museums, etc.	1.20
IV	Safety essential for the needs of the civil security, public order, defense and survival of the region: hospitals, barracks, ambulance stations, museums, libraries housing major works or irreplaceable collections, etc.	1.40

Table 2.9. *Coefficient of importance*

3) Damping correction coefficient, η.

Any element of a structure or partition that appears in the expression of the seismic acceleration calculation contributes to the damping of the whole building. The damping correction coefficient is expressed by:

$$\eta = \sqrt{\frac{10}{5+\xi}} \geq 0.55, \qquad [2.21]$$

where ξ is the critical percentage of viscous damping (expressed in %), which depends on the nature of the material used. In cases where several materials are used simultaneously, it is necessary to choose the most unfavorable damping. Table 2.10 illustrates the viscous damping values according to the nature of the construction material.

Material	Critical percentage of damping (%) ξ
Concrete	3.00
Reinforced concrete	4.00
Prestressed concrete	2.00
Reinforced masonry	6.00
Chained masonry	5.00
Bolted steel	4.00
Welded steel	2.00
Bolted wood	4.00
Nailed wood	5.00
Glued wood	4.00

Table 2.10. *Critical percentage of damping*

4) Acceleration calculation, a_g.

The acceleration calculation appearing in the expression of the spectrum is expressed by:

$$a_g = \gamma_1 . a_{gR} \qquad [2.22]$$

5) Vertical acceleration, a_{vg}.

The seismic acceleration that corresponds to the vertical acceleration of the masses is given in Table 2.11.

Seismicity zone	Very weak	Weak	Moderate	Medium	Strong
a_{vg} / a_g	0.80	0.80	0.80	0.80	0.90

Table 2.11. *Vertical acceleration*

Taking into account the importance of the building, the values of the acceleration a_{vg} can be deduced according to the seismicity area and importance of the buildings (Table 2.12).

Seismic zone	Very weak	Weak	Moderate	Medium	Strong
Coefficient of importance I	0.26	0.45	0.70	1.02	2.16
Coefficient of importance II	0.32	0.56	0.88	1.28	2.70
Coefficient of importance III	0.38	0.67	1.06	1.54	3.24
Coefficient of importance IV	0.45	0.78	1.23	1.79	3.78

Table 2.12. *Vertical acceleration*

6) Structural behavior coefficient, q.

The structural behavior coefficient is a reduction of forces. It is equal to the quotient of the seismic forces that the structure undergoes if its response is completely elastic with a viscous damping of 5%, and the seismic forces can be used during the design with a linear model while continuing to ensure the structure's satisfactory response.

In European rules, the values of the behavior coefficient, q, including the influence of viscous damping different to 5%, are indicated for various materials and structural systems according to various levels of ductility. The choice and justification for this coefficient is generally complex and its economic importance is fundamental. In this case, we identify that:

1) The structure has a dissipation capacity and a limited ductility. Such ductility concerns buildings of low seismicity, which corresponds to a seismic acceleration $a_g \leq 0.78$ m/s². In this case, it is possible to use a behavior coefficient $q \prec 1.5$.

2) For medium- or high-ductility classes, compliance with the rules of global and local ductility is required as well as with the construction features of European standards. In this case, it is acceptable to use the behavior coefficient $q \geq 1.5$.

7) Spectrum calculation, $S_d(T)$.

The spectrum calculation of an elastic analysis allows you to determine the acceleration that the seismic mass of the structure can acquire as a function of the vibration period. $S_d(T)$ is the response of the spectrum calculation for the elastic analysis in the period T. In this case, the horizontal spectrum calculation is defined by the expressions:

$$S_d(T) = a_g.S.[\frac{2}{3} + \frac{T}{T_B}.(\frac{2.5}{q} - \frac{2}{3})] \qquad 0 \leq T \leq T_B \qquad [2.23a]$$

$$S_d(T) = a_g.S.\frac{2.5}{q} \qquad T_B \leq T \leq T_C \qquad [2.23b]$$

$$S_d(T) = a_g.S.\frac{2.5}{q}[\frac{T_C}{T}] \geq \beta a_g \qquad T_C \leq T \leq T_D \qquad [2.23c]$$

$$S_d(T) = a_g.S.\frac{2.5}{q}[\frac{T_C.T_D}{T^2}] \geq \beta a_g \qquad T_D \leq T \leq 4s \qquad [2.23d]$$

Similarly, the spectrum calculation for vertical elastic analysis is defined by the expressions:

$$S_d(T) = a_{vg}.[\frac{2}{3} + \frac{T}{T_B}.(\frac{2.5}{q} - \frac{2}{3})] \qquad 0 \leq T \leq T_B \qquad [2.24a]$$

$$S_d(T) = a_{vg}.\frac{2.5}{q} \qquad T_B \leq T \leq T_C \qquad [2.24b]$$

$$S_d(T) = a_{vg}.\frac{2.5}{q}[\frac{T_C}{T}] \geq \beta a_g \qquad T_C \leq T \leq T_D \qquad [2.24c]$$

$$S_d(T) = a_{vg}.\frac{2.5}{q}[\frac{T_C.T_D}{T^2}] \geq \beta a_g \qquad T_D \leq T \leq 4s \qquad [2.24d]$$

β is the coefficient that corresponds to the lower limit of the horizontal spectrum calculation. The European code 8 sets it at 0.2 while S, T_C and T_b are the spectrum calculation parameters.

EXAMPLE 2.5.–

A two-level building is simulated by the branching model, with each level represented by its mass (Figure 2.5). For reasons of simplicity, $m = m_1 = m_2$ is adopted and flexural rigidity (EI) is also constant for both levels.

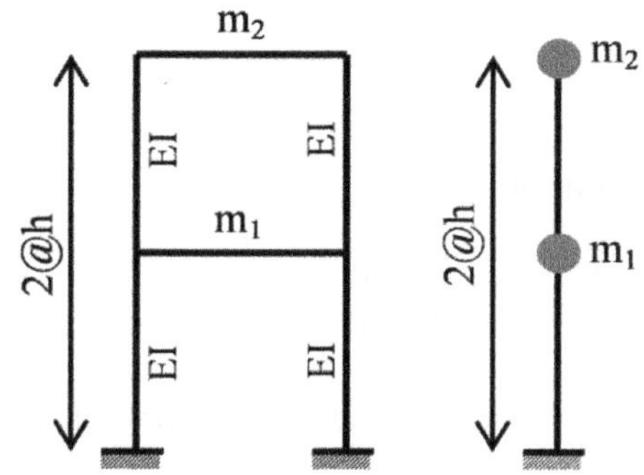

Figure 2.5. *Structure and its modeling*

The mass and rigidity matrices of the structure are:

$$[M] = m\begin{bmatrix} 1 & 0 \\ 0 & 1 \end{bmatrix}$$

$$[K] = \frac{6}{7}\frac{EI}{h^3}\begin{bmatrix} 16 & -5 \\ -5 & 2 \end{bmatrix}.$$

The dynamic equation of the non-damped free vibrations is given by:

$$\left\|[K] - \omega[M]\right\| = 0.$$

where ω are the values.

The values correspond to:

$$\left\|\begin{bmatrix} 16 & -5 \\ -5 & 2 \end{bmatrix} - \frac{7\omega mh^3}{6EI}\begin{bmatrix} 1 & 0 \\ 0 & 1 \end{bmatrix}\right\| = 0.$$

We write: $\alpha^2 = \dfrac{7\omega\ mh^3}{6EI}.$

The determinant of the matrix equation leads to:

$$\omega^4 - 18\omega^2 + 7 = 0.$$

The roots of this equation are, respectively:

$$\omega_1 = 0.583\sqrt{\frac{EI}{mh^3}}.$$

The first vibration mode period is:

$$T_1 = \frac{2\pi}{0.588}\sqrt{\frac{mh^3}{EI}} = 10.762\sqrt{\frac{mh^3}{EI}}.$$

In the same way, the pulsation and the second vibration mode period are:

$$\omega_2 = 3.884\sqrt{\frac{EI}{mh^3}}.$$

$$T_2 = 1.617\sqrt{\frac{mh^3}{EI}}.$$

The seismic shear force at the base of the construction is given by:

$$V_b = m.\lambda.S_d(T_1),$$

with λ being a correction coefficient:

$\lambda = 0.85$ if $T_1 \leq 2.T_c$ and if the building has more than two levels;

$\lambda = 1$ in other cases;

Equation [2.23] allows us to calculate the response spectrum according to the fundamental period. This value $S_d(T_1)$ gives us the shear force at the base of the construction.

Using the equivalent static method, the vertical distribution of seismic shear force at the different levels is:

$$V_i - V_b \frac{m_i.z_i}{\sum_{i=1}^{2} m_i.z_i},$$

from which:

$$V_1 = \frac{2}{3} V_b ;$$

$$V_2 = \frac{1}{3} V_b .$$

2.7. Thermal loads

In this section, the European standards present the principles for calculating thermal actions on constructions such as buildings, bridges, cooling towers, silos, reservoirs, hot and cold water storage facilities and hot and cold water distribution systems.

In addition, they describe the temperature variations on the structural elements and thus evaluate the characteristic values of thermal actions of structural dimensions exposed to daily or seasonal climatic variations. Evaluation of thermal actions on unexposed structures is not necessary.

In this section, we only present the standards of thermal actions on buildings. Therefore, the thermal actions resulting from climatic variations and operating temperature must be taken into account in building design.

The uniform temperature component ΔT of a structural element is given by the formula:

$$\Delta T = T - T_0 \qquad [2.25]$$

where T_0 is the original temperature; that is the temperature of a structural element on the date when it is completed.

The temperature assessment depends on the season and the environment in which it is calculated. Tables 2.13–2.15 show the temperature variation inside, outside and below ground level.

Season	Inside temperature in °C
Summer	20 °C
Winter	25 °C

Table 2.13. *Inside temperature*

Season	Significant factor		Outside temperature in °C
Summer	Relative absorbency in relation to the surface color	0.5 light surface	$T_{max} + 18$
		0.7 light surface	$T_{max} + 30$
		0.9 light surface	$T_{max} + 42$
Winter			T_{min}

Table 2.14. *Outside temperature above ground level*

Season	Underground depth	Outside temperature in °C
Summer	Less than 1 m	8
	More than 1 m	5
Winter	Less than 1 m	–5
	More than 1 m	–3

Table 2.15. *Outside temperature below ground level*

2.8. Hydrostatic loads and land thrust

When structures are designed to retain a fluid, soil or granular material, they are solicited by a normal pressure on their walls, which is the main design parameter. This category of construction includes reservoirs, dams, supporting walls, silos, etc. These structures are calculated to withstand hydrostatic pressure and/or land thrust (Figure 2.5).

The hydrostatic pressure of a fluid exerts a normal force on the contact wall with an intensity equal to:

$$P = \gamma . z , \qquad [2.26]$$

where γ is the density of the fluid.

Pressure reaches its maximum value at the base of the structure:

$$P = \gamma . h , \qquad [2.27]$$

where h is the height of the wall.

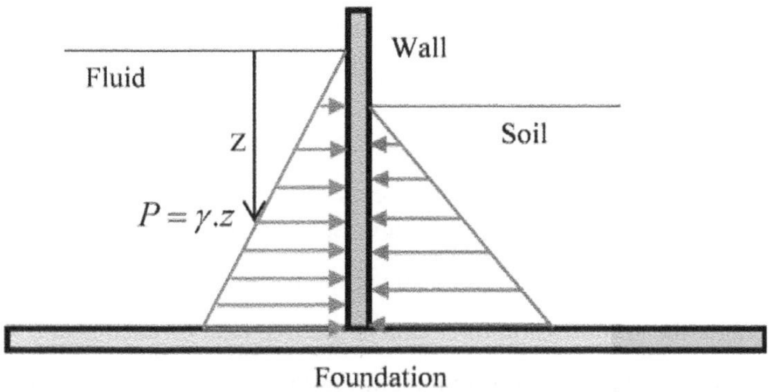

Figure 2.6. *Hydrostatic pressure and land thrust. For a color version of the figure, please see www.iste.co.uk/khalfallah/analysis1.zip*

2.9. Conclusion

In this chapter, we have illustrated the loads that can be applied to constructions and works. These loads can be grouped into three categories: (1) permanent (dead) loads, (2) live loads and (3) environmental loads.

Permanent loads are characterized by constant intensity and a fixed location. They act permanently during the life of the building.

Live loads have a variable magnitude and/or position and their valuation depends on the use of the construction. Each structural element must be designed according to the position of the operating load, which causes adverse internal actions.

Finally, environmental forces must be considered in the design load quantification. These forces are often wind, snow and earthquakes and their intensities are evaluated using the European standards.

2.10. Problems

For color versions of the figures in the following exercises, please see www.iste.co.uk/khalfallah/analysis1.zip.

Exercise 1

A slab in a residential building with dimensions of 6 × 4 m is subjected to a permanent load of 10 daN/m² (including dead weight).

1) Calculate the equivalent permanent load.

2) Calculate the equivalent operating load.

Exercise 2

The geometric dimensions of the elements of a residential building consisting of two levels are shown in the following figure. Each floor rests on four identical posts.

1) Calculate the compression force due to the permanent load on each post.

2) Calculate the compression force due to the live load on each post.

3) What is the combination of the operating load application that corresponds to the least favorable compression force?

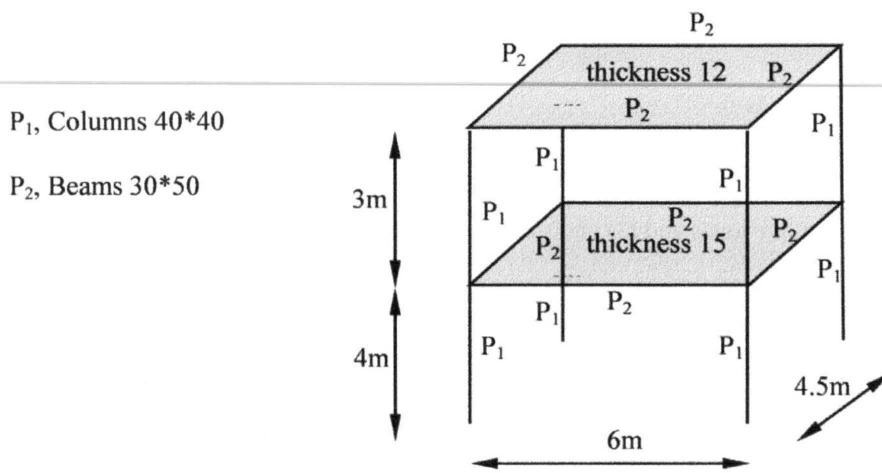

Exercise 3

Repeat Exercise 2 for an identical construction that is for hospital use.

Exercise 4

A beam is subjected to an operating load q_0.

1) Determine the bending moment when:

a) the live load runs along the entire length of the beam;

b) the live load is applied to a single span.

2) Comment on these results.

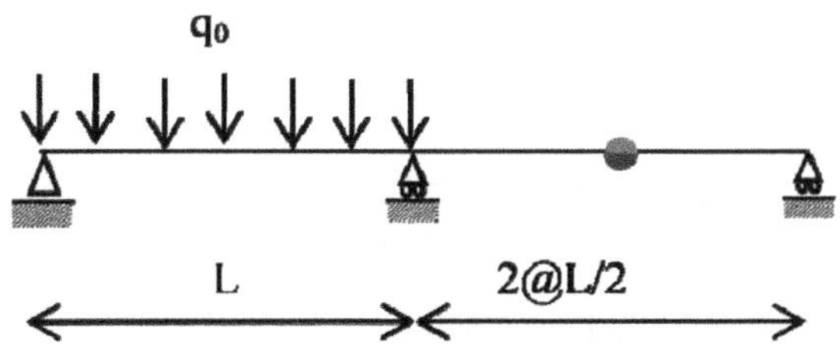

Exercise 5

A building is 20 m long, 15 m deep, 40 m high and has three levels. The building is located in wind zone 4 and on land type III.

1) Draw the wind force diagram.

2) Draw the overturning moment diagram.

Take the other data from Example 2.4.

Exercise 6

Repeat Exercise 5 using an opening area of 5 m² at each level.

Exercise 7

A structure is modeled using the branching model.

1) Write the differential equation of the movement of the mass m.

2) Calculate the pulsation, the frequency and the vibration period.

3) Calculate the shear force at the base.

4) Calculate the static displacement of the mass m.

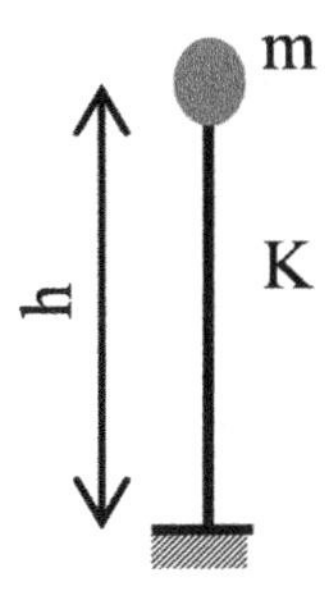

Exercise 8

To find the mathematical model of a structure:

– calculate the matrices of rigidity and mass;

– calculate the natural pulsations of vibration;

– deduce the corresponding periods;

– calculate the shear force at the base;

– deduce the vertical distribution of shear force.

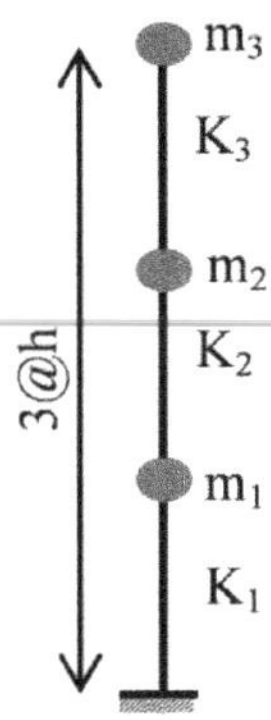

1) Study the following cases:

Case 1: $m_1 = m_2 = m_3$ and $k_1 = k_2 = k_3$

Case 2: $m_1 = 2m_2 = m_3$ and $k_1 = 2k_2 = k_3$

Case 3: $m_1 = 2m_2 = 2m_3$ and $k_1 = 2k_2 = 2k_3$

2) Comment on the results obtained.

Consider a damping coefficient equal to 7%.

Exercise 9

1) Calculate the seismic force acting on the faces of a hospital block consisting of two levels, 14 m in length, 12 m in depth and 6 m in height. The structure is located in an area of medium seismicity. The damping percentage is considered to be 5%.

2) Draw diagrams of the overturning moment and the shear force.

The density of the material used is γ, the sections of the columns and beams have identical dimensions $b \times 3b$ and the damping coefficient is equal to 5%.

Exercise 10

1) A dam retains two liquids with respective densities of ρ_1 and ρ_2 with $\rho_1 < \rho_2$ and heights of h_1 and h_2 (Figure (a)).

a) Draw the hydrostatic pressures diagram.

b) Calculate the resulting force and its application point (consider a 1 m wide band).

c) Deduce the particular case $\rho_1 = \rho_2$.

d) Calculate the overturning moment at the base.

2) Assume that the dam is buried at a depth h_3 and repeat the previous calculations (Figure ((b). Comment on the stability of the dam.

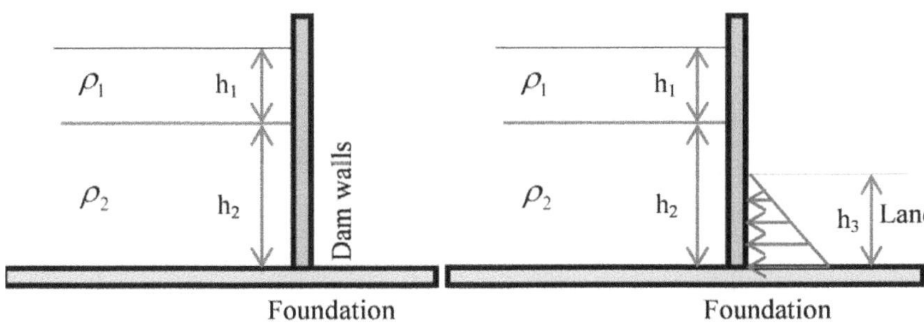

(a) (b)

3

External Analysis of Plane Structures

There are multiple teaching objectives for this chapter:

– the calculation of support reactions using equilibrium equations, matrix methods and the virtual work principle;

– the distinction between internal and external actions;

– differentiation between different types of support;

– the calculation of the degree of static indeterminacy of structures to distinguish between statically determinate, statically indeterminate and unstable structures;

– the advantages of using the superposition of effects principle.

First, we present the concept of structural equilibrium equations. The definition of internal and external actions is described in the second part. The different types of support structures are illustrated to calculate the unknowns of the problem. Matrix methods and the virtual work principle are used to calculate support reactions.

The last part of the chapter describes the superposition of effects principle.

3.1. External equilibrium of structures

A statically determinate structure is in equilibrium if it instantaneously maintains absolute equilibrium when subjected to a set of actions that may be forces and/or moments. In another way, a structure is in equilibrium if and only if the static equations are simultaneously verified. Therefore, for a structure subject to forces and at moments along three directions, the six static equations can be written as

$$\sum F_x = 0 \quad \sum F_y = 0 \quad \sum F_z = 0 \qquad [3.1]$$

$$\sum M_x = 0 \quad \sum M_y = 0 \quad \sum M_z = 0 \qquad [3.2]$$

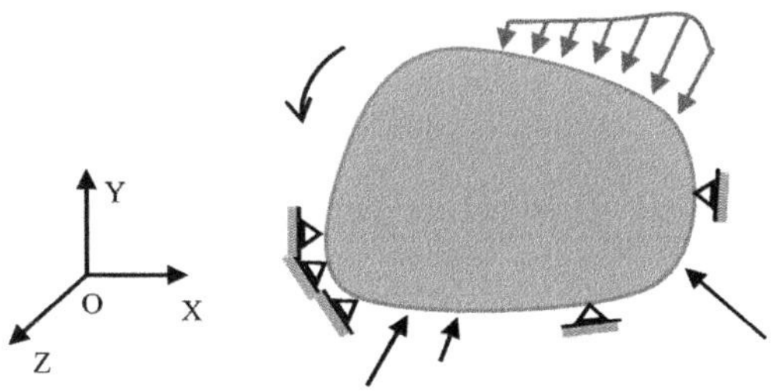

Figure 3.1. *Three-dimensional structure*[1]

If the structure is referred to a single plane (OXY), equations [3.1] and [3.2] become

$$\sum F_x = 0 \qquad [3.3a]$$

$$\sum F_y = 0 \qquad [3.3b]$$

$$\sum M_z = 0 \qquad [3.3c]$$

These relationships are called the static equations of plane structures and when satisfied, the corresponding structure is in equilibrium or stable.

In general, two categories of plane structures can be distinguished according to the directions of the applied forces.

1) The applied forces are concurrent and go through a singular point. In this case, the rotational equilibrium equation ($\sum M_z = 0$) is satisfied and equilibrium equations [3.3] are reduced to

$$\sum F_x = 0 \qquad [3.4a]$$

$$\sum F_y = 0 \qquad [3.4b]$$

1 All of the figures in this chapter are available to view in full color at www.iste.co.uk/khalfallah/analysis.zip.

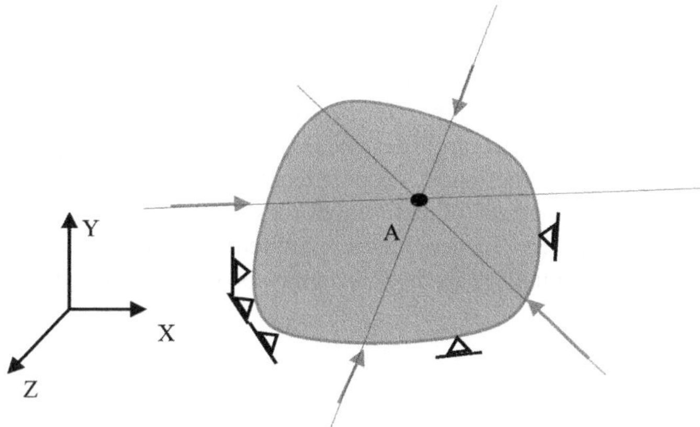

Figure 3.2. *Structure under concurrent forces*

2) When the applied forces are parallel, the sum of applied forces in the normal direction is zero (Figure 3.3). In this case, we write:

$$\sum F_n = 0 \quad [3.5]$$

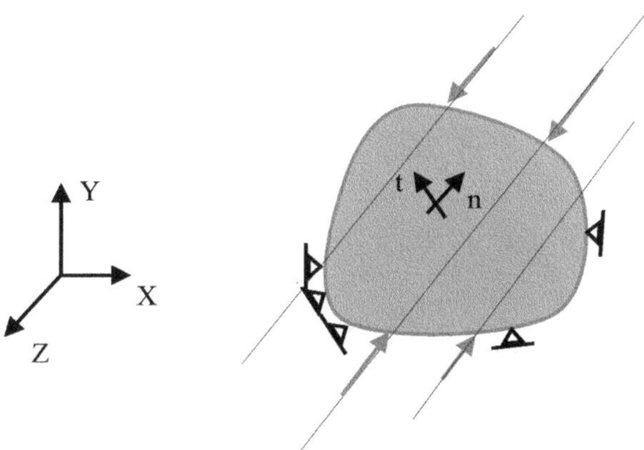

Figure 3.3. *Structure subject to parallel forces*

3.2. External and internal actions

Actions can be forces and/or moments to which a structure is subjected. They are grouped into two categories: external actions and internal actions.

3.2.1. *External actions*

External actions can be support reactions, forces and/or moments exerted by the external environment on the structure. These forces are the loads described in Chapter 2 that tend to deform and destabilize the structure from its stable equilibrium position. On the other hand, support reactions are the actions that occur at support level due to the effect of external forces, whose role is to block the structure's potential movement. In general, support reactions are the unknowns of the problem and their determination leads to the analysis of the proposed structure.

3.2.2. *Internal actions*

In the same way, internal actions are forces and/or moments that occur within the structure. The internal forces occur at any section with equal values but in the opposite direction to those of the adjacent part. Each part of the structure acts on the rest of the structure with the same actions and in the opposite direction (Newton's 3rd law) (Figure 3.4).

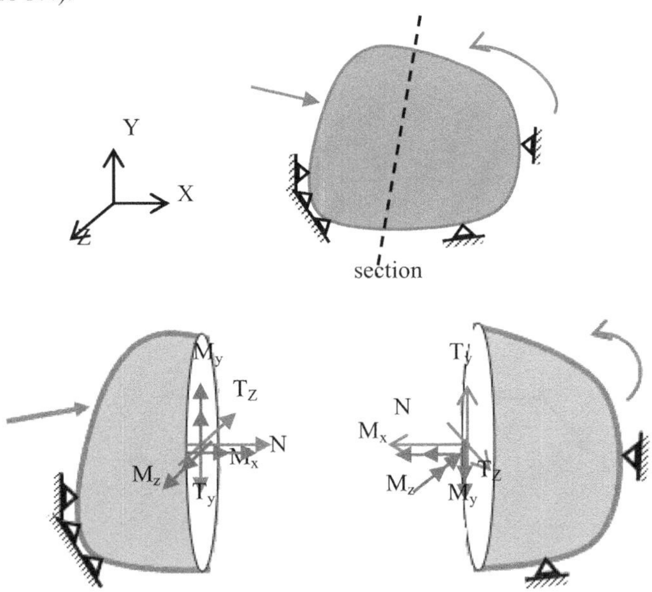

Figure 3.4. *Internal actions*

Internal actions are the unknowns of internal analysis and we use the equilibrium equations of a portion of the structure (the right part or the left part) to determine them.

3.3. Types of plane structure supports

The types of support grouped in Table 3.1 show the nature of the structure's bond with the external environment. Each support is characterized by the kinematic conditions (displacements and/or possible rotations) and at the same time by the components of the support reactions (Table 3.1).

Type of support	Graphic representation	Blocked displacements	Support reactions
Roller	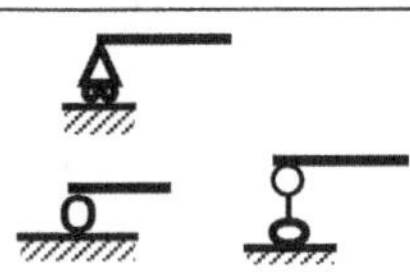	Vertical $\delta_y = 0$	Vertical
Hinge		Vertical and horizontal $\delta_x = \delta_y = 0$	Vertical and horizontal
Support on an inclined plane		Perpendicular and horizontal to the inclined plane $\delta_t = \delta_n = 0$	Perpendicular and horizontal to the inclined plane
Fixed	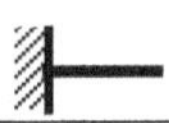	Vertical, horizontal and rotating	Vertical, horizontal and moment
Hinge		Rotation	/

Table 3.1. *Kinematic conditions and support reactions*

3.4. Static determinacy, static indeterminacy and structural stability

3.4.1. *External analysis*

The total number of unknowns of a problem is the number of independent support reactions. So, the staticity or external instability of a structure depends on the number of support reactions (r) and the number of equations that can be constructed (n_e).

To determine the nature of a structure, it is necessary to apply static equations [3.1]. In this case, $n_e = 3$. From this, we can distinguish the following cases:

– If $r = n_e$, the structure is statically determinate externally. The unknowns of the problem can be determined using equilibrium equations. Figure 3.5 shows some statically determinate structures.

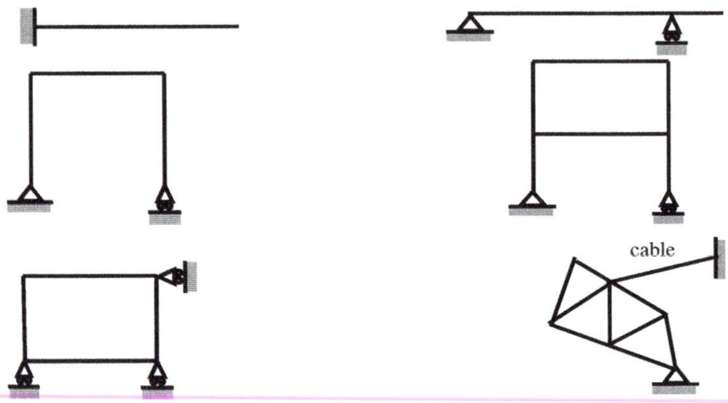

Figure 3.5. *Statically determinate externally of structures*

Each structure in Figure 3.5 shows three support reactions that can be determined using equilibrium equations.

– For $r > n_e$, the number of reactions exceeds the number of static equations, which means that the structure is based on a large number of supports. In this case, support reactions cannot be evaluated by the equilibrium equations and the structure is statically indeterminate externally. The degree of static indeterminacy (f) of a structure is defined by the difference between the number of support reactions and the number of static equations, $f = r - n_e$. Figure 3.6 shows some statically indeterminate plane structures.

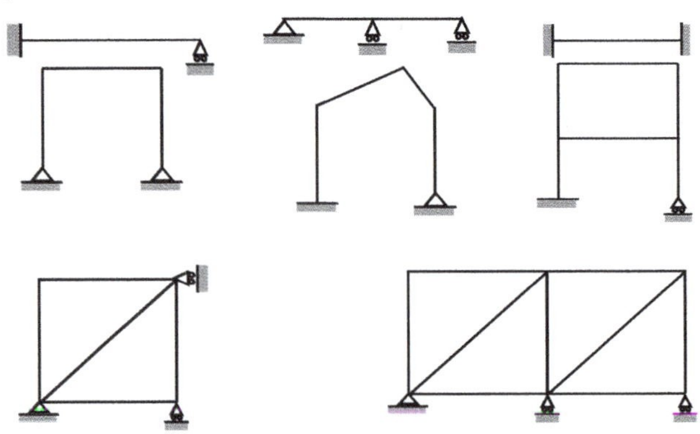

Figure 3.6. *Statically indeterminate externally of structures*

– If $r < n_e$, the number of unknowns is lower than that of equilibrium equations. The structure is supported by fewer supports than n_e. The support reactions are not sufficient to prevent potential structural deformations. Therefore, the structure cannot stay in equilibrium under the effect of the applied loading system. Figure 3.7 shows some unstable structures.

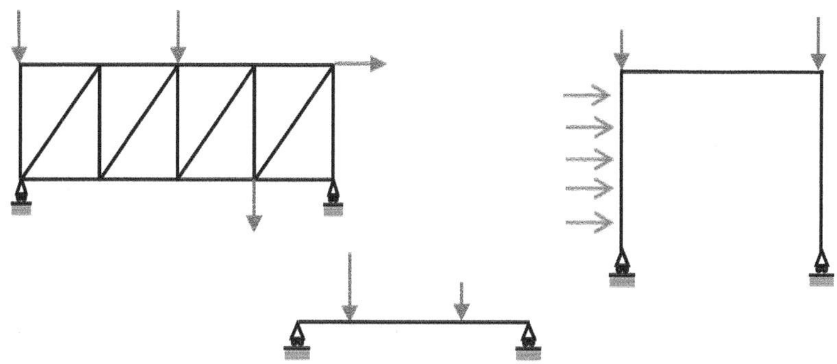

Figure 3.7. *Unstable structures*

The support reactions prevent transverse movement of the structure but the structure can move horizontally. For this reason, the structure is unstable externally.

3.4.2. *Internal analysis*

For beam or frames, the number of internal unknowns is equal to 3 if they are plane. These are a normal force, a tangential force and a bending moment. In this case, the three static equations suffice to determine the internal actions at any section of the structure.

The problem is quite the opposite for trusses. To determine the nature of a truss having b bars, n joints and r support reactions, it is necessary to compare the number of unknowns, with the number of independent equations.

The truss is statically determinate internally if

$$b-(2n-3)=0 \qquad [3.6]$$

It is statically indeterminate internally and has f times degrees of internal static indeterminacy if

$$f=b-(2n-3)\succ 0 \qquad [3.7]$$

And lastly, it is unstable when

$$f = b - (2n - 3) \prec 0 \qquad [3.8]$$

EXAMPLE 3.1.–

Specify the nature of the trusses (Figure 3.8).

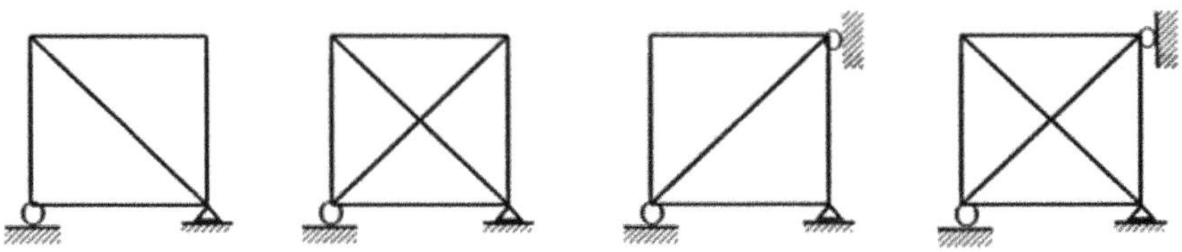

(a) (d) (c) (d)

Figure 3.8. *Trusses*

System	(a)	(b)	(c)	(d)
b	5	6	5	6
2n-3	5	5	5	5
r	3	3	4	4
f	0	1	0	1

System	Internal analysis	External analysis
(a)	Statically determinate	Statically determinate
(b)	Once statically indeterminate	Statically determinate
(c)	Statically determinate	Once statically indeterminate
(d)	Once statically indeterminate	Once statically indeterminate

EXAMPLE 3.2.–

Determine the nature of the trusses (Figure 3.9).

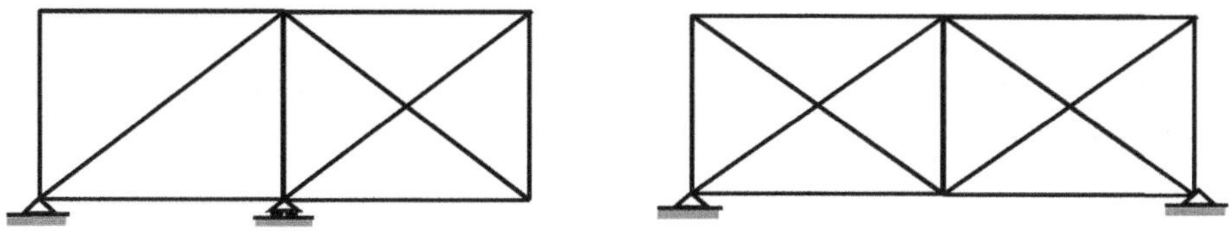

(i) (ii)

Figure 3.9. *Given structures*

– System (i)

b = 10, 2n – 3 = 9 and r = 3

Therefore, the system (i) is statically determinate externally and once statically indeterminate internally.

– System (ii)

b = 11, 2n – 3 = 9 and r = 4

The system (ii) is once statically indeterminate externally and twice statically indeterminate internally.

3.5. Calculation of support reactions

Solving the constructed equations is the only way of calculating support reactions. For structures that are statically determinate externally, static equations are applied to determine the support reactions. But for statically indeterminate structures, the equilibrium equations are not sufficient to determine the unknowns of the problem.

The following procedure can be applied successfully to calculate support reactions of a structure, which is statically determinate externally and subjected to a set of loads.

Representation of a free-body diagram

– The given structure has the plane reference (OXY).

– The structure is isolated from its supports or any attached structural body, which results in a free-body diagram.

– All applied forces and moments are positioned by indicating their directions and intensities.

– Instead of removing supports, we use equivalent actions. The actions considered depend on the nature of the support removed (Table 3.1).

Checking the structure's static determinacy

Before beginning the analysis, it is necessary to check that the structure is statically determinate externally. If this is not the case, we must to stop the analysis at this phase.

Application of equilibrium equations

The three static equations are used to determine the support reactions. If the analysis leads to a negative intensity of a reaction or a moment, it is necessary to change the direction. In the opposite case, the selected direction is maintained.

EXAMPLE 3.3.– Beam

Calculate the structural support reactions (Figure 3.10):

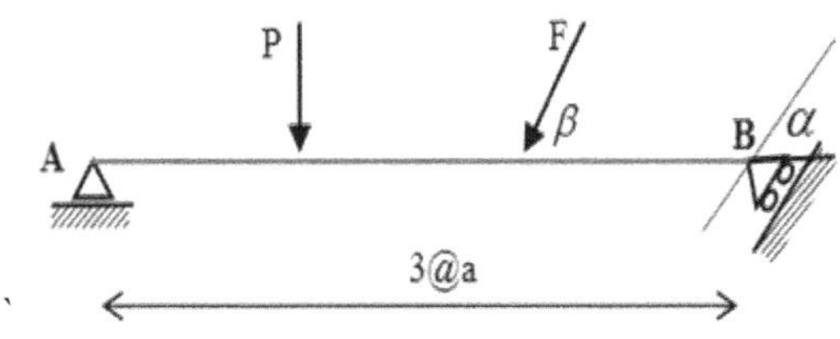

Figure 3.10. *Given beam*

– Stage 1: Construction of the free-body diagram (Figure 3.11).

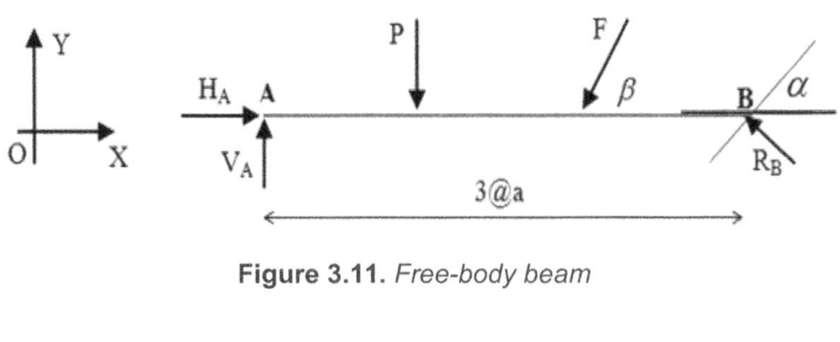

Figure 3.11. *Free-body beam*

– Stage 2: Calculation of reactions

The unknowns of the problem are H_A, V_A and R_B.

The beam is in equilibrium if and only if:

$$\sum \vec{F} = 0$$

$$\overrightarrow{H_A} + \vec{V}_A + \vec{R}_B + \vec{P} + \vec{F} = \vec{0}$$

The projection of forces according to the axis (XX):

$$H_A - R_B \sin\alpha - F\cos\beta = 0$$

The projection of forces according to the axis (YY):

$$V_A + R_B \cos\alpha - F\sin\beta - P = 0$$

We have two equations with three unknowns. To solve the problem, it is necessary to formulate a third equation. It is

$$\sum \vec{M}_A = 0 \ :$$

$$\overrightarrow{AA} \wedge \vec{V}_A + \overrightarrow{AA_1} \wedge \vec{P} + \overrightarrow{AA_2} \wedge \vec{F} + \overrightarrow{AB} \wedge \vec{R}_B = 0$$

$$0 + a\vec{i} \wedge (-P.\vec{j}) + 2a.\vec{i} \wedge (-F\sin\beta.\vec{j}) + 3a.\vec{i} \wedge R_B \cos\alpha.\vec{j} = 0$$

$$(-Pa - 2F.a.\sin\beta + 3R_B a\cos\alpha).\vec{k} = \vec{0}$$

or

$$-Pa - 2F.a.sin\beta + 2R_B acos\alpha = 0$$

From the last equation, the support reaction at joint B is

$$R_B = \frac{P + 2F.\sin\beta}{3.\cos\alpha}$$

The horizontal reaction at support A can be calculated by

$$H_A = \frac{P + 2F.\sin\beta}{3.\cos\alpha}.\sin\alpha + F.\cos\beta$$

Similarly, the vertical reaction of support A is

$$V_A = P + F.\sin\beta - \frac{P + 2F.\sin\beta}{3.\cos\alpha}$$

In this particular case, for F = P and $\alpha = \beta$, the previous equations give

$$R_B = \frac{1 + 2\sin\alpha}{3.\cos\alpha}P$$

$$H_A = \frac{2 + \sin\alpha + \cos^2\alpha}{3.\cos\alpha}P$$

$$V_A = \frac{(3.\cos\alpha - 1).(1 + \sin\alpha) - \sin\alpha}{3.\cos\alpha}P$$

EXAMPLE 3.4.– Frame

Calculate support reactions of the frame (Figure 3.12).

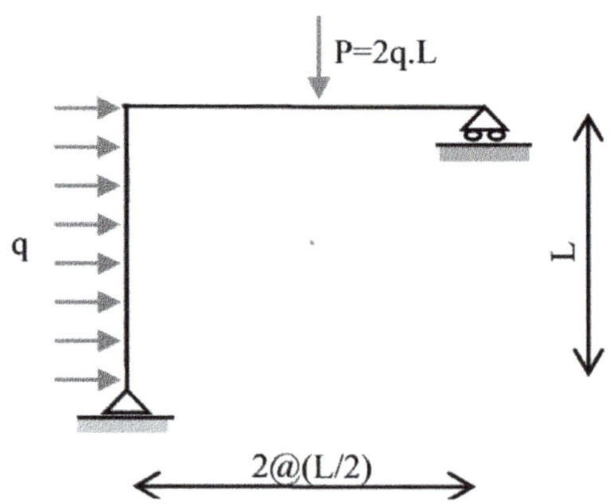

Figure 3.12. *Given frame*

– Stage 1: The free-body diagram is shown in Figure 3.13.

– Stage 2: Calculate the support reactions.

The unknowns of the problem are H_A, V_A and V_C.

The beam is in balance, and we can write:

$$\sum \vec{F} = 0 \ : \ \overrightarrow{H_A} + \vec{V}_A + \vec{V}_c + \vec{P} + \vec{q}L = \vec{0}$$

The projection of forces according to the axis (XX):

$H_A + qL = 0 \quad : H_A = -qL$

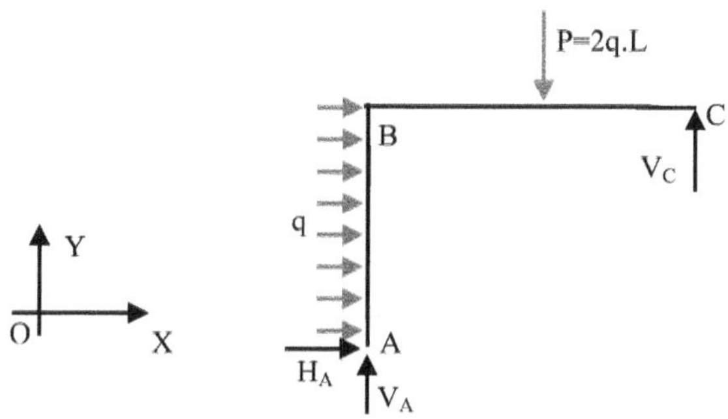

Figure 3.13. *Free-body diagram*

The projection of forces in relation to the axis (YY):

$V_A + V_c - P = 0$

$$\sum \overrightarrow{M}_A = 0 \ : \ V_C L - P.\frac{L}{2} - q.L \ \frac{L}{2} = 0$$

The use of these equations leads to the calculation of the support reactions:

$$V_C = \frac{3}{2} q.L$$

$$V_A = P - V_C = \frac{1}{2} q.L$$

EXAMPLE 3.5.– Statically indeterminate frame

Calculate support reactions of the frame (Figure 3.14): the frame is subjected to a uniformly distributed load of magnitude q and a concentrated force is applied at the middle point on the inclined bar.

In this example, there are four unknowns with three static equations. An additional equation can be added. It is $\sum \vec{M}_{hinge} = \vec{0}$. So, the problem becomes solvable. A joint is placed at point B.

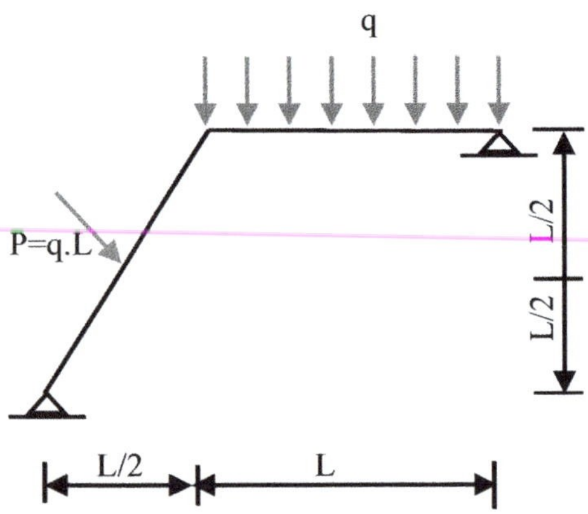

Figure 3.14. *Statically indeterminate frame*

– Stage 1: Construction of the free-body diagram (Figure 3.15).

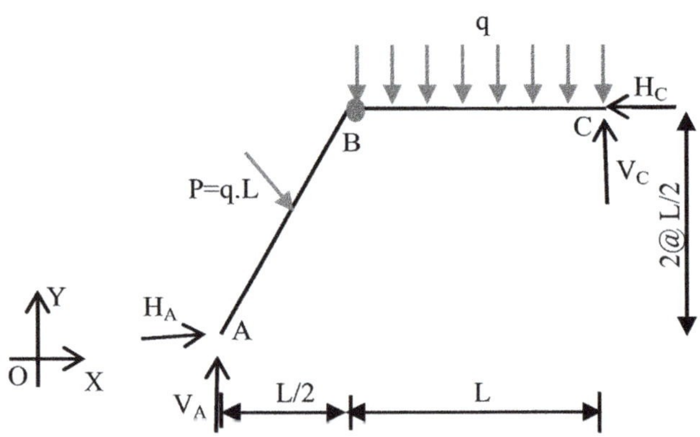

Figure 3.15. *Free-body frame*

We apply the equilibrium equations:

$$\sum \overrightarrow{M}_B = 0 \;:\; V_c\, L - qL.(\frac{L}{2})=0$$

$$V_c = \frac{1}{2}\, qL$$

$$\sum \overrightarrow{F}_y = 0 \;:\; V_A + V_c - qL- \frac{\sqrt{5}}{5}\, P =0$$

$$V_A = \frac{5+2\sqrt{5}}{10}\, qL$$

$$\sum \overrightarrow{M}_A = 0 \;:\; H_c\, L- qL^2- \frac{\sqrt{5}}{4}\, PL =0$$

$$H_c = \frac{\sqrt{5}+4}{4}\, qL$$

$$\sum \overrightarrow{F}_X = \vec{0} \;:\; H_A - H_C + \frac{2\sqrt{5}}{5}\, P =0$$

$$H_A = \frac{20-3\sqrt{5}}{20}\, qL$$

EXAMPLE 3.6.– Truss

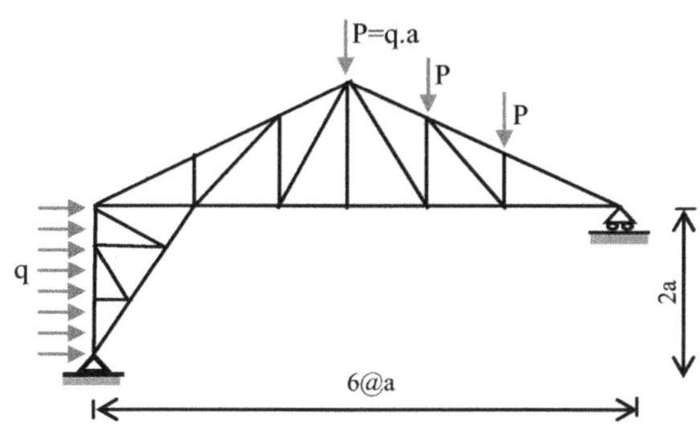

Figure 3.16. *Truss*

Calculate support reactions of the truss (Figure 3.14):

– Stage 1: Construction of the free-body diagram (Figure 3.17).

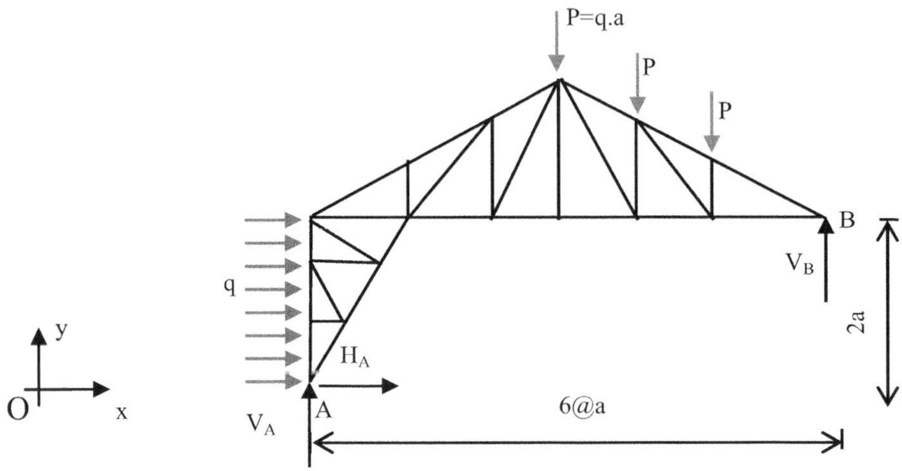

Figure 3.17. *Free-body diagram*

The rotational equilibrium of the truss in relation to point A leads to

$$\sum \vec{M}_A = 0 \ : \ V_B = \frac{5}{6} qa$$

$$\sum \vec{F}_y = 0 \ : \ V_A + V_B = 3P = 3qa$$

By substituting V_A in the last relation, we obtain

$$V_A = \frac{13}{6} qa$$

$$\sum \vec{F}_x = 0 \ : \ H_A + 2qa = 0$$

$$H_A = -2qa$$

3.6. Superposition principle

The superposition principle of forces is among the most used hypotheses in structural analysis and strength of materials. In linear elasticity, the effect caused by a set of actions simultaneously applied to a structure is equal to the sum of the effect of each action. The superposition principle facilitates greatly the analysis of structures subject to multiple external actions. This principle can be used for structures that satisfy the following conditions: (1) the deflections of the structures are considered to be small and the equilibrium equations can be established on the non-deformed geometry of the structure, and (2) the material constituting the structure has linear elastic behavior.

Consider a beam subjected to a uniformly distributed load of magnitude q and a concentrated force P as shown in Figure 3.18.

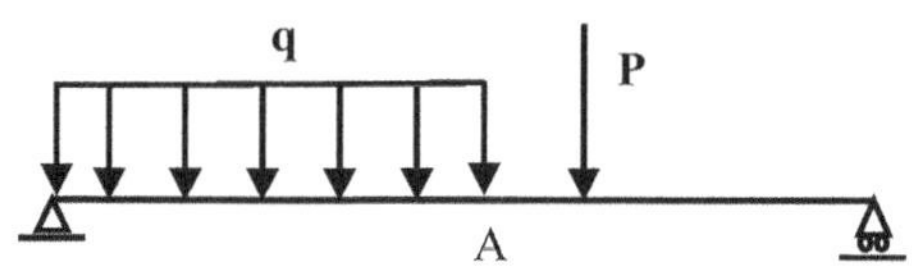

Figure 3.18. *Beam under multiple loads*

For example, to calculate the displacement at point A, the superposition principle can be used by decomposing the effect of each force (Figure 3.19).

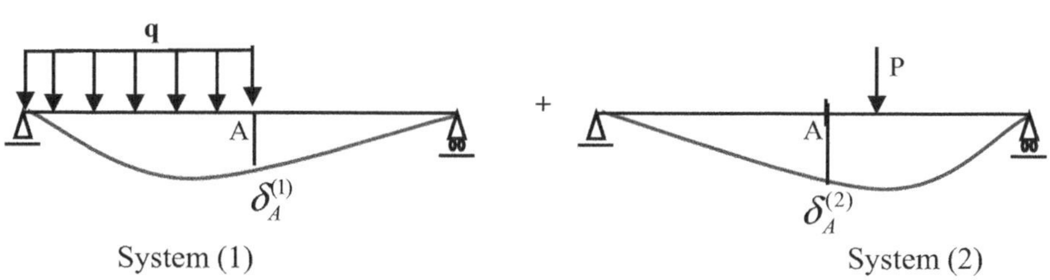

Figure 3.19. *Beam deflections*

The vertical displacement at point A is the algebraic sum of the displacements of two systems.

$$\delta_A = \delta_A^{(1)} + \delta_A^{(2)} \quad [3.9]$$

$\delta_A^{(1)}$ and $\delta_A^{(2)}$ are the vertical displacements due to the uniformly distributed load q and to the concentrated force P, respectively (Figure 3.19).

3.7. Matrix formulation

Due to the availability of computing machines, the matrix analysis of structures has become a popular task. It may be noted that the matrix method cannot offer any advantage to simple problems. It is very successful in the case of large or complex structures.

Consider a beam (Figure 3.20) in which there are three unknowns.

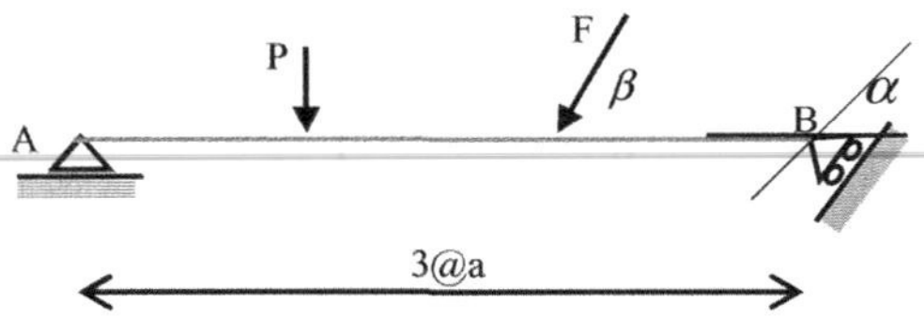

Figure 3.20. *Statically determinate beam*

The applied forces are divided and the structure relates to an axis system (OXY) (Figure 3.21).

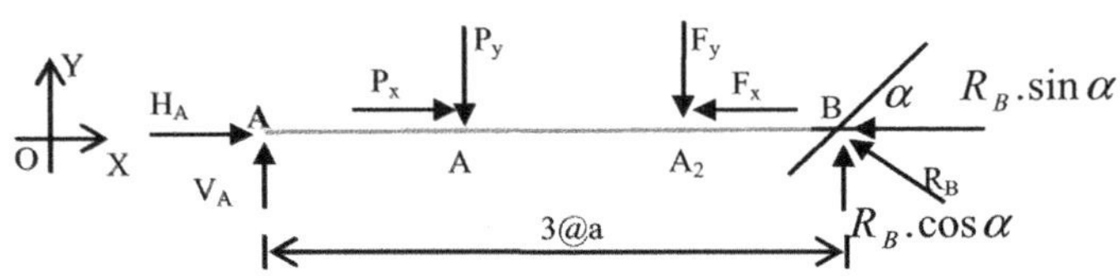

Figure 3.21. *Free-body beam*

The equilibrium equations are:

$$\sum \vec{F}_x = \vec{0} : H_A + P_x - F_x - R_B \sin\alpha = 0$$

$$\sum \vec{F}_y = \vec{0} : V_A + R_B \cos\alpha - F_y - P_y = 0$$

$$\sum \overrightarrow{M}_A = \vec{0} : - P_y a \ - \ 2F_y .a + 3R_B a\cos\alpha = 0$$

In matrix form, these equations can be written as

$$\begin{bmatrix} 1 & 0 & -\sin\alpha \\ 0 & 1 & \cos\alpha \\ 0 & 0 & 3\cos\alpha \end{bmatrix} \begin{Bmatrix} H_A \\ V_A \\ R_B \end{Bmatrix} = \begin{bmatrix} -1 & 0 & 1 & 0 \\ 0 & 1 & 0 & 1 \\ 0 & 1 & 0 & 2 \end{bmatrix} . \begin{Bmatrix} P_x \\ P_y \\ F_x \\ F_y \end{Bmatrix}$$

In a compacted form, the previous equation is written as

$$[A]\{R\} = [S].\{F\},$$

where $\{R\}$ and $\{F\}$ are the support reactions vector and the applied forces, respectively.

The support reactions vector can be deduced by:

$$\{R\} = [A]^{-1}[S].\{F\}$$

We put:

$$\{B\} = [A]^{-1}[S]$$

$$\{R\} = [B].\{F\}$$

The support reactions vector is:

$$\begin{Bmatrix} H_A \\ V_A \\ R_B \end{Bmatrix} = \begin{bmatrix} 1 & 0 & -\sin\alpha \\ 0 & 1 & \cos\alpha \\ 0 & 0 & 3.\cos\alpha \end{bmatrix}^{-1} \begin{bmatrix} -1 & 0 & 1 & 0 \\ 0 & 1 & 0 & 1 \\ 0 & 1 & 0 & 2 \end{bmatrix} . \begin{Bmatrix} P_x \\ P_y \\ F_x \\ F_y \end{Bmatrix}$$

In the particular case $\alpha = \beta = \dfrac{\pi}{4}$, the support reactions vector becomes:

$$\begin{Bmatrix} H_A \\ V_A \\ R_B \end{Bmatrix} = \begin{bmatrix} -1.0 & 0.334 & 1.0 & 0.667 \\ 0 & 0.667 & 0 & 0.334 \\ 0 & 0.471 & 0 & 0.943 \end{bmatrix} . \begin{Bmatrix} P_x \\ P_y \\ F_x \\ F_y \end{Bmatrix}$$

We substitute the values of the external forces in this relation to get:

$$\begin{Bmatrix} H_A \\ V_A \\ R_B \end{Bmatrix} = \begin{bmatrix} -1.0 & 0.334 & 1.0 & 0.667 \\ 0 & 0.667 & 0 & 0.334 \\ 0 & 0.471 & 0 & 0.943 \end{bmatrix} . \begin{Bmatrix} 0 \\ 1 \\ 0.707 \\ 0.707 \end{Bmatrix} p = \begin{Bmatrix} 1.5117 \\ 0.9023 \\ 1.1381 \end{Bmatrix} p$$

From which, the support reactions are:

$$H_A = 1.512P, \quad V_A = 0.902P \quad \text{and} \quad R_B = 1.38P$$

EXAMPLE 3.7.– Statically indeterminate beam

Using the matrix method, calculate the structural support reactions (Figure 3.22).

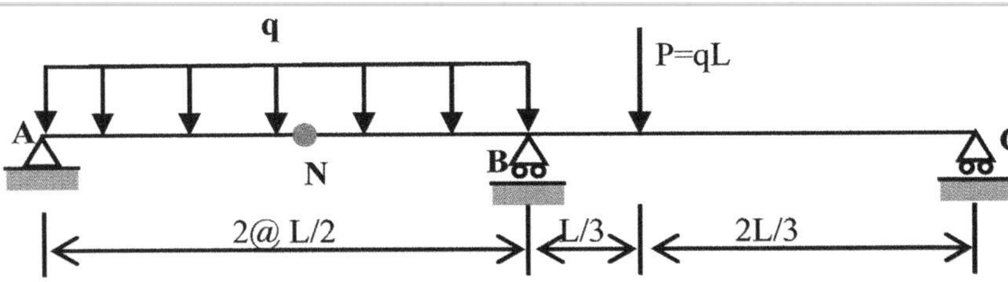

Figure 3.22. *Given beam*

The unknowns of the problem are V_A, H_A, V_B and V_c.

In this case, we have three static equations and a hinge joint allows us to add an additional equation ($\sum \overrightarrow{M}_N = \vec{0}$). In this case, the number of unknowns equals that of the equations.

The vector of external loads is:

$$\{F\} = \begin{Bmatrix} q \\ P_x \\ P_y \end{Bmatrix}$$

The free-body diagram is shown in Figure 3.23.

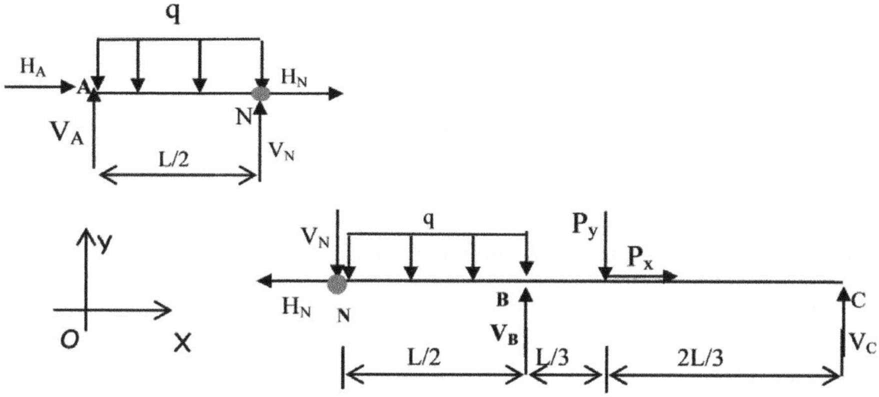

Figure 3.23. *Beam decomposition*

The application of equilibrium equations, $\sum \overrightarrow{M}_C = \vec{0}$, $\sum \vec{F}_x = \vec{0}$, $\sum \vec{F}_y = \vec{0}$ and $\sum \overrightarrow{M}_N = \vec{0}$ allows writing the following matrix

$$\begin{bmatrix} 0 & 2L & L & 0 \\ 1 & 0 & 0 & 0 \\ 0 & 1 & 1 & 1 \\ 0 & \frac{L}{2} & 0 & 0 \end{bmatrix} \begin{Bmatrix} H_A \\ V_A \\ V_B \\ V_C \end{Bmatrix} = \begin{bmatrix} 0 & \frac{2}{3}L & \frac{3}{2}L^2 \\ 0 & 0 & 0 \\ 0 & 1 & L \\ 0 & 0 & \frac{1}{8}L^2 \end{bmatrix} \cdot \begin{Bmatrix} P_{1x} \\ P_{1y} \\ q \end{Bmatrix}$$

The matrix relationship is written as

$$\begin{Bmatrix} H_A \\ V_A \\ V_B \\ V_C \end{Bmatrix} = \begin{bmatrix} 0 & 2L & L & 0 \\ 1 & 0 & 0 & 0 \\ 0 & 1 & 1 & 1 \\ 0 & \frac{L}{2} & 0 & 0 \end{bmatrix}^{-1} \begin{bmatrix} 0 & \frac{2}{3}L & \frac{3}{2}L^2 \\ 0 & 0 & 0 \\ 0 & 1 & L \\ 0 & 0 & \frac{1}{8}L^2 \end{bmatrix} \cdot \begin{Bmatrix} P_{1x} \\ P_{1y} \\ q \end{Bmatrix}$$

For L = 3 m, P_{1x} = 0, P_{1y} = 70 KN and q = 20 KN/mL, this relationship becomes:

$$\begin{Bmatrix} H_A \\ V_A \\ V_B \\ V_C \end{Bmatrix} = \begin{bmatrix} 0 & 0 & 0 \\ 0 & 0 & 0.75 \\ 0 & 0.667 & 3 \\ 0 & 0.334 & -0.75 \end{bmatrix} \cdot \begin{Bmatrix} P_{1x} \\ P_{1y} \\ q \end{Bmatrix}$$

The support reactions have the following values:

$H_A = 0$

$V_A = 15$ KN

$V_B = 106.667$ KN

$V_c = 8.334$ KN

3.8. Concept of displacement

All structures deform under the effect of applied forces according to the law governing the behavior of the structure's material. So, every point on the structure moves from its initial position to another position that differs from the first position. The distance between the position points is called "displacement", which generates a deflection. It can be an axial deflection or an angular deflection (Figure 3.24).

In this context, the behavior of the structures is linear elastic and the structure can be modeled by a stiffening spring equal to that of the structure itself (Figure 3.25).

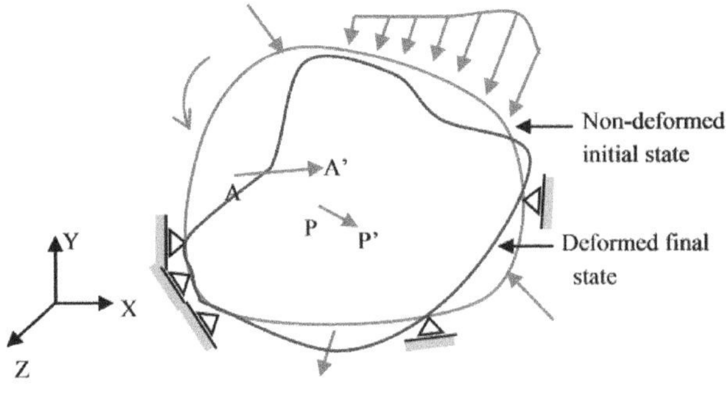

Figure 3.24. *Deformation of a structure*

In linear elasticity, the one-dimensional behavior of the spring is given by the relation [3.10].

$$F = K_s.\ \Delta \quad [3.10]$$

For three-dimensional problems, the relation [3.10] is written in matrix form by

$$\{F\} = [K].\{q\} \quad [3.11]$$

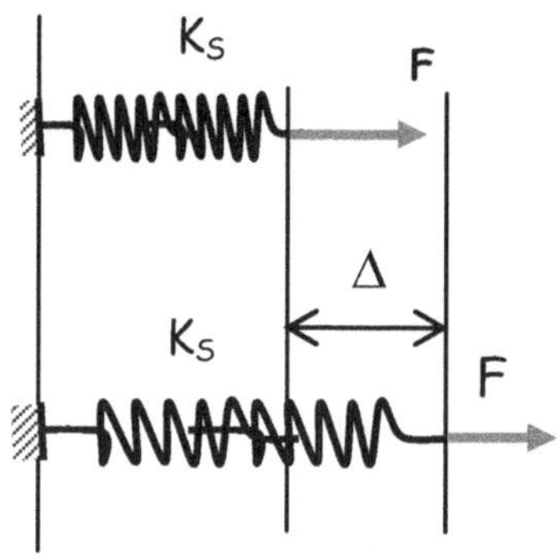

Figure 3.25. *"Spring" mechanical model*

where $\{F\}$ and $\{q\}$ are, respectively, the vector of the applied forces and the vector of the corresponding displacements, and $[K]$ is the rigidity matrix of the structure. The displacements are calculated by

$$\{q\} = [K]^{-1}\{F\} \quad [3.12]$$

3.9. Concept of work

The "work of actions" concept is particularly important in structural analysis, often using energy methods. With in this concept, we define the elementary work of a force F which generates an elementary displacement $d\Delta$ by the scalar product.

$$dW = \vec{F}.\overrightarrow{d\Delta} \quad [3.13]$$

Knowing that $\vec{F}$ and $\overrightarrow{d\Delta}$ have the same direction, the total work of the force causing a displacement Δ is

$$W = \int_0^{\Delta} F.d\Delta \qquad [3.14]$$

where F depends on the displacement Δ. In this case, the total work to move the structure of a quantity Δ is the hatched portion between the curve $F = f(\Delta)$ along $[0, \Delta]$ (Figure 3.26).

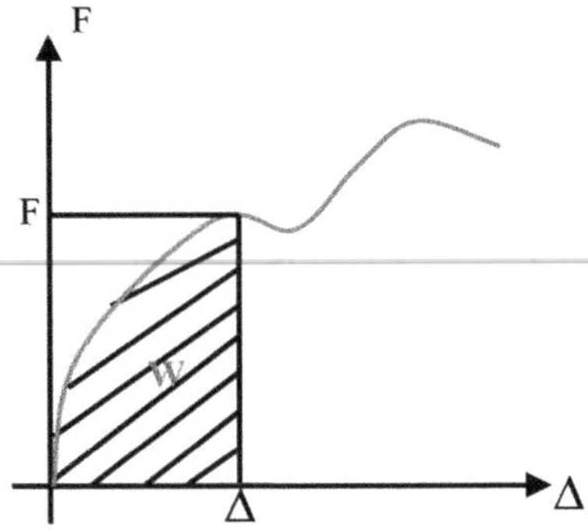

Figure 3.26. *Force–displacement relationship*

If the force is independent of the displacement generated (Figure 3.27), the total work of the force F is written as

$$W = F.\Delta \qquad [3.15]$$

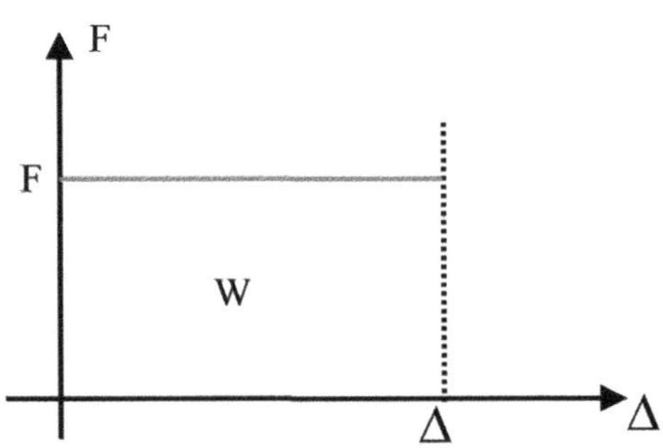

Figure 3.27. *Force and displacement are independent*

The applied force is usually proportional to the displacement (Figure 3.28); the total work of the force F is:

$$W = \int_0^{\Delta} F.d\Delta \qquad [3.16]$$

Knowing that F = k. Δ, the relationship [3.16] is written as:

$$W = \frac{1}{2} K\Delta^2 \qquad [3.17]$$

or:

$$W = \frac{1}{2} F\Delta \qquad [3.18]$$

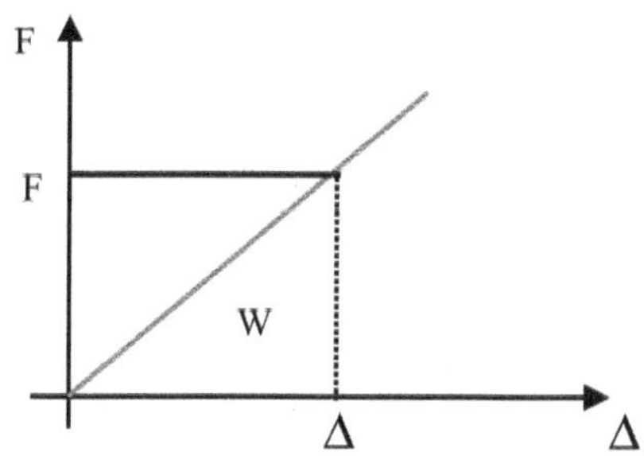

Figure 3.28. *Linear behavior law*

Generally, the structure is subjected to n forces generating n independent displacements; the total work of the forces applied using the superposition principle is written as:

$$W = \sum_{i=1}^{n} \int_0^{\Delta_i} F_i .d\Delta_i = \frac{1}{2}\sum_{i=1}^{n} F_i .\Delta_i = \frac{1}{2}\{F\}^t .\{\Delta\} \qquad [3.19]$$

When the forces are independent of the generated displacements, the work of the forces is:

$$W = \sum_{i=1}^{n} F_i .\Delta_i = \{F\}^t .\{\Delta\} \qquad [3.20]$$

3.10. Principle of virtual work

The principle of virtual work is described for the first time by John Bernouilli in 1717. It can be stated as follows: when a rigid body is in equilibrium under a set of forces and if it is subjected to virtual displacement, the virtual work of external forces is zero.

To clarify this principle, consider a beam subjected to an external force inclined at an angle α (Figure 3.29(a)). The beam is associated with virtual displacement that allows it to move from its initial position (AB) to another position (A'B') (Figure 3.29(c)). The total virtual displacement of the beam can be divided into a horizontal displacement δ_{vx}, a vertical displacement δ_{vy} and a rotation around point A, ω_{vA}. As the beam (AB) has undergone virtual displacement from the initial configuration to another final position, the forces applied therein produce virtual work.

The total virtual work, W_{vt}, can be expressed as the sum of the virtual work W_{vx} and W_{vy} during the displacements δ_{vx} and δ_{vy} and the virtual work W_{vr} during the beam's rotation.

$$W_{vt} = W_{vx} + W_{vy} + W_{vr} \qquad [3.21]$$

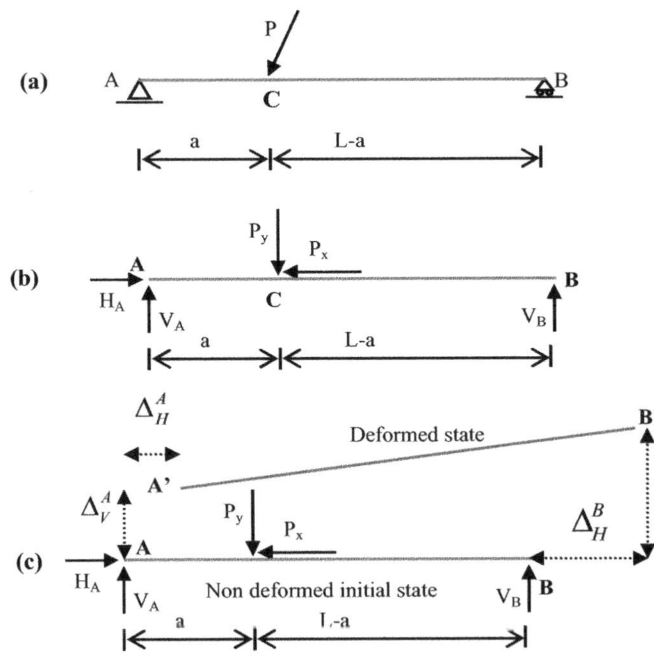

Figure 3.29. *Displacements of the beam*

During virtual displacements δ_{vx} and δ_{vy}, the virtual work of forces is given:

– the virtual work of horizontal forces:

$$W_{vx} = H_A.\ \delta_{vx} - P_x.\ \delta_{vx} = (H_A - P_x)\ \delta_{vx} = (\sum F_x)\delta_{vx} \qquad [3.22]$$

– the virtual work of vertical forces:

$$W_{vy} = V_A\delta_{vy} - P_y\delta_{vy} + V_B\delta_{vy} = (V_A - P_y + V_B)\delta_{vy} \qquad [3.23]$$

– the virtual work of moments:

$$W_{yr} = V_B L\omega_{vA} - P_y a\omega_{vA} = (V_B L - P_y a)\omega_{vA} = (\sum M_A)\omega_{vA} \qquad [3.24]$$

Using the superposition principle, the total virtual work is written as:

$$W_{vt} = (\sum F_x)\delta_{vx} + (\sum F_y)\delta_{vy} + (\sum M_A)\omega_{vA} \qquad [3.25]$$

The beam is in equilibrium, $(\sum F_x) = 0$, $(\sum F_y) = 0$ and $(\sum M_A) = 0$, this means that the total virtual work is zero.

$$W_{vt} = 0 \qquad [3.26]$$

3.11. Calculation of reactions by the virtual work method

We have already mentioned above that any structure is in equilibrium if the virtual work of applied forces is zero. This statement can be used to determine the support reactions of statically determinate structures.

The beam is under the effect of the external load system (Figure 3.30).

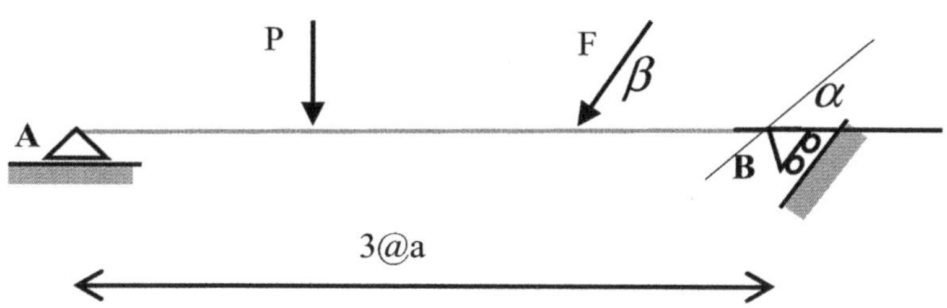

Figure 3.30. *Given beam*

The beam is isolated from its supports; the support reactions are shown schematically in Figure 3.31.

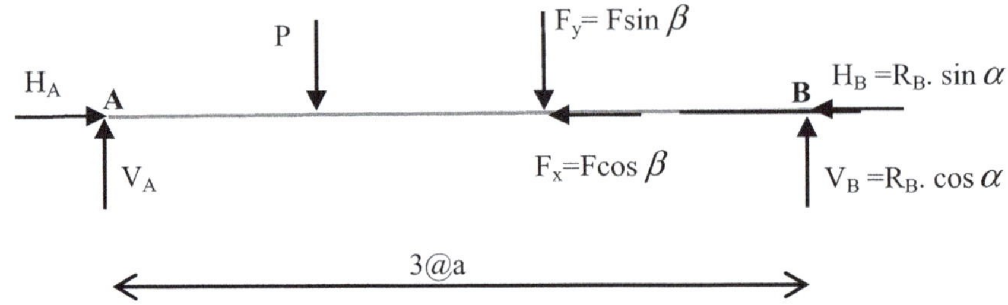

Figure 3.31. *Free-body beam*

Using the superposition principle allows us to break down the forces on the beam.

1) Horizontal movement: we consider the effect of horizontal forces producing a rigid body movement accompanied by virtual displacement δ_{vx} (Figure 3.32).

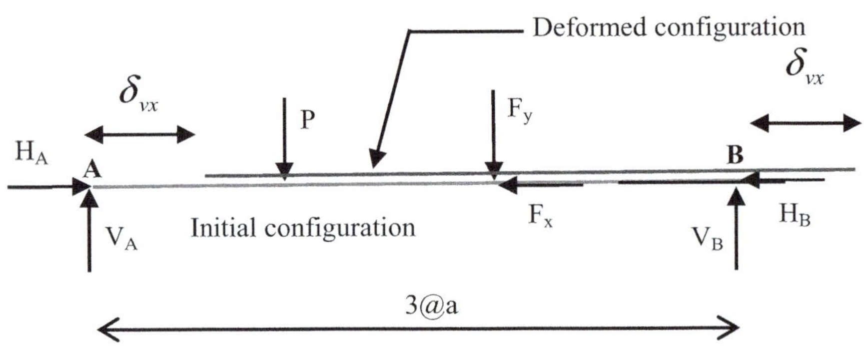

Figure 3.32. *Effect of horizontal forces*

$$W_{vx} = (\sum F_x)\delta_{vx} = (H_A - F_x - H_B)\ \delta_{vx} = 0$$

$$H_A - F_x - H_B = 0$$

2) This time we consider the effect of vertical forces (Figure 3.33).

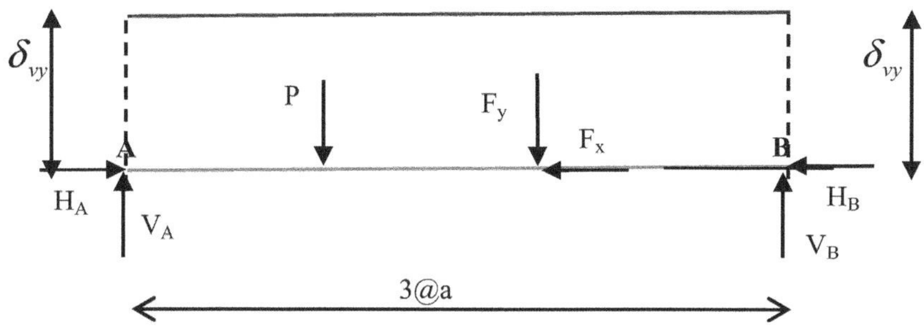

Figure 3.33. *Effect of vertical forces*

$$W_{vy} = (\sum F_y)\delta_{vy} = (V_A - P_y - F_y + V_B)\ \delta_{vy} = 0$$

$$V_A - P_y - F_y + V_B = 0$$

3) In the same way, the effect of moments is represented in Figure 3.34.

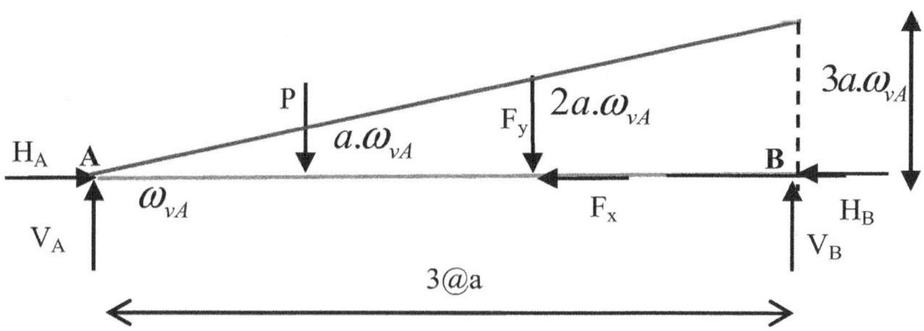

Figure 3.34. *Effect of moments*

$$W_{vr} = (\sum M_A).\omega_{vA} = V_B.3a.\ \omega_{vA} - F_y.2a.\ \omega_{vA} - P_y.a.\ \omega_{vA} = (3a.V_B - 2a.F_y - a.P_y).\ \omega_{vA} = 0$$

or:

$$3V_B - 2F_y - P_y = 0$$

Solving the equations allows us to calculate the support reactions.

$$V_A = \frac{2P_y + F_y}{3}$$

$$H_A = F_x + \frac{2F_y + P_y}{3} tg\alpha$$

$$R_B = \frac{2F_y + P_y}{3} tg\alpha$$

3.12. Conclusion

In this chapter, we have seen that any structure is in equilibrium under applied actions when the equilibrium equations [3.1] and [3.2] are fulfilled.

Especially for plane structures, equations [3.3] are sufficient to calculate the unknowns of this structure.

A structure is said to be statically indeterminate externally if the number of support reactions r is greater than 3. It is said to be statically determinate externally if $r = 3$. Lastly, it is unstable externally if $r < 3$.

The degree of static indeterminacy externally can be deduced by the formula $f = r - 3$.

The procedure for determining support reactions is presented using structural equilibrium techniques.

Similarly, the superposition principle can be used to evaluate the response of a structure submitted to a set of actions. The final structural solution is obtained by the algebraic summation of each force.

Using the static equations, the matrix method and the virtual work method, we can calculate structural support reactions.

3.13. Problems

Exercise 1

Determine the nature of the following structures:

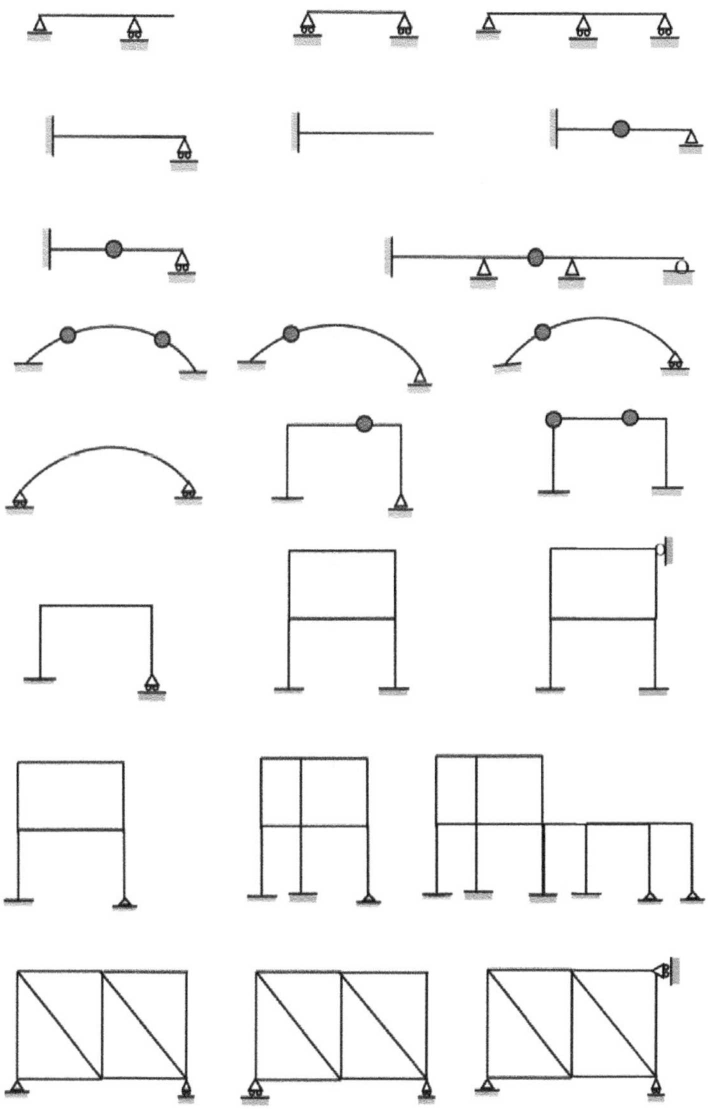

Exercise 2

Calculate the support reactions of the following beams:

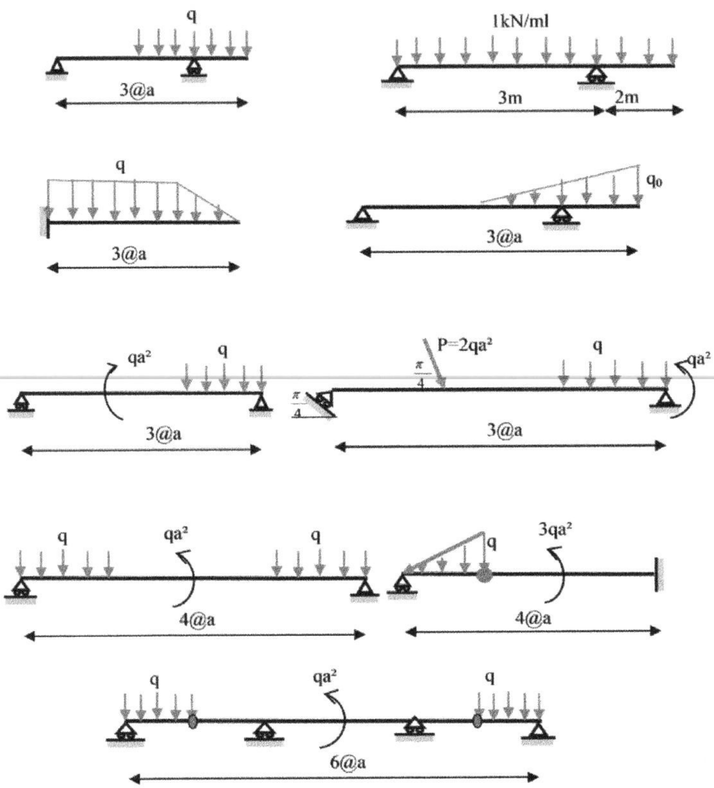

Exercise 3

A vehicle weighing $W = 20$ kN crosses a bridge 15 m long at a constant speed. The vehicle is represented by a constant force W located at a distance x from the left side of the bridge.

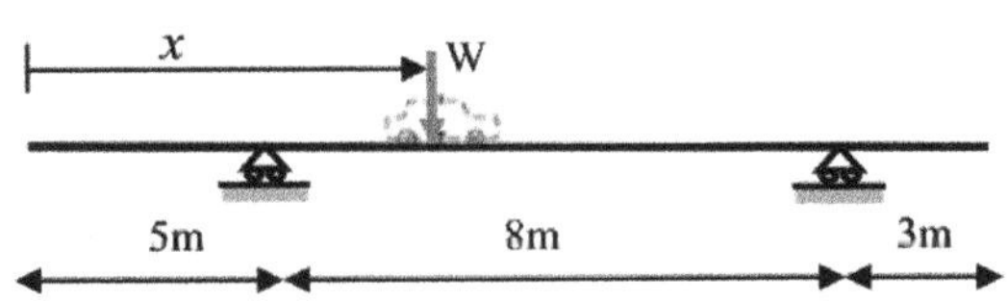

Exercise 4

Calculate the support reactions of the following frames:

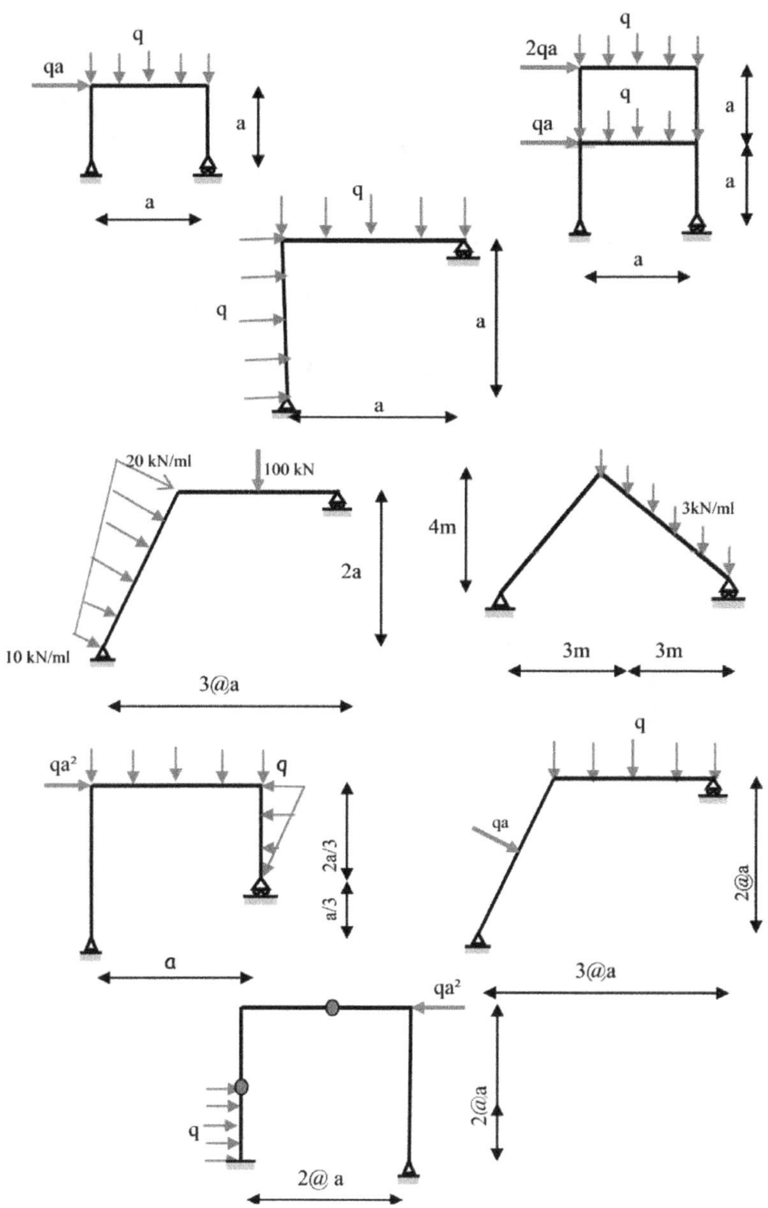

Exercise 5

Determine the support reactions of the following trusses:

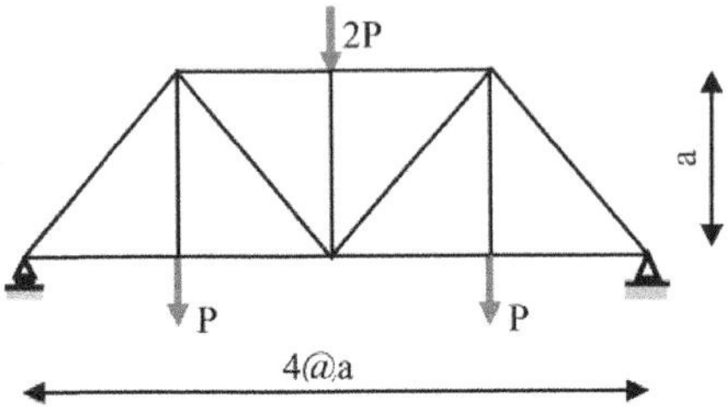

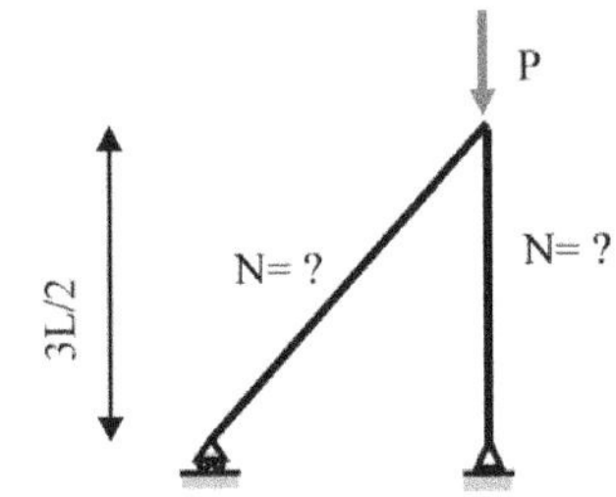

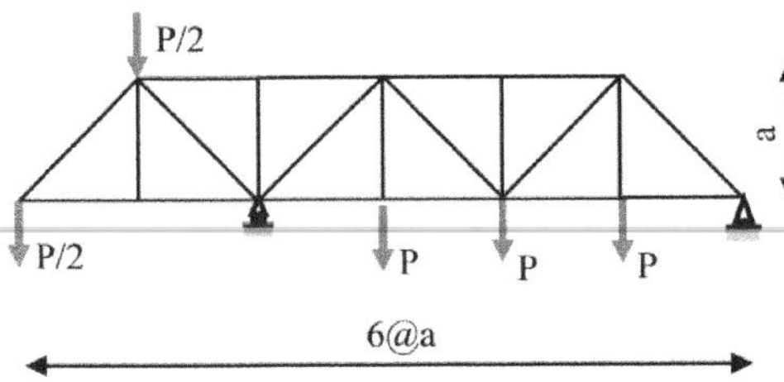

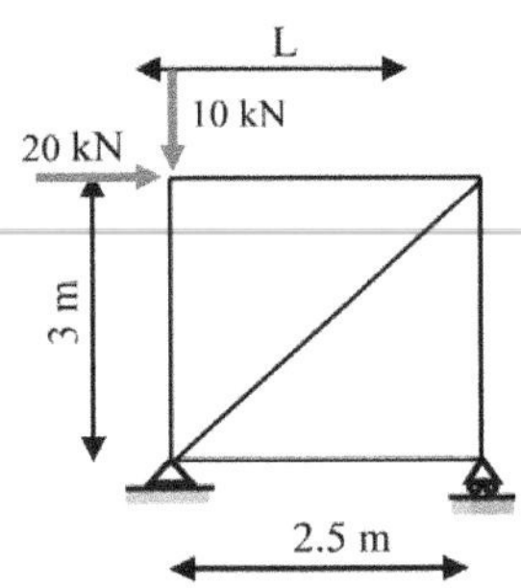

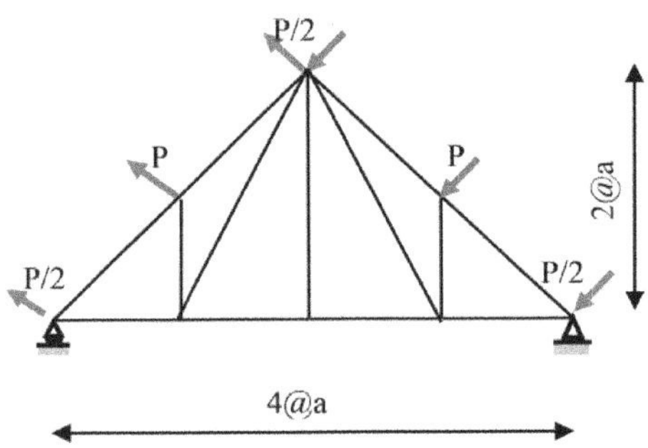

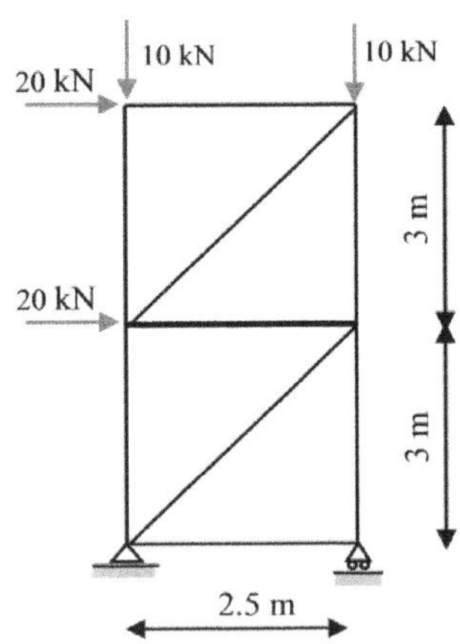

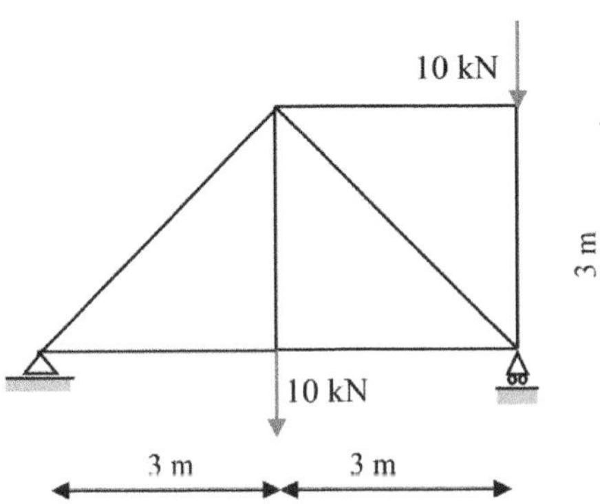

Exercise 6

Determine the support reactions of the following arches:

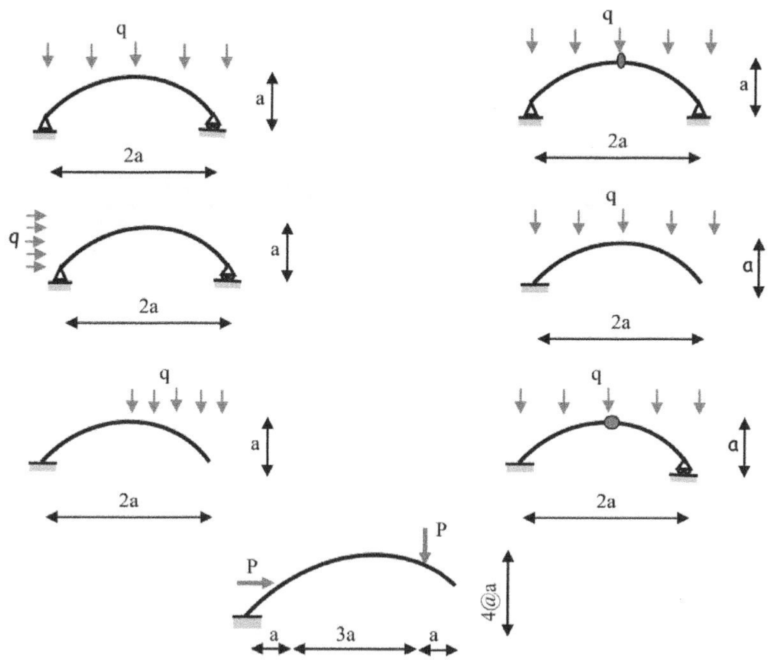

Exercise 7

Determine the unknowns by using hinges in the case of non-determinate structures.

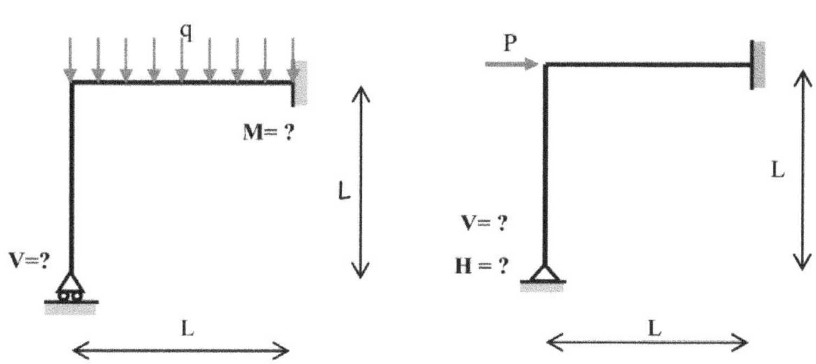

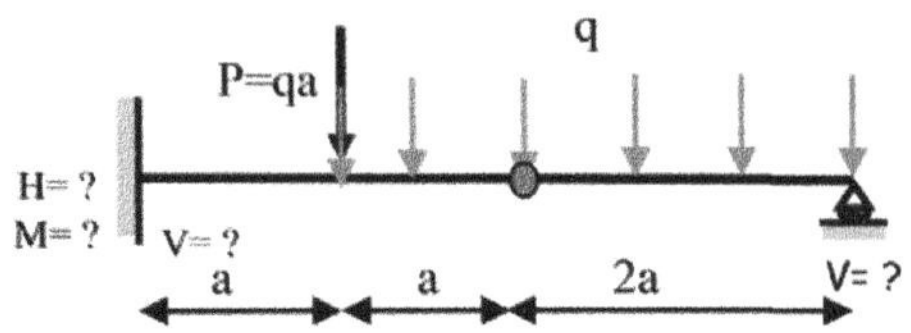
q
P=qa
H= ?
M= ?
V= ?
a
a
2a
V= ?

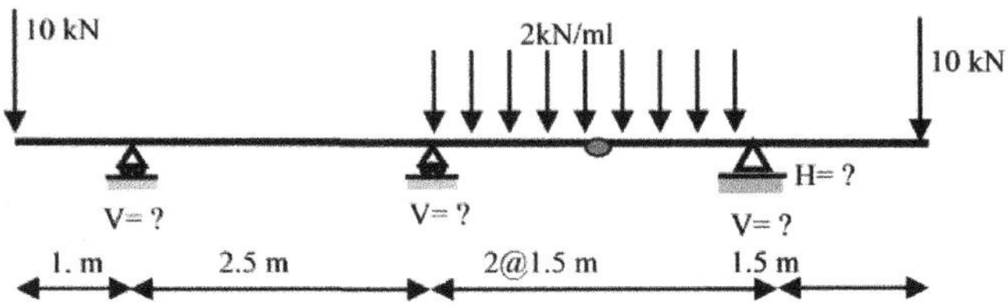
10 kN
2kN/ml
10 kN
H= ?
V= ?
V= ?
V= ?
1. m
2.5 m
2@1.5 m
1.5 m

4

Analysis of Trusses

The teaching objectives for this chapter are as follows:

– mastery and implementation of truss analysis methods;

– distinction between truss analysis methods and their limitations;

– analysis of the static indeterminacy of trusses;

– analysis of composite and space trusses;

– implementation of the matrix method.

In the first part of this chapter, we present the static indeterminacy of trusses. In these structures, we distinguish between external and internal static indeterminacy. The second part describes truss analysis methods. In this part, the joint equilibrium method, the method of sections, the graphical method and the matrix method are described. Lastly, the third part presents the analysis of composite and space trusses.

4.1. Definition of a truss

A truss is an assembly of straight or curved bars biarticulated at their ends, which forms a stable structure. For their lightness and strength, trusses are widely used to solve the problems of range, resistance and aesthetics. Trusses are used in bridges, floors, factories, large sports and conference halls, and domes, etc. The bars of a truss are joined by bolts or rivets through a plate called a "gusset" (Figure 4.1).

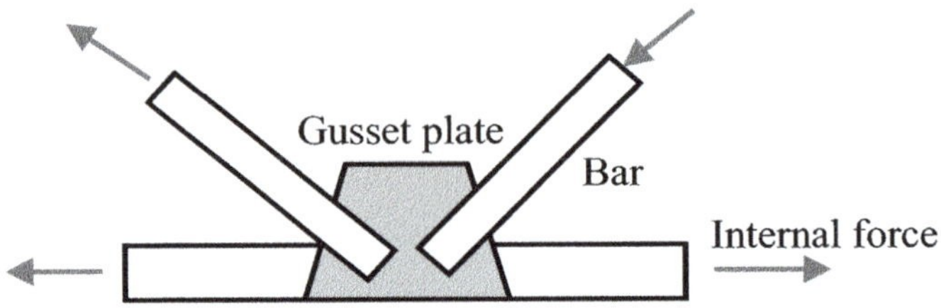

Figure 4.1. *Connecting gusset*[1]

When the bars of a structure are on a single plane, the truss is said to be plane (Figure 4.2). In general, trusses are spatial and have a perfect (rigid) connection between their bars (Figure 4.3). In many cases, bridge or firm structures are considered to be plane structures, meaning they can be analyzed without influencing the accuracy of the results obtained.

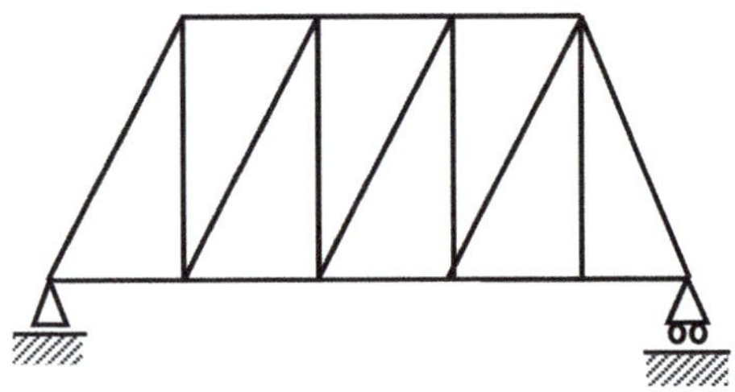

Figure 4.2. *Plane truss*

Trusses may be statically indeterminate internally or externally. If the number of support reactions is greater than the number of static equations, the truss is said to be statically indeterminate externally. Similarly, the truss is statically indeterminate internally if the static equations are insufficient to calculate the internal forces on the bars.

4.2. Hypothesis of analysis

The analysis of trusses is based on the following simplifying assumptions:

1) all the bars of the plane or space structures are biarticulated at their ends;

1 All of the figures in this chapter are available to view in full color at www.iste.co.uk/khalfallah/analysis1.zip

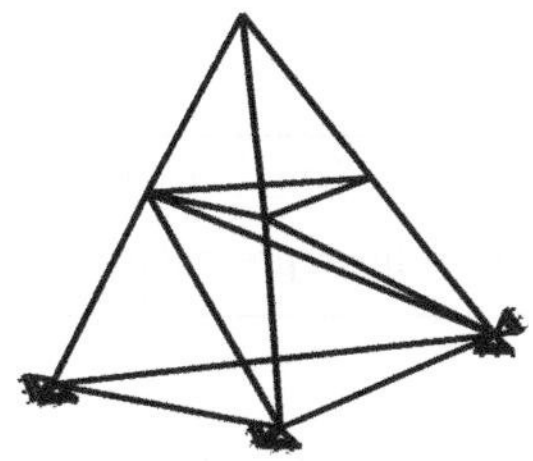

Figure 4.3. *Space truss*

2) the external forces are exclusively applied to the joints of the truss;

3) the axis of each bar coincides with the line between the centers of two adjacent joints (Figure 4.4).

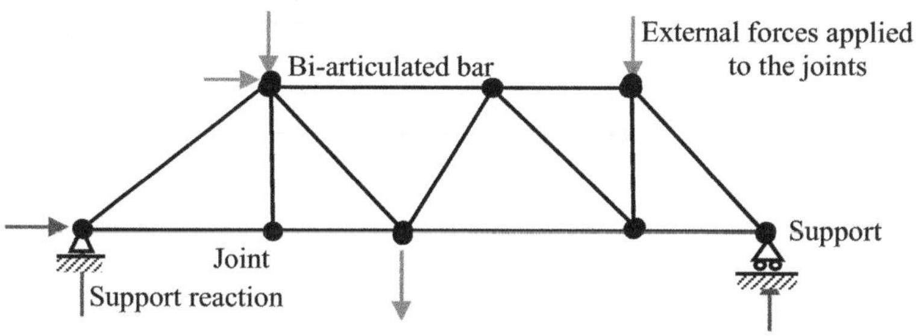

Figure 4.4. *Model of a truss*

In trusses, the loading system only generates forces on the structure's bars. Each internal force can be a traction, a compression or a zero force (neutral).

4.3. Sign convention and representation of internal forces

For consistent notation, a sign convention is necessary for truss analysis. In reality, there is no universal sign convention. In truss analysis, we will use the positive sign for traction force and the negative sign for compression force. This convention even conveys the physical direction of each force. In addition, the internal traction force lengthens the bar as the compressive force compresses it (Figure 4.5).

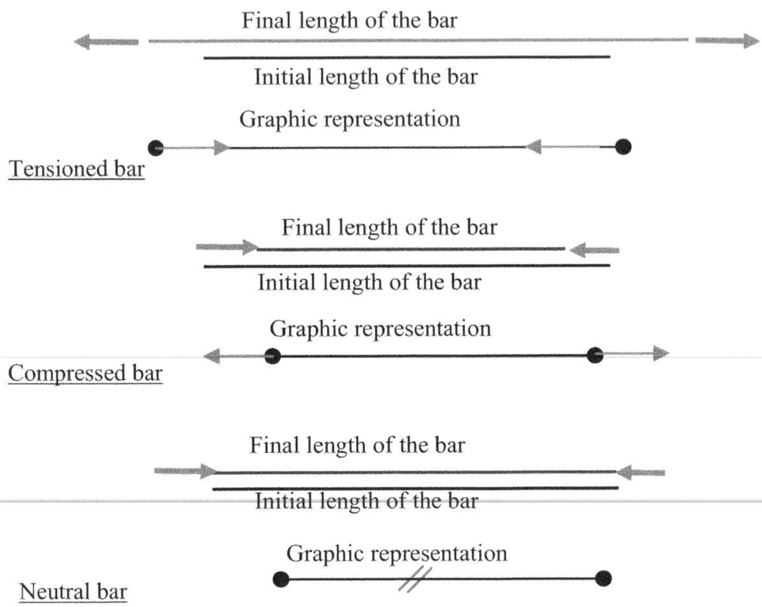

Figure 4.5. *Sign convention*

4.4. Degree of static indeterminacy and stability of trusses

We have already covered the static indeterminacy of trusses in section 1.8. The static indeterminacy of trusses is distinguished on two different levels; it is an internal analysis and an external analysis.

For a truss with b bars, n joints and r support reactions, the degrees of external and internal static indeterminacy can be determined by the formulas discussed in the following sections.

4.4.1. *External analysis*

The static indeterminacy of a truss is determined by comparing both the number of unknowns of the problem and the number of static equations. Figure 4.6 shows some cases of static indeterminacy of a truss.

$r = n_e$: the structure is statically determinate externally.

$r > n_e$: the structure is statically indeterminate externally and the degree of static indeterminacy is $f = r - 3$.

$r < n_e$: the structure is unstable externally.

n_e is the number of static equations, $n_e = 3$.

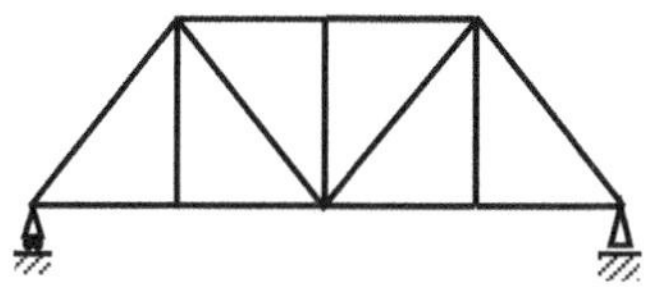

Number of unknowns, $r = 3$.
Number of equations, $n_e = 3$.
The structure is statically determinate externally.

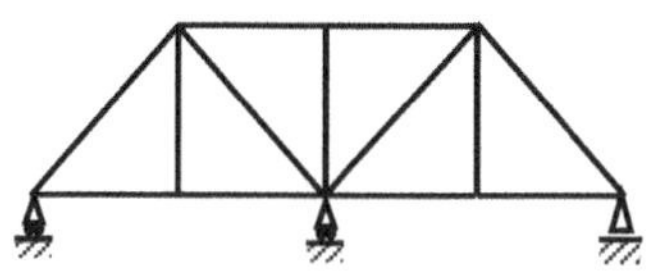

Number of unknowns, $r = 4$.
Number of equations, $n_e = 3$.
The structure is statically indeterminate externally.

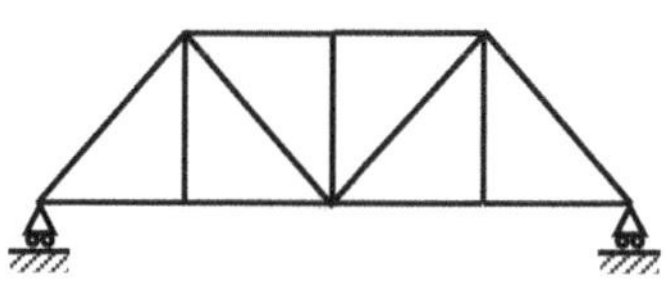

Number of unknowns, $r = 2$.
Number of equations, $n_e = 3$.
The structure is unstable externally.

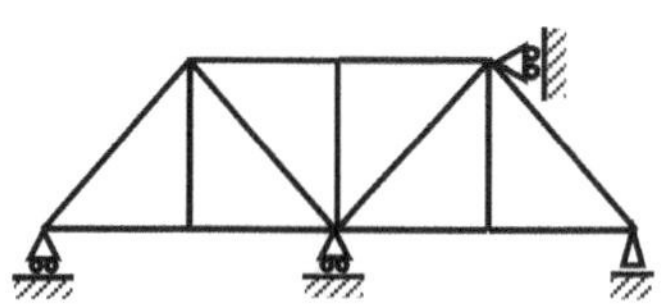

Number of unknowns, $r = 5$.
Number of equations, $n_e = 3$.
The structure is twice statically indeterminate externally.

Figure 4.6. *External analysis of trusses*

4.4.2. *Internal analysis*

Using the same concept as in section 4.4.1, we define the internal static indeterminacy of a truss by the difference between the number of unknowns of the problem and the number of independent equations.

– The truss is statically determinate internally if:

$$b = 2n - 3 \qquad [4.1]$$

– The truss is statically indeterminate internally if:

$$b > 2n - 3 \quad [4.2]$$

– The truss is unstable if:

$$b < 2n - 3 \quad [4.3]$$

Therefore, the degree of internal static indeterminacy can be deduced by:

$$f = b - (2n - 3) \quad [4.4]$$

Figure 4.7 shows the internal static indeterminacy of some trusses.

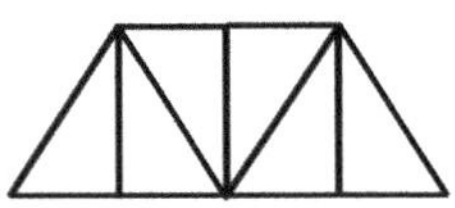

Number of bars, b = 13.
Number of joints, n = 8.
$b = 2n - 3$, the structure is statically determinate internally.

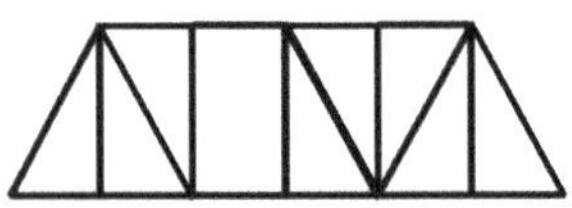

Number of bars, b = 20.
Number of nodes, n = 12.
$b < 2n - 3$, the structure is unstable internally.

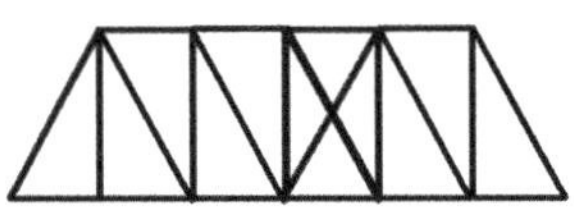

Number of bars, b = 22.
Number of nodes, n = 12.
$b > 2n - 3$, the structure is once statically indeterminate internally.

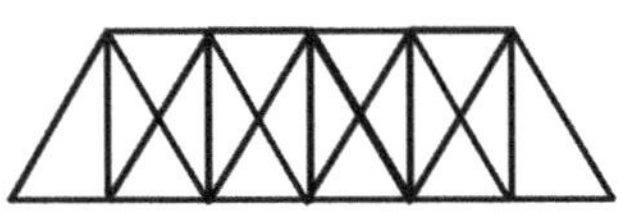

Number of bars, b = 25.
Number of nodes, n = 12.
$b > 2n - 3$, the structure is four times statically indeterminate internally.

Figure 4.7. *Internal analysis of trusses*

4.5. Analysis methods of trusses

The primary purpose of truss analysis is to determine the internal forces on the bars. As seen in [4.4], the number of unknowns in a truss is the number of bars in the free-body diagram. So, to analyze it, it is necessary to provide at least *b* independent equations. In the literature, there are several analysis methods. In this subject, we can identify the following.

4.5.1. *Method of joint equilibrium*

In the method of joint equilibrium, the internal forces on the bars of a truss can be determined by considering the equilibrium of each joint. The equilibrium of the whole structure makes it possible to check the overall equilibrium. At each joint, the forces on the bars and the nodal forces and/or the support reactions must satisfy the equilibrium equations.

$$\sum F_x = 0 \qquad [4.5]$$

$$\sum F_y = 0 \qquad [4.6]$$

Equations [4.5] and [4.6] are checked at each joint. For this reason, it is recommended to start the analysis with the joint having a maximum of two bars. By successively applying the equilibrium equations at each joint, we can then determine the forces on the bars.

To enrich the analysis using this method, consider the truss (Figure 4.8).

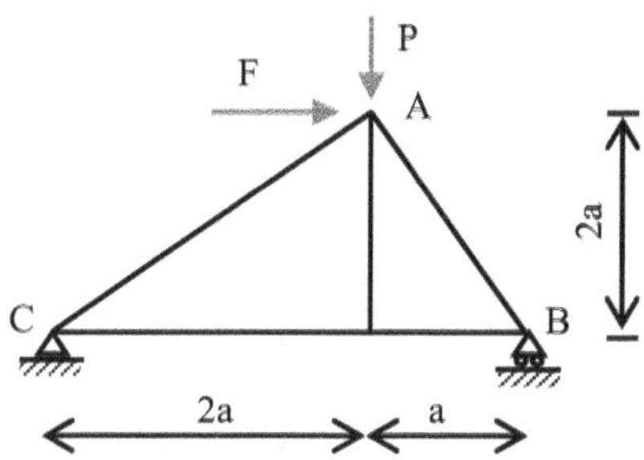

Figure 4.8. *Truss*

External analysis

The structure is linked to the external environment by two supports; a roller support in B (V_B) and a hinged support in C (V_C and H_c). If the number of support reactions is greater than the number of the static equations, the structure is said to be statically indeterminate externally.

Internal analysis

Number of bars, b = 5.

The number of independent equations is equal to $2 \times n - 3 = 5$.

$b = 2 \times n - 3$, the structure is statically determinate internally and externally, the method of joint equilibrium can be applied to calculate internal forces on the bars.

The procedure for analyzing trusses using the method of joint equilibrium is as follows.

Calculation of support reactions

We construct the free-body diagram (Figure 4.9).

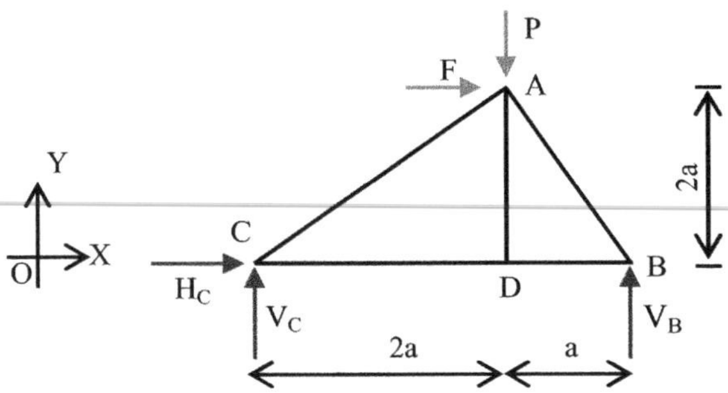

Figure 4.9. *Free-body diagram*

$$\sum F_x = 0.: F + H_c = 0.$$

$$H_c = -F.$$

$$\sum F_y = 0.: V_B + V_c - P = 0.$$

$$\sum M_C = 0.: V_B\,(3a) + -P.(2a) + F.(2a) = 0.$$

$$V_B = \frac{2}{3}P + \frac{2}{3}F$$

$$V_C = \frac{1}{3}P - \frac{2}{3}F$$

The equilibrium of joint C makes it possible to calculate the forces on bars N_{CD} and N_{CA}.

$\sum F_x = 0.: H_c + N_{CD} + N_{CA}.\cos\alpha = 0.$

$\sum F_y = 0.: V_c + N_{cA}.\sin\alpha = 0.$

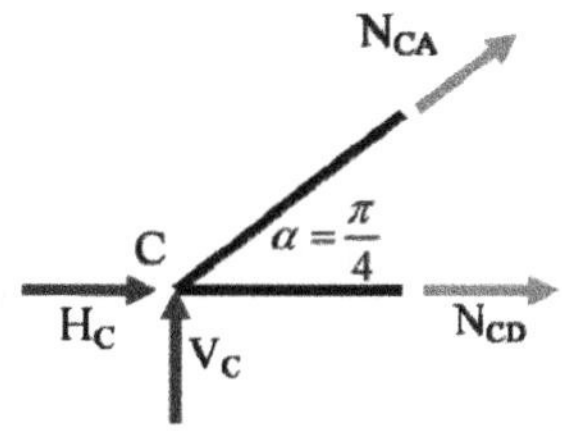

$$N_{CA} = -\frac{\sqrt{2}}{3}P + \frac{2\sqrt{2}}{3}F$$

$$N_{CD} = -H_C - N_{CA}.\cos\alpha = \frac{1}{3}P + \frac{1}{3}F$$

Joint D:

$\sum F_y = 0.: N_{DA} = 0.$

$\sum F_x = 0.: N_{DB} - N_{DC} = 0.$

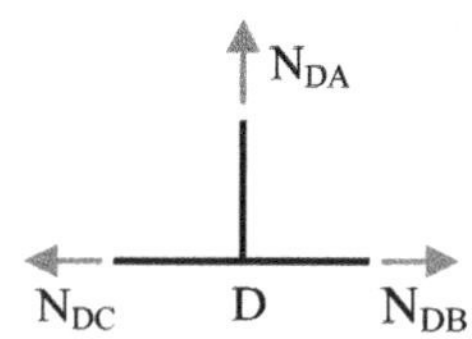

$$N_{DB} = N_{DC} = \frac{1}{3}P + \frac{1}{3}F$$

Finally, the equilibrium of joint B makes it possible to determine the force in the bar (BA).

The vertical equilibrium is sufficient to determine the force in the bar (BA):

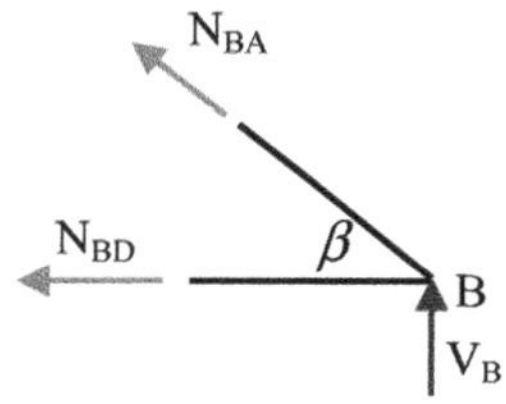

$\sum F_y = 0.: V_B + N_{BA}.\sin\beta = 0.$

$$N_{BA} = -\frac{V_B}{\sin\beta} = -\frac{\sqrt{5}}{3}P - \frac{\sqrt{5}}{3}F$$

At this stage, all forces on the bars are determined and the equilibrium of joint A can be used to verify the calculation.

The horizontal and vertical equilibrium of joint A is:

$F - N_{AC}.\cos\alpha + N_{AB}.\cos\beta = 0.$

$$-P- N_{AD} - N_{AB}.\sin\beta - N_{AC}.\sin\alpha = 0.$$

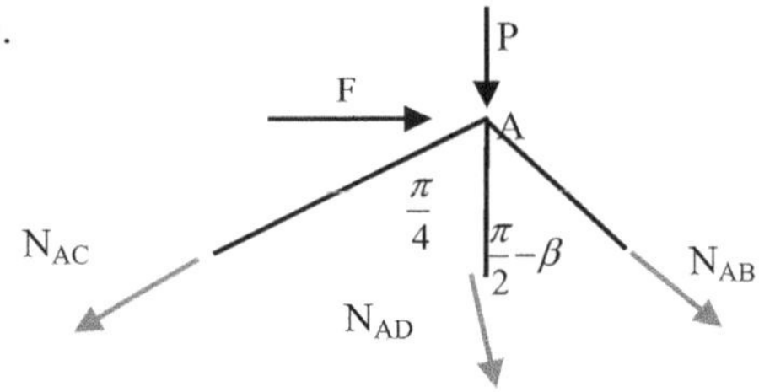

$$F + \frac{1}{3}P - \frac{2}{3}F - \frac{1}{3}P - \frac{1}{3}F = 0.$$

$$-P + \frac{2}{3}P + \frac{2}{3}F + \frac{1}{3}P - \frac{2}{3}F = 0.$$

The above equalities are verified by substituting the expressions of the normal forces N_{AC}, N_{AD} and N_{AB}.

The forces on the bars are calculated using the equilibrium equations. To distinguish the nature of the internal force and to draw the diagram of normal forces (Figure 4.10), we can look at case P = F. The forces on the bars and their natures are grouped in Table 4.1.

Bar	CD	CA	DA	DB	BA
Normal force	$\frac{2}{3}P$	$\frac{\sqrt{2}}{3}P$	0	$\frac{2}{3}P$	$-\frac{2\sqrt{5}}{3}P$
Nature of force	Traction	Traction	Neutral	Traction	Compression

Table 4.1. *Internal forces on the bars*

4.5.2. *Method of sections*

The method of joint equilibrium presented in section 4.5.1 is effective for calculating the forces for all the bars. However, when we limit which forces to calculate, the method of sections is the preferred option.

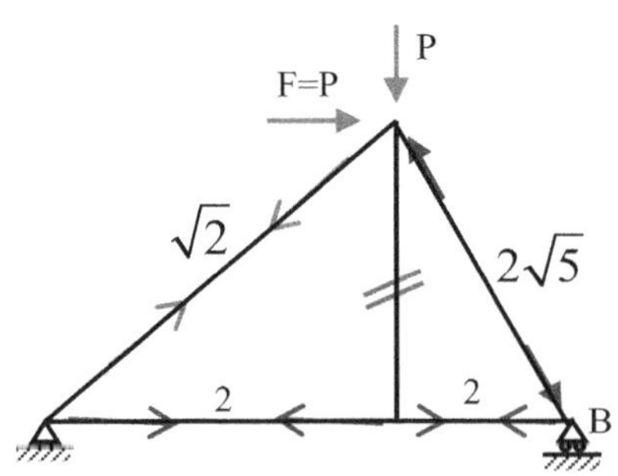

Figure 4.10. *Normal force*($\frac{P}{3}$)*

The method of sections consists of cutting the bars by a section, whose internal forces are the unknowns of the problem. The section divides the truss into two portions – each of which must satisfy the equilibrium conditions.

$$\sum F_x = 0. \quad [4.7]$$

$$\sum F_y = 0. \quad [4.8]$$

$$\sum M_i = 0. \quad [4.9]$$

EXAMPLE 4.1.–

Determine the internal forces on bars 1, 2 and 3 of the truss (Figure 4.11).

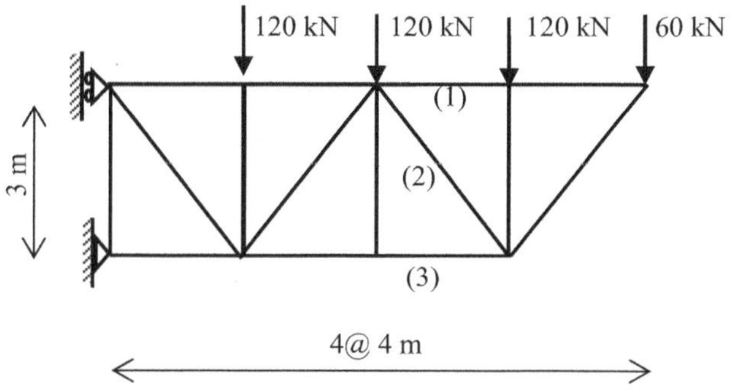

Figure 4.11. *Given truss*

The section divides bars 1, 2 and 3 and considers the right part of the structure (Figure 4.12).

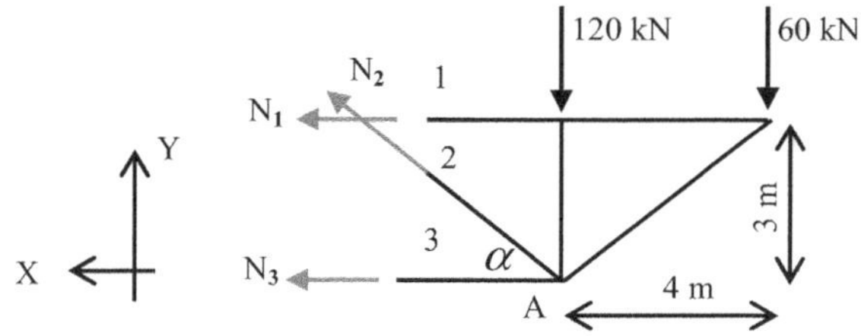

Figure 4.12. *Part of the structure proposed*

The equilibrium equations of the right portion (Figure 4.12) are:

$$\sum F_x = 0.: \; N_1 + N_2 \cos\alpha + N_3 = 0$$

$$\sum F_y = 0.: \; N_2 \sin\alpha - 120 - 60 = 0$$

Knowing that: $\sin\alpha = \dfrac{3}{5}: \; N_2 = 300\,\text{kN}$

$$\sum M_A = 0.: \; N_1 = 80\,\text{kN} \text{ and } N_3 = -320\,\text{kN}$$

Bars (1) and (2) are tensioned while bar (3) is compressed.

EXAMPLE 4.2.–

Determine the internal forces on bars 1, 2 and 3 (Figure 4.13).

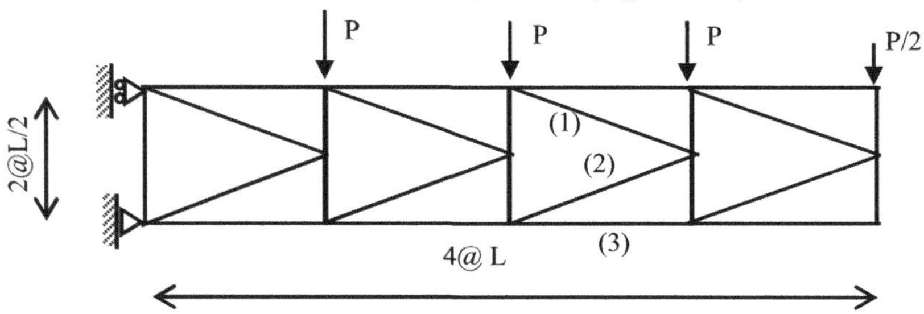

Figure 4.13. *Given structure*

We cut the truss by a curve as shown in Figure 4.14. The right part is considered for analysis.

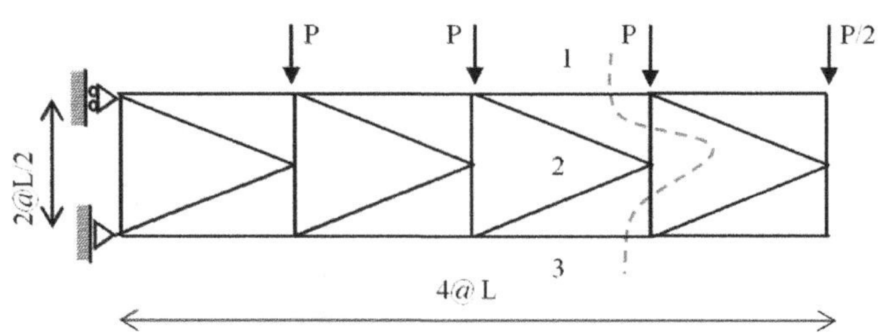

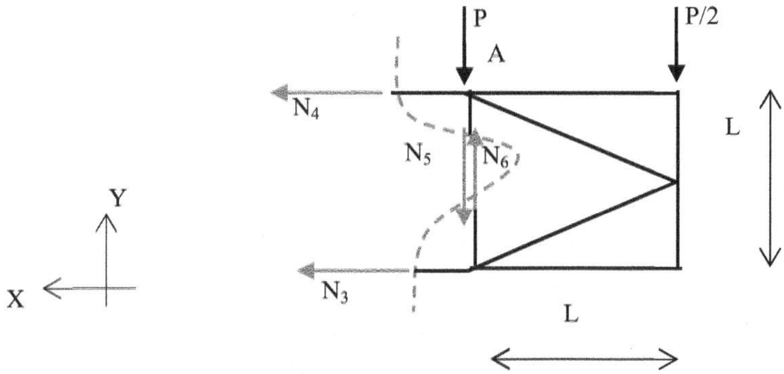

Figure 4.14. *Part considered*

$$\sum M_A = 0.: \ N_3.L + \frac{P}{2}L = 0$$

$$N_3 = -\frac{P}{2}$$

$$\sum F_x = 0.: \ N_3 + N_4 = 0$$

$$N_4 = \frac{P}{2}$$

The equilibrium equations of the right portion (Figure 4.15) are:

$$\sum F_x = 0 \ : \ N_1 \cos\alpha + N_2 \cos\alpha = 0$$

$$\sum F_y = 0 \ : \ N_1 \sin\alpha - N_2 \sin\alpha - P - \frac{P}{2} = 0$$

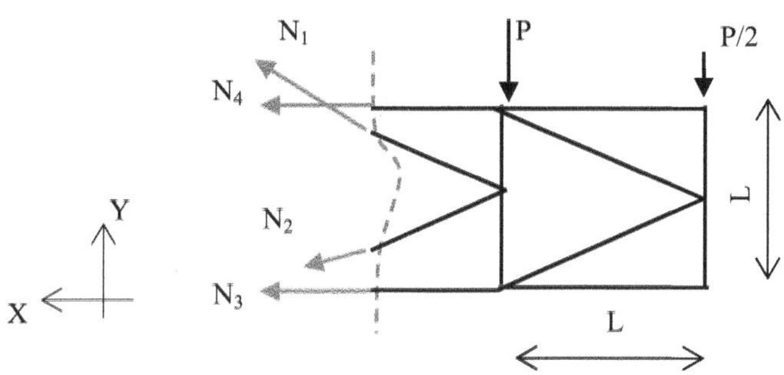

Figure 4.15. *Forces on bars 1, 2, and 3*

Hence, $N_1 = -N_2 = \dfrac{3P}{4\sin\alpha}$

Knowing that $\sin\alpha = \dfrac{\sqrt{5}}{5}$, $N_1 = -N_2 = \dfrac{3\sqrt{5}P}{4}$

Table 4.2 groups the forces on the bars of the truss (Figure 4.13).

Bar	1	2	3
Normal force	$\dfrac{3\sqrt{5}P}{4}$	$-\dfrac{3\sqrt{5}P}{4}$	$-\dfrac{P}{2}$
Nature of force	Traction	Compression	Compression

Table 4.2. *Forces on bars 1, 2 and 3*

4.5.3. *Matrix method*

We consider a truss with several bars in Figure 4.16.

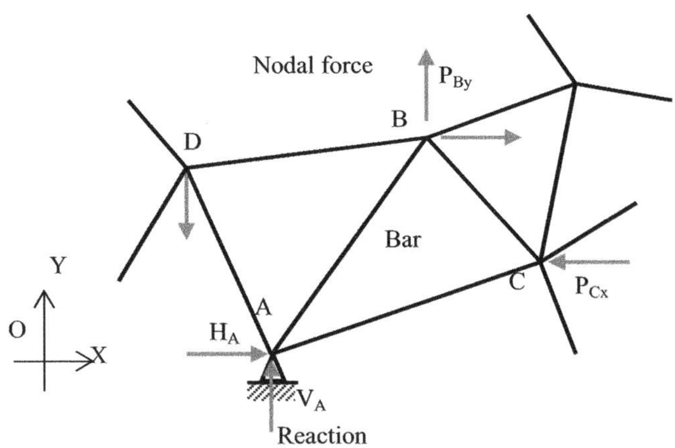

Figure 4.16. *Portion of a truss*

We choose bar (AB) (Figure 4.17).

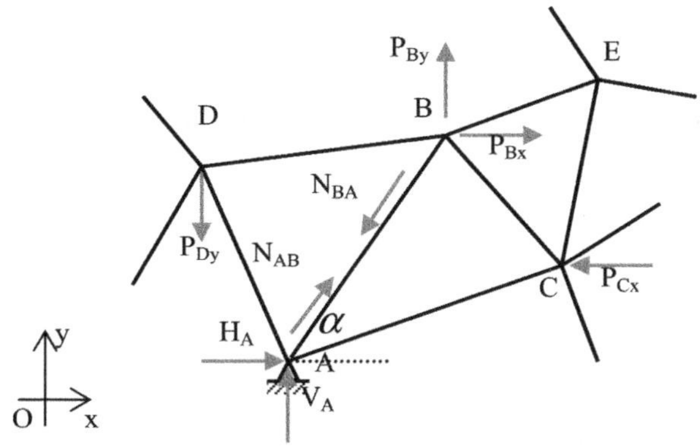

Figure 4.17. *Study of bar (AB)*

The force in the bar (AB) can be divided into:

$$X_{AB} = N_{AB} \cos\alpha = N_{AB} \frac{x_B - x_A}{L_{AB}} = N_{AB} l_{AB} \quad [4.10]$$

$$Y_{AB} = N_{AB} \sin\alpha = N_{AB} \frac{y_B - y_A}{L_{AB}} = N_{AB} m_{AB} \quad [4.11]$$

l_{AB}, m_{AB} are the components of the unit vector of bar (AB).

$$l_{AB} = \cos\alpha$$

$$m_{AB} = \sin\alpha$$

Similarly, we can deduce that:

$$X_{BA} = N_{BA} l_{BA} \quad [4.12]$$

$$Y_{BA} = N_{BA} m_{BA} \quad [4.13]$$

with:

$$l_{BA} = -l_{AB}$$

$$m_{BA} = -m_{AB}$$

Applying the equilibrium equations to joints A and B allows us to write:

$$X_{AB} + X_{AC} + X_{AD} + H_A = 0 \quad [4.14a]$$

$$Y_{AB} + Y_{AC} + Y_{AD} + V_A = 0 \quad [4.14b]$$

$$X_{BD} + X_{BA} + X_{BC} + X_{BE} + P_{Bx} = 0 \quad [4.14c]$$

$$Y_{BD} + Y_{BA} + Y_{BC} + Y_{BE} + P_{By} = 0 \quad [4.14d]$$

We introduce the equations [4.10] – [4.13] in the relationship [4.14]; the internal forces of the bars can be deduced by:

$$N_{AB}l_{AB} + N_{AC}l_{AC} + N_{AD}l_{AD} + H_A = 0 \quad [4.15a]$$

$$N_{AB}m_{AB} + N_{AC}m_{AC} + N_{AD}m_{AD} + V_A = 0 \quad [4.15b]$$

$$N_{BD}l_{BD} + N_{BA}l_{BA} + N_{BC}l_{BC} + N_{BE}l_{BE} + P_{Bx} = 0 \quad [4.15c]$$

$$N_{BD}m_{BD} + N_{BA}m_{BA} + N_{BC}m_{BC} + N_{BE}m_{BE} + P_{By} = 0 \quad [4.15d]$$

Knowing that:

$$l_{BA} = -l_{AB}$$

$$m_{BA} = -m_{AB}$$

The system of equations [4.15] is written as:

$$N_{AB}l_{AB} + N_{AC}l_{AC} + N_{AD}l_{AD} + H_A = 0 \quad [4.16a]$$

$$N_{AB}m_{AB} + N_{AC}m_{AC} + N_{AD}m_{AD} + V_A = 0 \quad [4.16b]$$

$$N_{BD}l_{BD} - N_{BA}l_{AB} + N_{BC}l_{BC} + N_{BE}l_{BE} + P_{Bx} = 0 \quad [4.16c]$$

$$N_{BD}m_{BD} - N_{AB}m_{AB} + N_{BC}m_{BC} + N_{BE}m_{BE} + P_{By} = 0 \quad [4.16d]$$

The matrix equation of equilibrium of the bars in the structure is written as:

$$\begin{bmatrix} .. & .. & .. & .. & .. & .. & .. & .. & .. & .. \\ .. & .. & l_{AD} & l_{AB} & l_{AC} & 0 & 0 & 1 & 0 & 0 \\ .. & .. & m_{AD} & m_{AB} & m_{AC} & 0 & 0 & 0 & 1 & 0 \\ .. & .. & 0 & -l_{AB} & 0 & 0 & l_{BC} & l_{BE} & 0 & 0 \\ .. & .. & 0.. & -m_{AB} & 0 & 0 & m_{BC} & m_{BE} & 0 & 0 \\ .. & .. & .. & .. & .. & .. & .. & .. & .. & .. \\ .. & .. & .. & .. & .. & .. & .. & .. & .. & .. \\ .. & .. & .. & .. & .. & .. & .. & .. & .. & .. \\ .. & .. & .. & .. & .. & .. & .. & .. & .. & .. \\ .. & ... & .. & .. & .. & .. & .. & .. & .. & .. \end{bmatrix} \begin{Bmatrix} .. \\ N_{AD} \\ N_{AB} \\ N_{AC} \\ N_{BD} \\ N_{BC} \\ N_{BE} \\ H_A \\ .V_A. \\ .. \end{Bmatrix} = \begin{Bmatrix} .. \\ 0 \\ 0 \\ -P_{Bx} \\ -P_{By} \\ .. \\ .. \\ .. \\ .. \\ .. \end{Bmatrix} \quad [4.17]$$

EXAMPLE 4.3.–

Analyze the structure (Figure 4.18) using the matrix method.

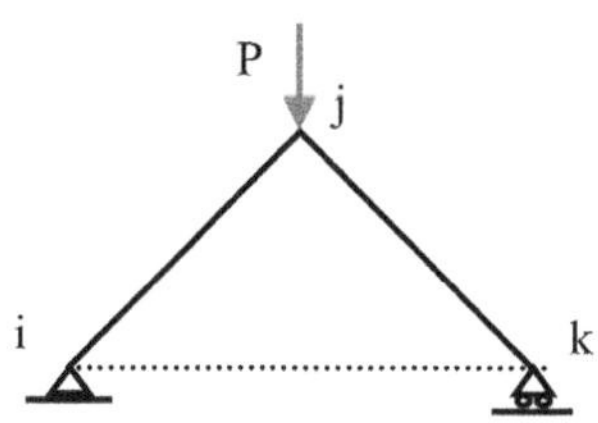

Figure 4.18. *Given structure*

Joint i

$$R_{ix} + X_{ij} = 0$$

$$R_{iy} + Y_{ij} = 0$$

Joint j

$$X_{ji} + X_{jk} = 0$$

$$Y_{ji} + Y_{jk} - P = 0$$

Joint k

$$R_{ky} + Y_{kj} = 0$$

We can write:

$$X_{ij} = N_{ij} l_{ij}$$

$$X_{ji} = N_{ji} l_{ji} = -N_{ij} l_{ij}$$

and:

$$Y_{ij} = N_{ij} m_{ij}$$

$$Y_{ji} = N_{ji} m_{ji} = -N_{ij} m_{ij}$$

In the matrix form, the previous equations are grouped into:

$$\begin{bmatrix} l_{ij} & 0 & 1 & 0 & 0 \\ m_{ij} & 0 & 0 & 1 & 0 \\ -l_{ij} & l_{jk} & 0 & 0 & 0 \\ -m_{ij} & m_{jk} & 0 & 0 & 0 \\ 0 & -m_{jk} & 0 & 0 & 1 \end{bmatrix} \begin{Bmatrix} N_{ij} \\ N_{jk} \\ R_{ix} \\ R_{iy} \\ R_{ky} \end{Bmatrix} = \begin{Bmatrix} 0 \\ 0 \\ 0 \\ P \\ 0 \end{Bmatrix}$$

Resolving the matrix equation leads to determine the internal forces on bars (ij) and (jk), and the support reactions.

$$N_{ij} = N_{jk} = -\frac{P}{2m_{ij}} = -\frac{P}{2\sin\alpha}$$

The first equation leads to deduce the vertical reaction of support A.

$$R_{ix} = -l_{ij} N_{ij} = +\frac{Pl_{ij}}{2m_{ij}} = \frac{P}{2tg\alpha}$$

Similarly, the second equation leads to:

$$R_{iy} = -m_{ij} N_{ij} = +\frac{P}{2}$$

Finally, the previous equations lead to determine the vertical reaction.

$$R_{ky} = m_{jk} N_{jk} = -\frac{P m_{jk}}{2 m_{ij}} = \frac{P}{2}$$

4.5.4. *Graphic or Cremona method*

The graphic or Cremona method applies only to internally and externally statically determinate trusses. It describes the method of joint equilibrium in the graphical form.

Method steps

1) The intervals between the external forces are numbered in a clockwise direction.

2) The intervals between the bars are numbered from left to right.

3) The polygon of external forces is established according to a scale of forces. The polygon of external forces forms a closed cycle and its construction is carried out by following the forces in a clockwise direction.

4) Likewise, the polygon of internal forces is built around each joint starting with the joint that contains no more than two bars (two unknowns).

5) The measurement of the distance between two successive points represents the internal force of the bar defined by these points.

EXAMPLE 4.4.–

Determine the forces on the bars of the truss (Figure 4.19).

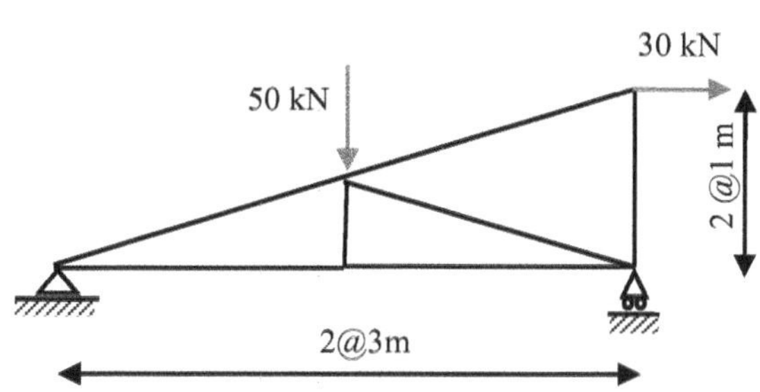

Figure 4.19. *Given structure*

Numbering the intervals between the external forces and the network is shown in Figure 4.20.

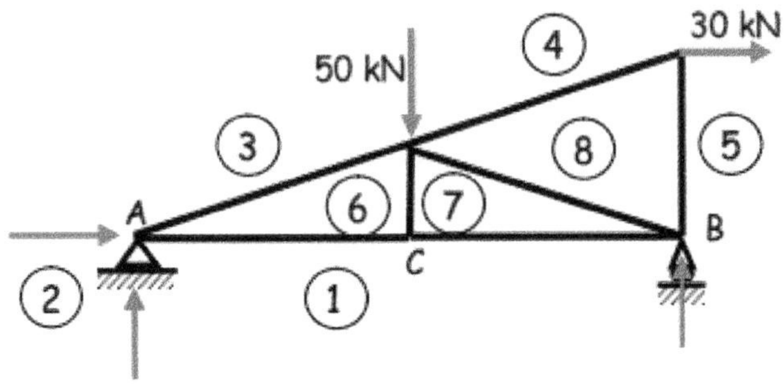

Figure 4.20. *Numbering of joints and bars*

The structure's external equilibrium leads to the determination of support reactions:

$$V_A = 15 \text{ kN}, \quad H_A = -30 \text{ kN} \quad \text{and } V_B = 35 \text{ kN}$$

To represent the polygons of external and internal forces, it is necessary to choose a scale of forces; 1 cm on the Cremona diagram is worth 10 kN of force.

Figure 4.21 shows the Cremona diagram, in which the distance between two numbers represents the normal force on the bar separating the regions numbered by these two numbers.

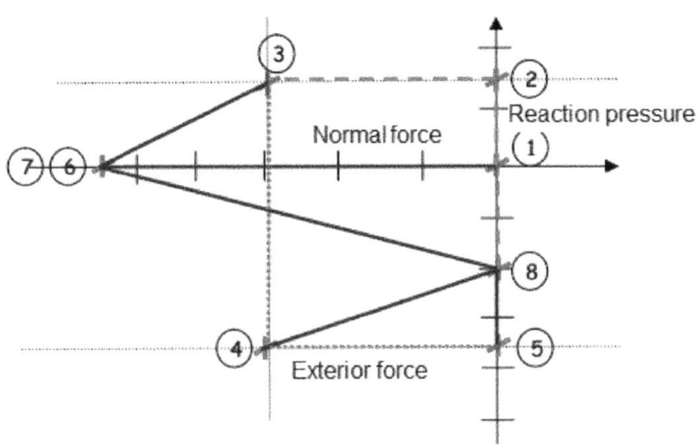

Figure 4.21. *Cremona diagram*

For example, the internal force in the bar (CB) is the distance between ① and ⑦.

To identify the nature of the internal force, we turn in a clockwise direction around a joint at the end of the bar. Then, the direction of travel from the first joint to the second joint shows the direction of the force relative to the considered joint. We can use joint A as an example (Figure 4.22).

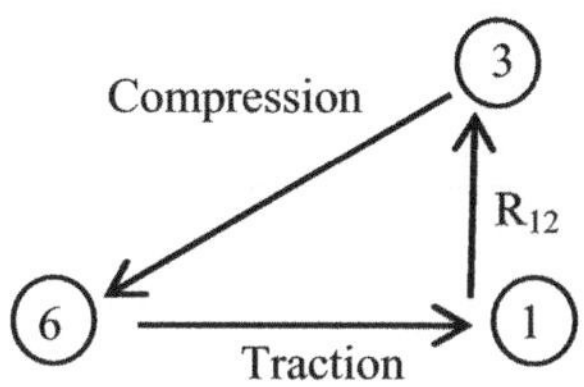

Figure 4.22. *Forces around joint 1*

The forces on the bars (the force and direction values) are grouped in Table 4.3.

Bar	AD	AC	DC	DB	BE	CE	BC
Distance	6-1	3-6	6-7	7-1	5-8	4-8	7-8
Force	75	47.43	0	75	10	31.62	79
Direction	+	–	/	+	–	+	–
Nature	Trac.	Comp.	Neutral	Trac.	Comp.	Trac.	Comp.

Table 4.3. *Forces on the bars*

4.6. Composite trusses

The method of joint equilibrium or the method of sections can be used to calculate the forces on the bars of a composite truss. In this case, the method of sections is used to determine only a few internal forces. The other forces are calculated by the method of joint equilibrium.

EXAMPLE 4.5.–

Determine the internal forces on the structure (Figure 4.23).

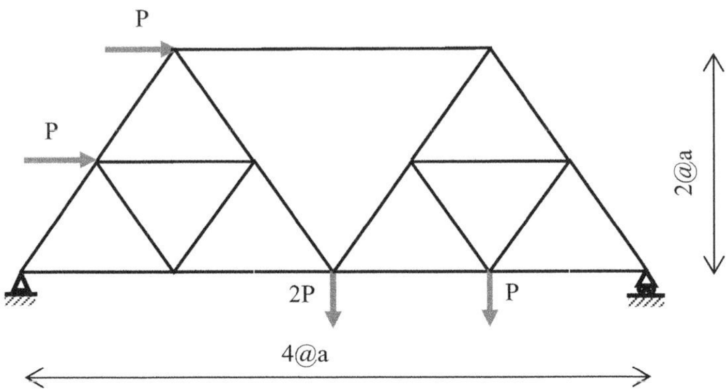

Figure 4.23. *Composite structure*

Determination of static determinacy

Before calculating support reactions, it is necessary to verify that the truss is statically determinate internally and externally (Figure 4.24).

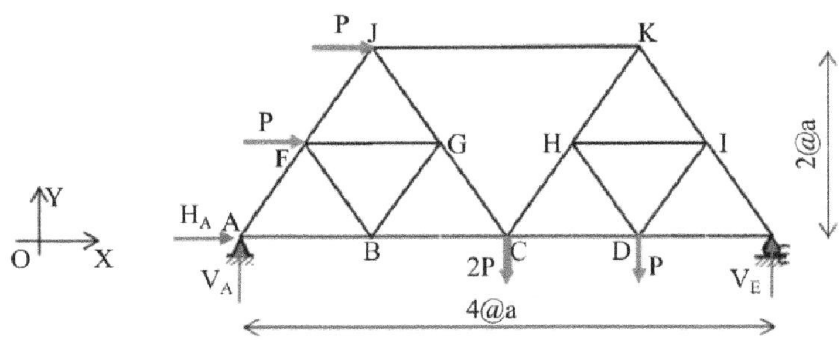

Figure 4.24. *Free-body diagram*

The number of support reactions is equal to 3.

The number of equations is also equal to 3.

The structure is statically determinate externally.

The number of bars is b = 19.

The number of equations is 2n – 3 = 2 × 11 – 3 = 19.

The structure is also statically determinate internally.

Finally, the truss (Figure 4.24) is statically determinate internally and externally.

External analysis

In this case, we consider the equilibrium of the whole structure (Figure 4.24).

$$\sum F_x = 0:\ H_A + P + P = 0$$

$$\sum F_y = 0:\ V_A + V_E - P - 2P = 0$$

$$\sum M_A = 0:\ V_E(4a) - Pa - P(2a) - 2P(2a) - \mathrm{P}(3\,\mathrm{a}) = 0$$

$$H_A = -2P$$

$$V_A = \frac{1}{2}P$$

$$V_E = \frac{5}{2}P$$

Calculation of forces on the bars

We cut the bars by section (JK), (GC) and (BC) (Figure 4.25).

We apply the static equations:

$$\sum M_J = 0:\ N_{BC}(2a) + Pa = \frac{Pa}{2} + 2P(2a)$$

$$N_{BC} = \frac{7}{4}P$$

$$\sum F_y = 0:\ -N_{GC}\sin\alpha = \frac{P}{2}$$

$$\sum F_x = 0:\ N_{GK} + \mathrm{N}_{GC}\cos\alpha + N_{BC} = 0$$

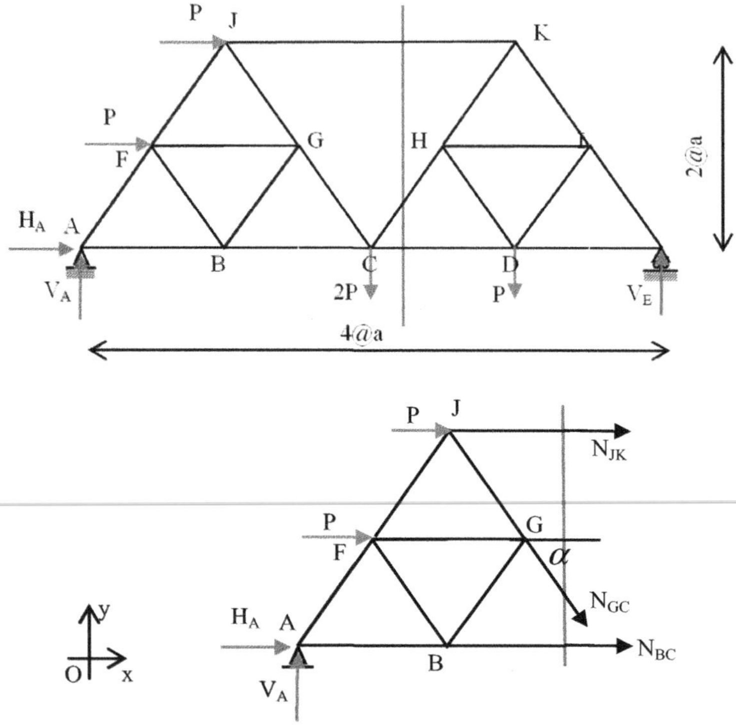

Figure 4.25. *Method of sections*

$$\sin\alpha = \frac{2}{\sqrt{5}} \text{ and } \cos\alpha = \frac{1}{\sqrt{5}}$$

$$N_{GC} = \frac{\sqrt{5}}{4}P$$

$$N_{JK} = -2P$$

Joint J

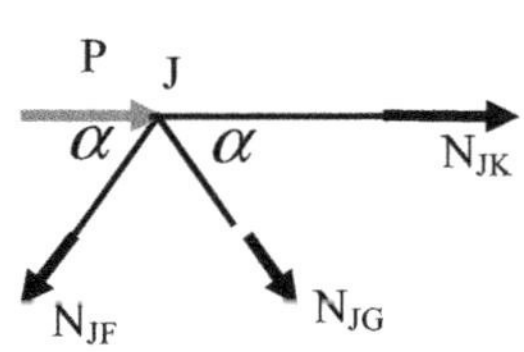

$$\sum F_x = 0: \ N_{JG}\cos\alpha - N_{JF}\cos\alpha + P + N_{JK} = 0$$

$$\sum F_y = 0: \ -N_{JF}\sin\alpha - N_{JG}\sin\alpha = 0$$

$$N_{JF} = -\frac{\sqrt{5}}{2}P$$

$$N_{JG} = \frac{\sqrt{5}}{2}P$$

Joint G

$$\sum F_x = 0: \ N_{GC}\cos\alpha - N_{GF} - N_{GJ}\cos\alpha = 0$$

$$\sum F_y = 0: \ N_{GJ}\sin\alpha - N_{GB}\sin\alpha - N_{GC}\sin\alpha = 0$$

$$N_{GF} = -\frac{1}{2}P$$

$$N_{GB} = \frac{\sqrt{5}}{4}P$$

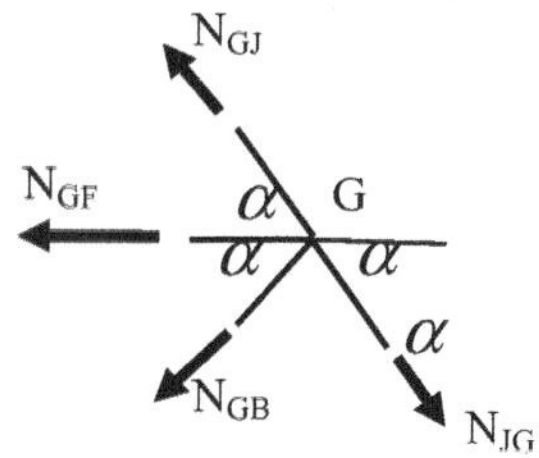

Joint B

$$\sum F_x = 0: \ N_{BC} + N_{BG}\cos\alpha - N_{BF}\cos\alpha - N_{BA} = 0$$

$$\sum F_y = 0: \ N_{BG}\sin\alpha + N_{BF}\sin\alpha = 0$$

$$N_{BA} = \frac{9}{4}P$$

$$N_{BF} = -\frac{\sqrt{5}}{4}P$$

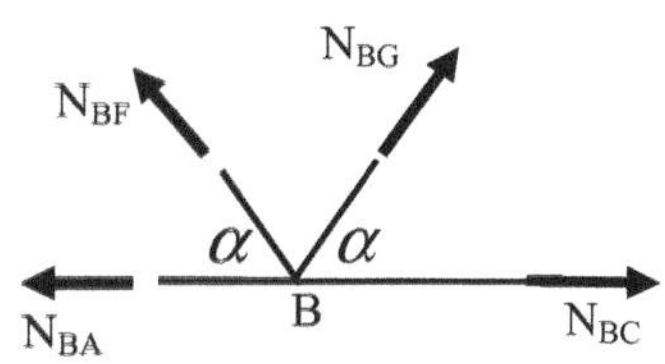

Joint A

$$\sum F_x = 0: \ N_{AB} + N_{AF}\cos\alpha - 2P = 0$$

$$N_{AF} = -\frac{\sqrt{5}}{4}P$$

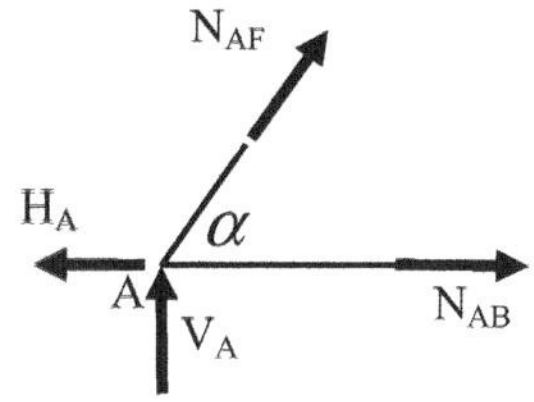

Joint E

$$\sum F_x = 0 : -N_{ED} - N_{EI}\cos\alpha = 0$$

$$\sum F_y = 0 : N_{EI}\sin\alpha + V_E = 0$$

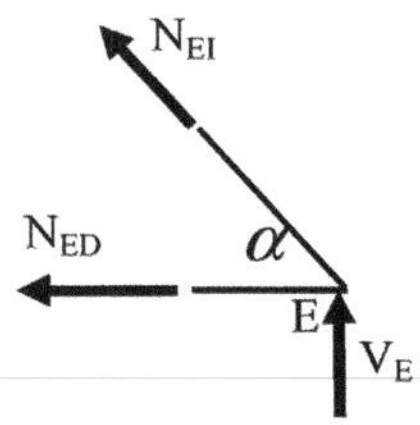

$$N_{EI} = -\frac{5\sqrt{5}}{4}P$$

$$N_{ED} = \frac{5}{4}P$$

Joint K

$$\sum F_x = 0 : N_{KI}\cos\alpha - N_{KH}\cos\alpha + N_{KJ} = 0$$

$$\sum F_y = 0 : -N_{KI}\sin\alpha - N_{KH}\sin\alpha = 0$$

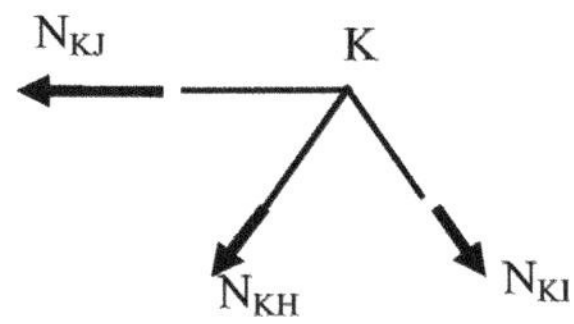

$$N_{KH} = -\sqrt{5}P$$

$$N_{KI} = +\sqrt{5}P$$

Joint I

$$\sum F_x = 0 : N_{IE}\cos\alpha - N_{IK}\cos\alpha - N_{IH} - N_{ID}\cos\alpha = 0$$

$$\sum F_y = 0 : N_{IK}\sin\alpha - N_{ID}\sin\alpha - N_{IH}\sin\alpha = 0$$

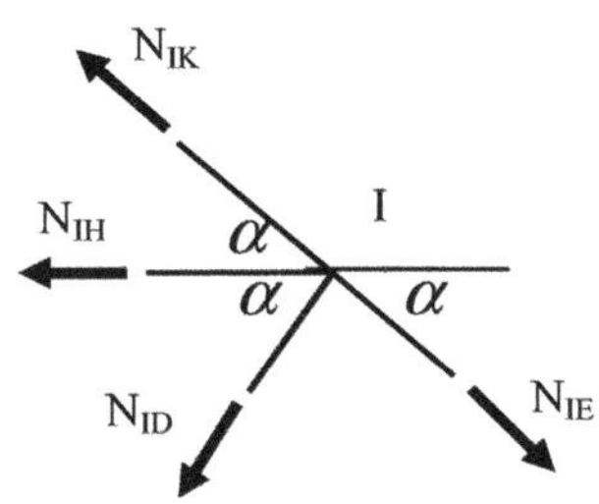

$$N_{IH} = -\frac{1}{2}P$$

$$N_{ID} = +\frac{\sqrt{5}}{4}P$$

Joint D

$$\sum F_x = 0: \; N_{DE} - N_{DC} + N_{DI}\cos\alpha - N_{DH}\cos\alpha = 0$$

$$\sum F_y = 0: \; N_{DI}\sin\alpha + N_{DH}\sin\alpha - P = 0$$

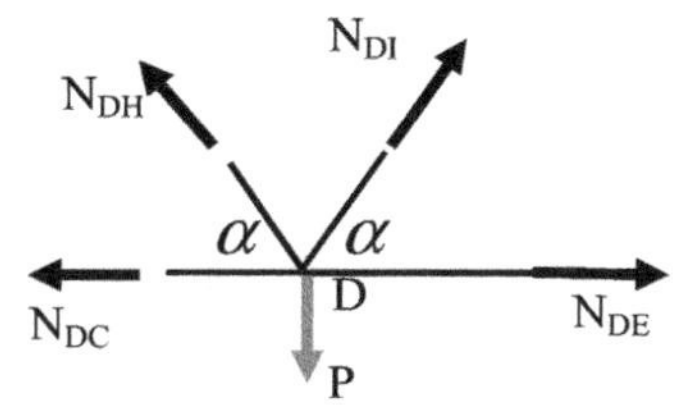

$$N_{DH} = \frac{\sqrt{5}}{4}P$$

$$N_{DC} = \frac{5}{4}P$$

Joint C

$$\sum F_y = 0:$$

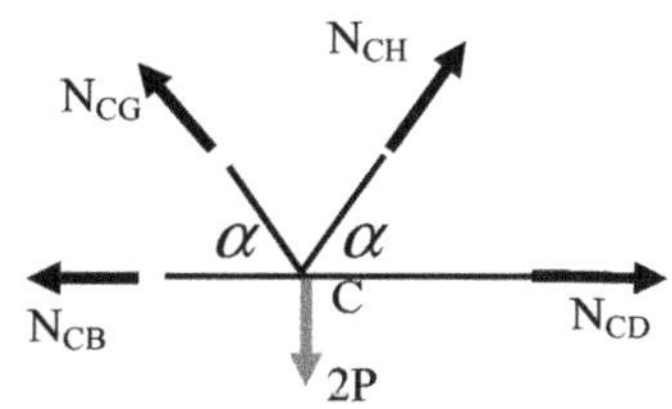

$$N_{CH}\sin\alpha + N_{CG}\sin\alpha - 2P = 0$$

$$N_{CH} = \frac{3\sqrt{5}}{4}P$$

The normal force diagram is shown in Figure 4.26.

Similarly, we have developed a numerical program that calculates the forces on the bars of a statically determinate or statically indeterminate truss. The given structure is modeled (Figure 4.27). The numerical calculation is carried out for the values a = 2 m and P = 1 kN; the numerical results are grouped in Table 4.4.

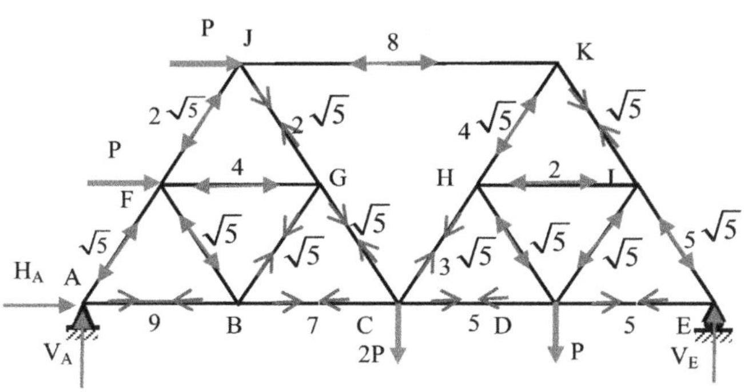

Figure 4.26. *Normal force diagram* $*(\frac{P}{4})$

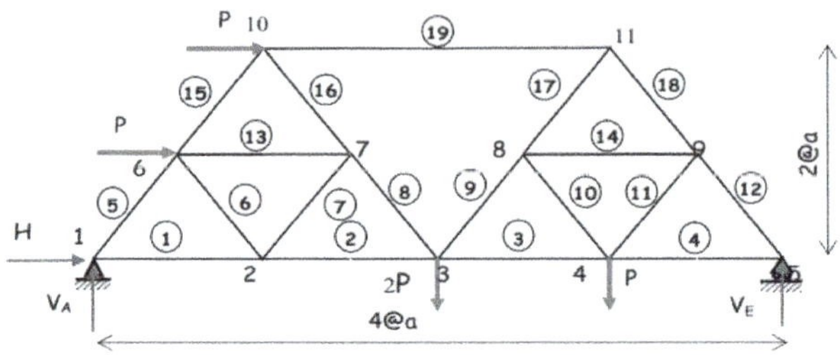

Figure 4.27. *Numbering of joints and bars*

Element	Joints	Normal force	Element	Joints	Normal force
1	1 2	–22,500E+01 0.22500E+01	11	4 9	–0.55902E+00 0.55902E+00
2	2 3	–0.17500E+01 0.17500E+01	12	5 9	0.27951E+01 –0.27951E+01
3	3 4	–0.12500E+01 0.12500E+01	13	6 7	0.50000E+00 –0.50000E+00
4	4 5	–0.12500E+01 0.12500E+01	14	8 9	0.50000E+00 –0.50000E+00
5	1 6	0.55902E+00 –0.55902E+00	15	6 10	0.11180E+01 –0.11180E+01
6	2 6	0.55902E+00 –0.55902E+00	16	7 10	–0.11180E+01 0.11180E+01
7	2 7	–0.55902E+00 0.55902E+00	17	8 11	–0.22361E+01 0.22361E+01
8	3 7	–0.55902E+00 0.55902E+00	18	9 11	0.22361E+01 –0.22361E+01
9	3 8	–0.16771E+01 16,771E+01	19	10 11	0.20000E+01 –0.20000E+01
10	4 8	–0.55902E+00 0.55902E+00			

Table 4.4. *Normal force in the bar structure (Figure 4.24)*

4.7. Complex trusses

Complex trusses are those which are not mentioned in the categories of simple or composite trusses. Complex trusses are statically determinate internally and externally, but they cannot be analyzed by the method of joint equilibrium or the method of sections. Figure 4.28 shows this category of trusses.

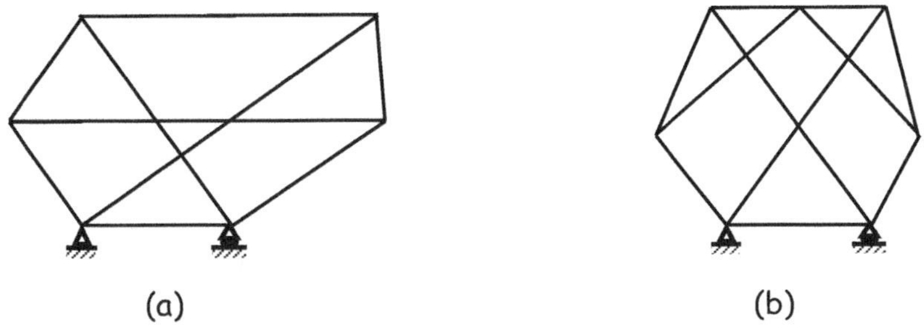

Figure 4.28. *Complex structures*

Truss (a)

Number of bars, b = 9.

Number of joints, n = 6.

b = 2n – 3, the structure is statically determinate internally.

Number of unknowns, r = 3.

Number of equations, n_e = 3.

The structure is externally statically determinate.

Truss (a) is statically determinate internally and externally.

Truss (b)

Number of bars, b = 11.

Number of joints, n = 7.

b = 2n – 3, the structure is statically determinate internally.

Number of unknowns, r = 3.

Number of equations, n_e = 3.

The structure is externally statically determinate.

Truss (b) is statically determinate internally and externally.

In trusses (a) and (b), there is no joint containing two bars, which makes it impossible to use the method of joint equilibrium or the method of sections.

As a result, matrix methods offer a very considerable advantage to the analysis of this category of structures. Matrix methods can be used interchangeably with the analysis of trusses without taking into account the static determinacy or static indeterminacy of the given truss.

4.8. Space trusses

A space truss is composed of bars interconnected at the ends, constituting a three-dimensional truss. External forces are always applied to the joints in all three directions. The simplest space truss is a tetrahedron consisting of six bars and four joints (Figure 4.29).

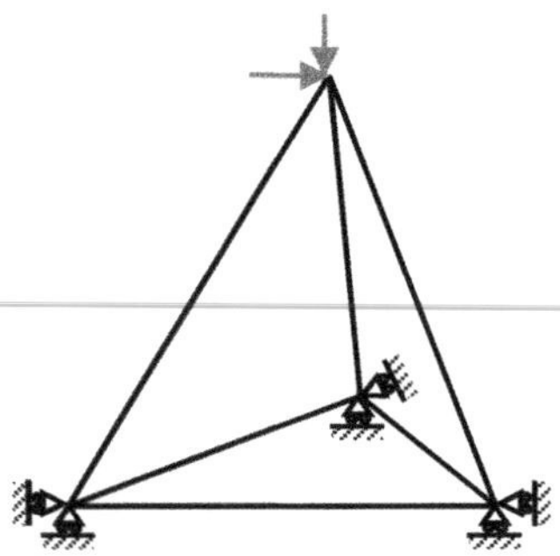

Figure 4.29. *Space truss*

Internal analysis

In a space truss with n joints, b bars and r support reactions, we can say that the structure is unstable internally if:

$$b+r-3n \prec 0 \tag{4.18a}$$

The structure is statically determinate internally if:

$$b+r-3n=0 \tag{4.18b}$$

The structure is statically indeterminate internally if:

$$b+r-3n \succ 0 \tag{4.19}$$

The static indeterminacy of a truss is defined by:

$$f=(b+r)-3n \tag{4.20}$$

External analysis

The equilibrium equations in the three directions are:

$$\sum \vec{F}_x = \vec{0} \quad [4.21a]$$

$$\sum \vec{F}_y = \vec{0} \quad [4.21b]$$

$$\sum \vec{F}_z = \vec{0} \quad [4.21c]$$

Similarly, the bending moments in the direction of the axis are:

$$\sum \vec{M}_x = \vec{0} \quad [4.22a]$$

$$\sum \vec{M}_y = \vec{0} \quad [4.22b]$$

$$\sum \vec{M}_z = \vec{0} \quad [4.22c]$$

The degree of external static indeterminacy can be deduced by:

$$f = r - 6 \quad [4.23]$$

We can distinguish that the space structure is unstable externally if:

$$f < 0 \quad [4.24]$$

The space structure is statically determinate externally if:

$$f = 0 \quad [4.25]$$

The space structure is statically indeterminate externally if:

$$f > 0 \quad [4.26]$$

4.8.1. *Space truss analysis*

Analysis methods for plane trusses have been discussed in section 4.5. These methods can be effectively extended to the analysis of space trusses.

For example, the method of joint equilibrium applies in the same way as in the analysis of plane trusses. Equations [4.21] must be satisfied at each joint of the space truss. In general, each joint has three forces or three independent reactions. For this reason, analyzing space trusses starts from the joint having, at most, three unknowns that can be deduced by applying static equations. This approach applies sequentially to each joint of the space truss until all the forces on the bars are determined.

EXAMPLE 4.6.–

The space truss (Figure 4.30) is composed of nine identical bars and is stressed by a nodal force F applied to the free joint in direction (ZZ):

– calculate support reactions:

– deduce the forces on the bars.

We isolate the space truss from its supports (Figure 4.31).

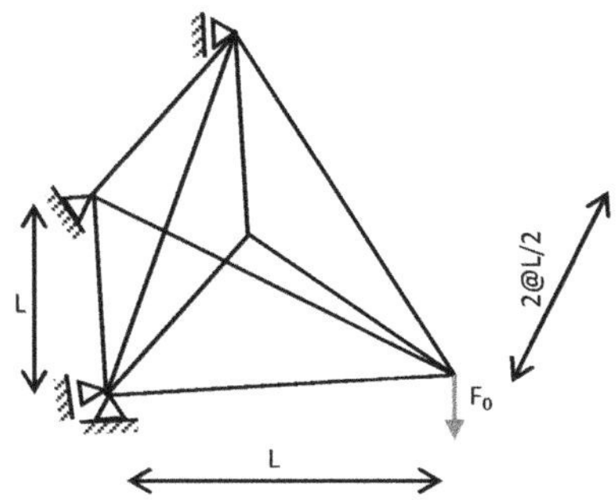

Figure 4.30. *Spatial truss*

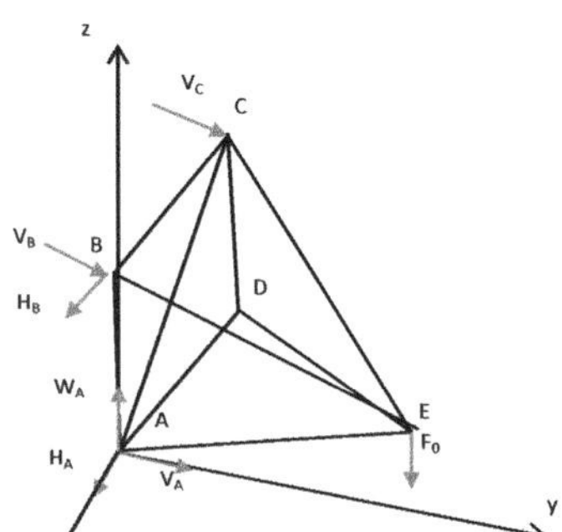

Figure 4.31. *Free-body diagram*

External analysis

The number of support reactions is equal to 6, which are H_A, V_A, W_A, H_B, V_B and V_C.

The number of static equations is equal to 6. The structure is therefore statically determinate externally.

Internal analysis

Number of bars, $b = 9$.

Number of support reactions, $r = 6$.

Number of equations, $3n = 3 \times 5 = 15$.

It is clear that $b + r - 3n = 0$.

The space truss is statically determinate internally and externally.

Calculation of support reactions

Conventionally, we associate the direction of any action along the positive direction of the axes of reference.

The application of relationships [4.22] and [4.24] can be written as:

$$\sum \vec{M}_x = \vec{0} : V_B.L - F_0\, L = 0$$

$$V_B = F_0$$

$$\sum \vec{M}_z = \vec{0} : V_C.L = 0$$

$$V_C = 0.$$

$$\sum \vec{M}_y = \vec{0} : H_B.L - F_0\, \frac{L}{2} = 0$$

$$H_B = \frac{F_0}{2}$$

The application of relationships [4.21] leads to:

$$\sum \vec{F}_x = \vec{0} : H_A + H_B = 0$$

$$H_A = -H_B = -\frac{F_0}{2}$$

$$\sum \vec{F}_y = \vec{0} : V_A + V_B = 0$$

$$V_A = -V_B = F_0$$

$$\sum \vec{F}_z = \vec{0} : W_A - F_0 = 0$$

$$W_A = F_0$$

Calculation of forces on the bars

It is necessary to apply the method of joint equilibrium starting with joint B.

Joint B

The equilibrium joint B allows us to write:

$$N_{BE}.\cos\psi.\cos\alpha + V_B = 0.$$

$$\cos\psi = \frac{\sqrt{5}}{3}, \ \cos\alpha = \frac{2}{\sqrt{5}}$$

$$N_{BE} = \frac{3}{2}F_0$$

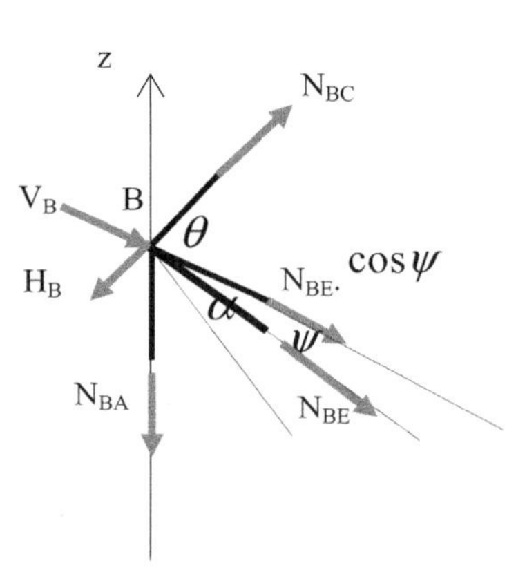

$$N_{BC} = H_B - N_{BE}.\cos\psi.\cos\theta$$

$$N_{BC} = 0.$$

$$N_{BA} + N_{BE}\sin\psi = 0.$$

$$N_{BA} = -F_0$$

Joint A

The equilibrium equation of joint C leads to:

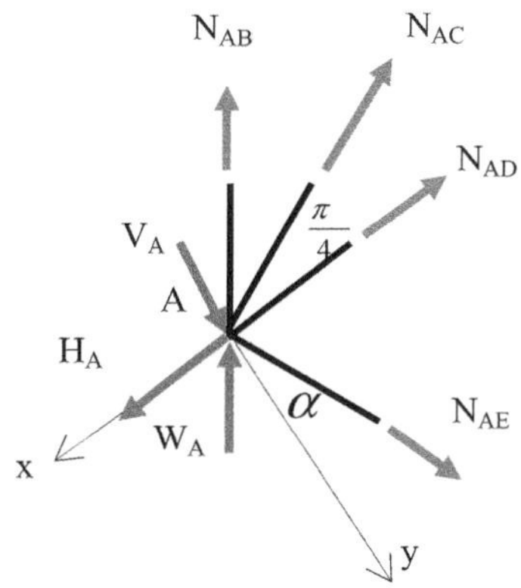

$$N_{AE} = -\frac{V_A}{\cos\alpha} = -\frac{\sqrt{5}}{2}F_0$$

$$N_{AC} = -\frac{(W_A + N_{AB})}{\cos\frac{\pi}{4}} = 0.$$

$$N_{AD} = H_A - N_{AE}.\sin\alpha = 0.$$

Joint D

Given that $N_{DA} = 0$, applying the equilibrium of joint D allows us to write:

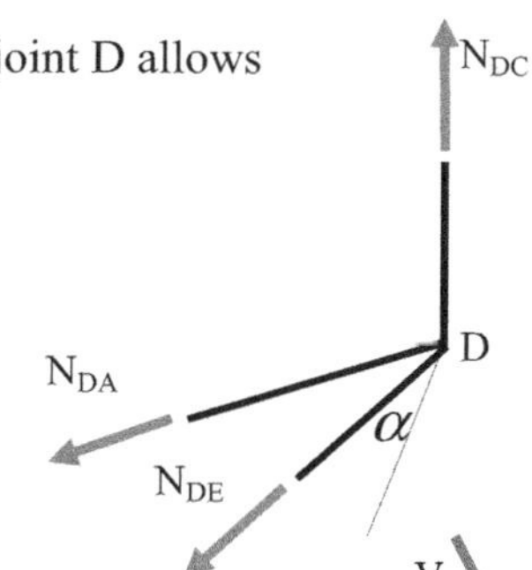

$$N_{DC} = 0.$$

$$N_{DE} = \frac{N_{DA}}{\cos\alpha} = 0.$$

Joint C

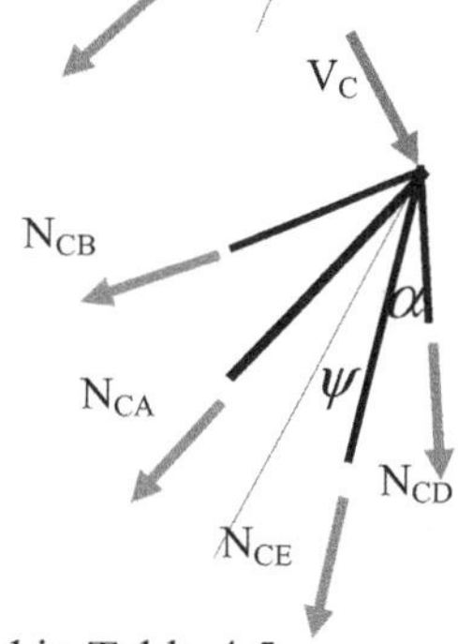

$$N_{CE}\cos\psi.\cos\alpha + V_C = 0.$$

$$N_{CE} = 0.$$

The forces on the bars of the space structure are grouped in Table 4.5.

Bar	Force	Nature of force
AB	$-F_0$	Compression
AC	0.	Neutral
AD	0.	neutral
AE	$-\frac{\sqrt{5}}{2}F_0$	Compression
DE	0.	Neutral
DC	0.	Neutral
BE	$\frac{3}{2}F_0$	Traction
BC	0.	Neutral
CE	0.	Neutral

Table 4.5. *Forces in the bars*

The normal force diagram is given in Figure 4.32.

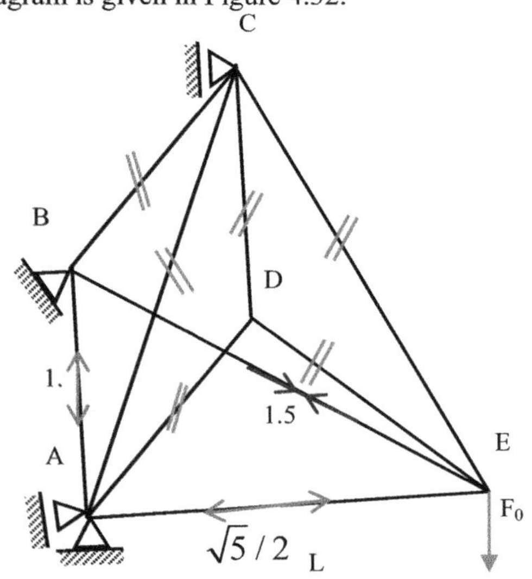

Figure 4.32. *Normal force on the bars *(F_0)*

4.9. Conclusion

A truss is a chain of bars connected to each other at the joints. Its analysis is based on the following assumptions:

– the bars are biarticulated;

– the loading system is a set of concentrated forces applied only to the joints;

– the bars of a truss are stressed by axial forces that can be traction, compression or neutral forces.

The external static indeterminacy of a truss is considered in the same way as that of beams and frames (Chapter 3), whereas the internal static indeterminacy is defined by $f = b - (2n - 3)$, hence:

$b > (2n - 3)$: the truss is statically indeterminate internally;

$b = (2n - 3)$: the truss is statically determinate internally;

$b < (2n - 3)$: the truss is unstable internally.

The forces on the bars of a truss can be deduced by (1) the method of joint equilibrium, (2) the methods of sections, (3) the matrix method, or (4) the graphical method.

In order to analyze composite structures, a combination of the method of joint equilibrium and the methods of sections is generally used.

The principle of the space structure analysis method is applied in the same way as in the analysis of plane trusses. The purpose of analysis methods of trusses is to determine support reactions and internal forces on the bars.

4.10. Problems

Exercise 1

Distinguish statically determinate, statically indeterminate and unstable structures among the following structures:

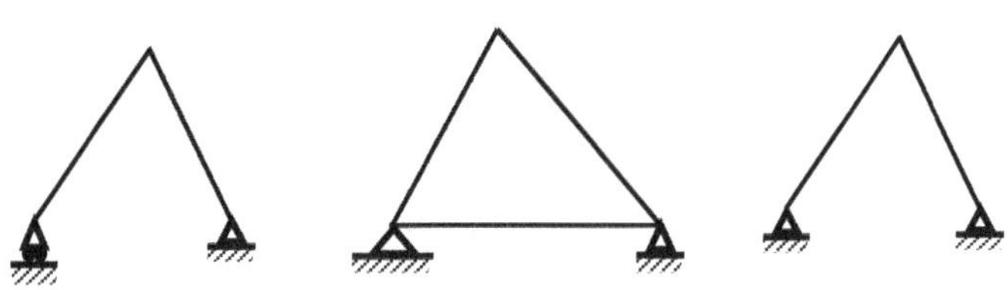

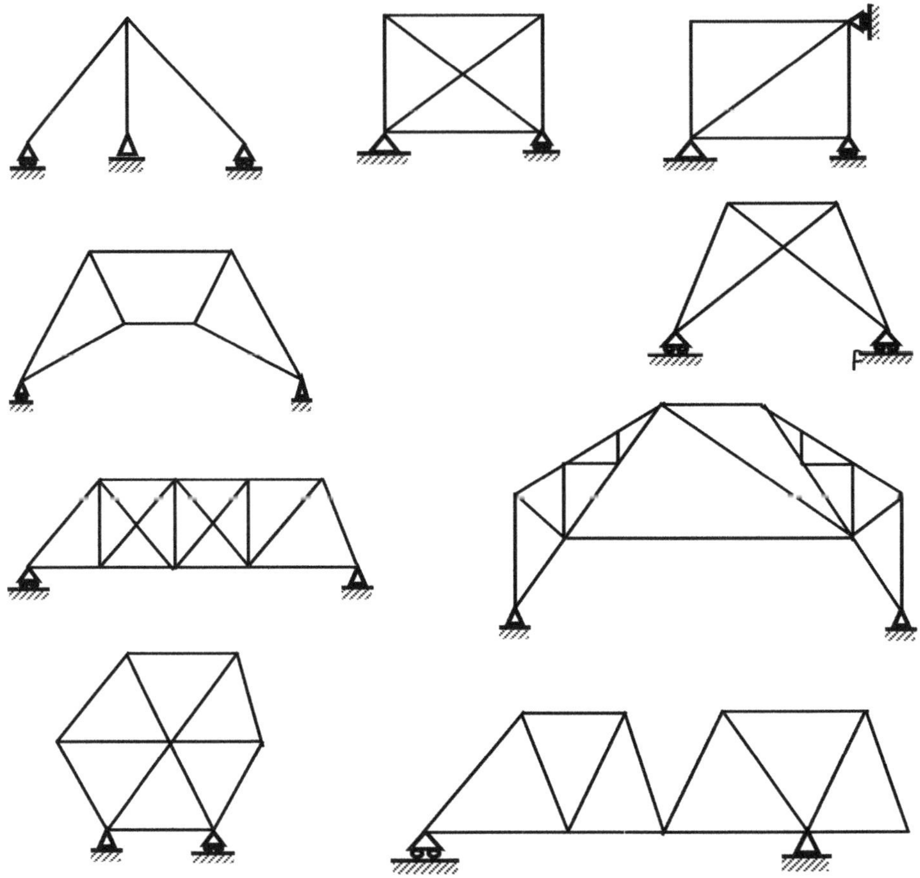

Exercise 2

Use the method of joint equilibrium to determine the forces on the bars of the following trusses and draw a diagram of internal forces:

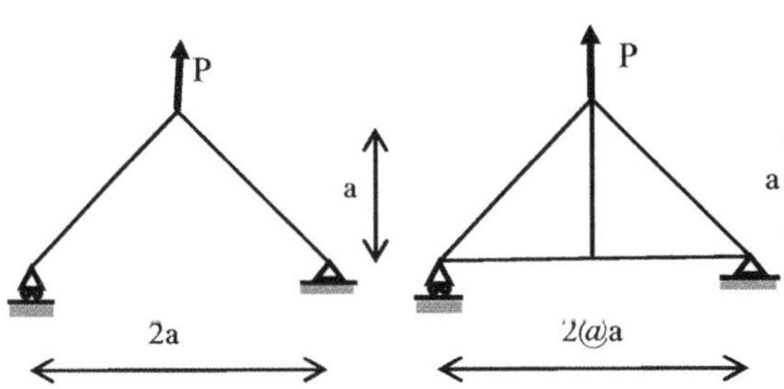

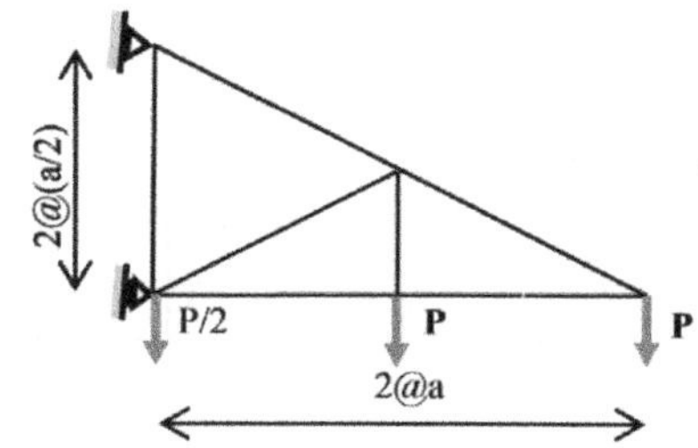

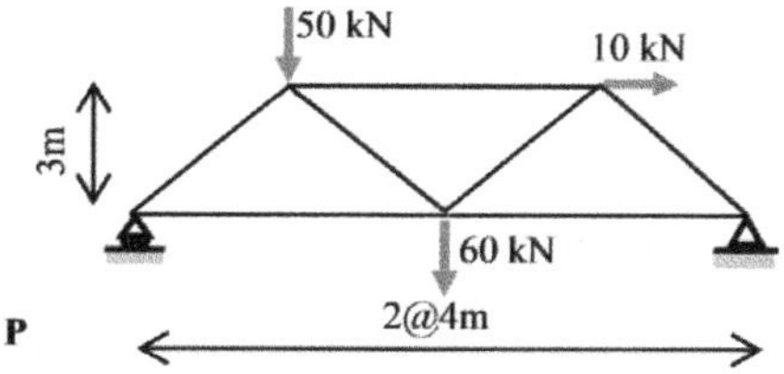

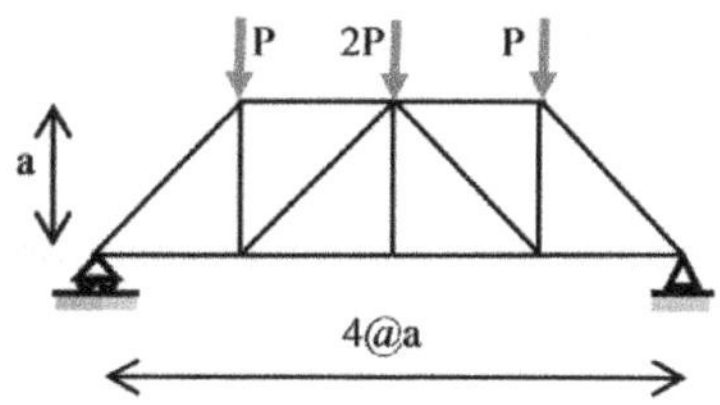

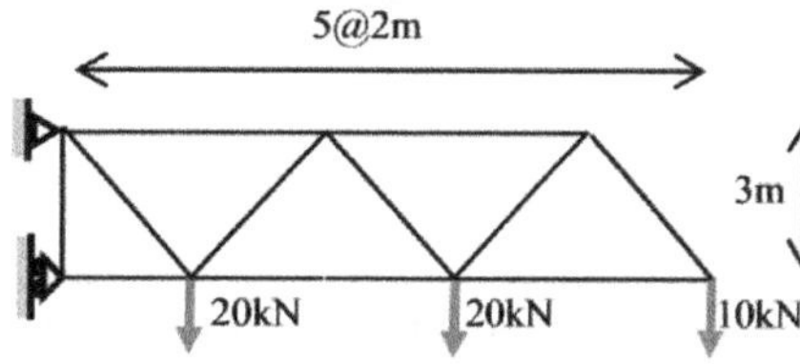

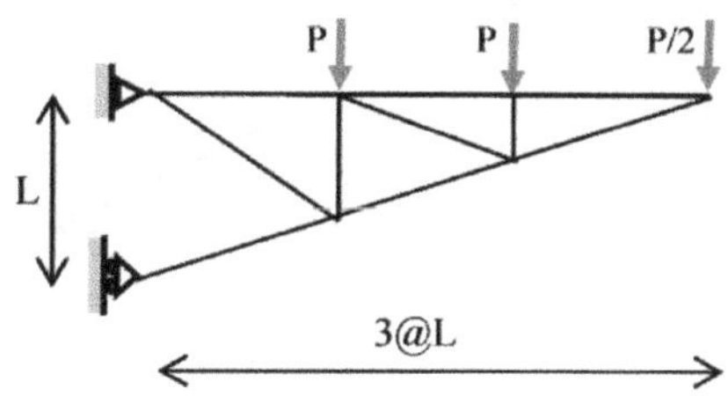

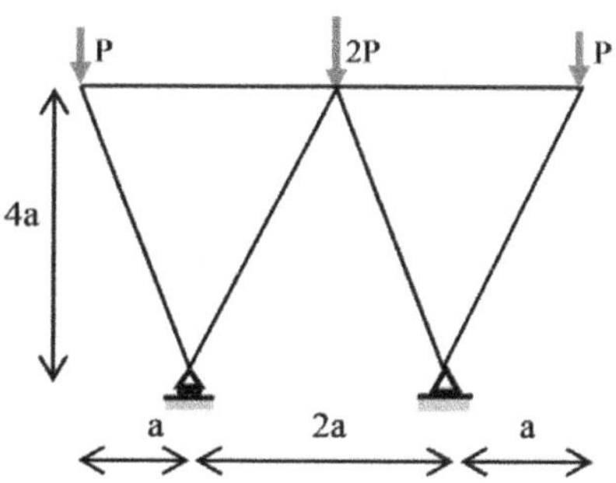

Exercise 3

Using the method of sections, determine the forces on bars (1), (2) and (3) of the following trusses:

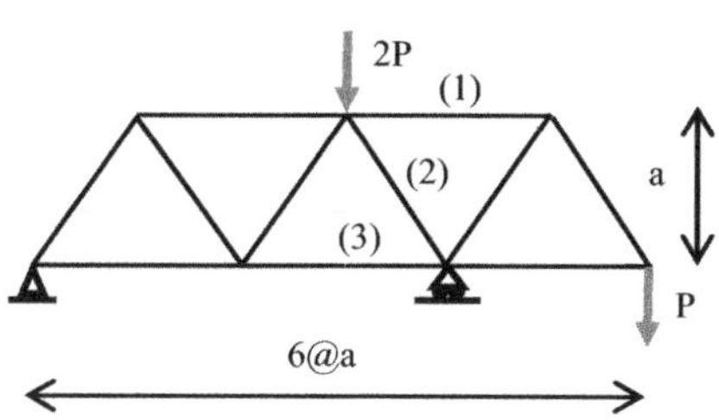

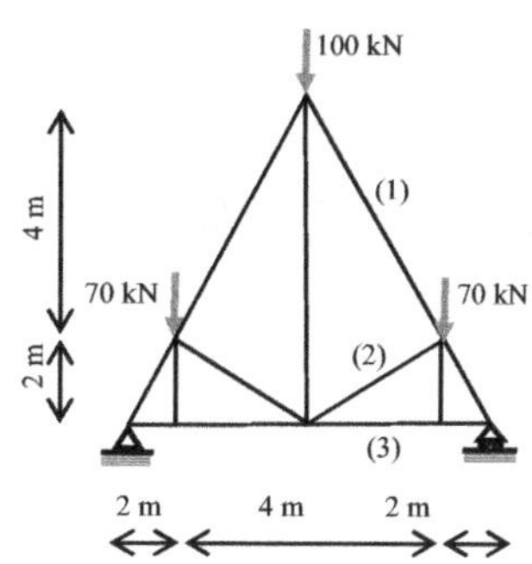

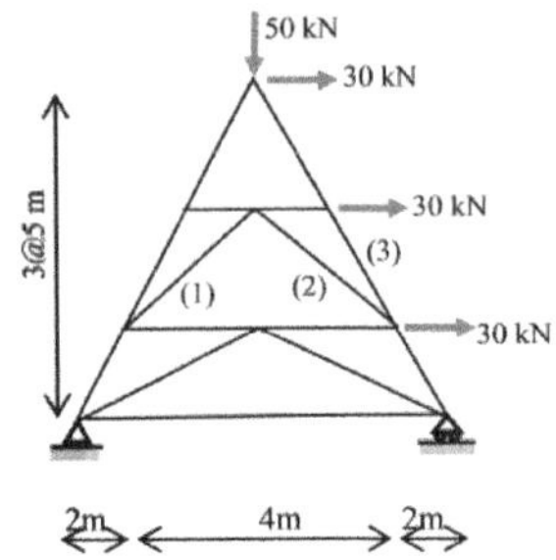

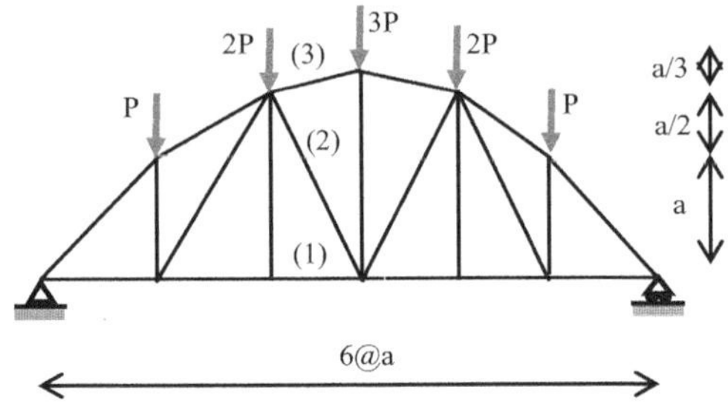

Exercise 4

Using the graphical method, evaluate the value and the direction of each force in the bars of the following trusses:

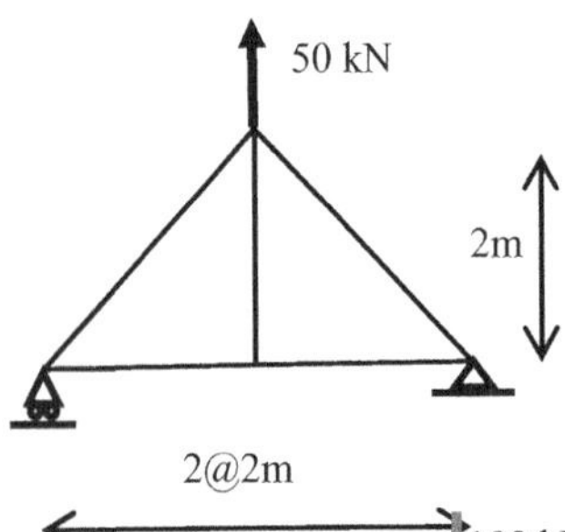

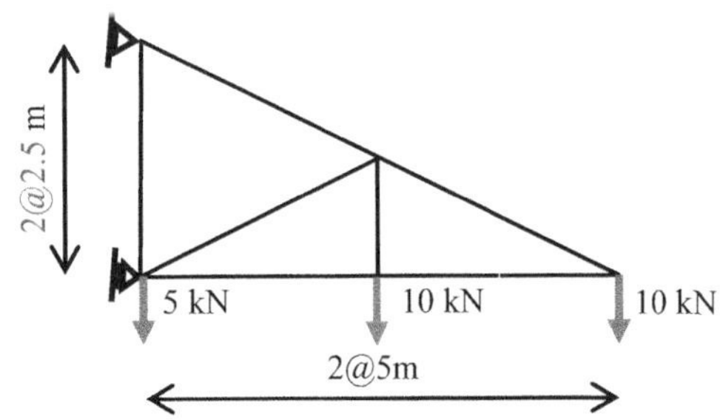

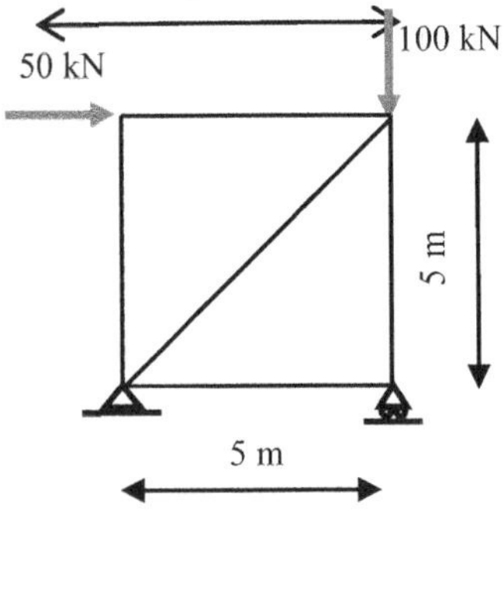

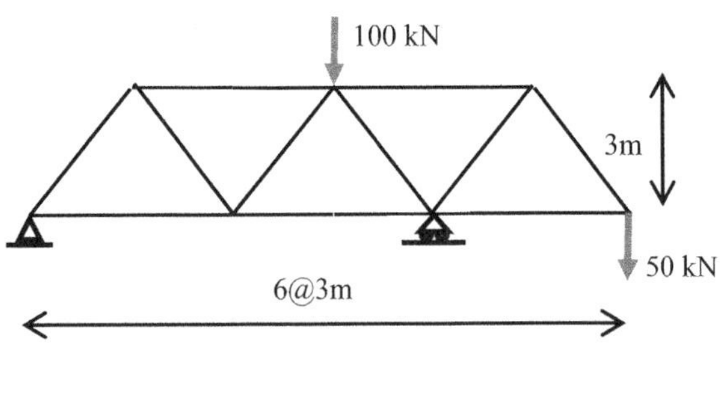

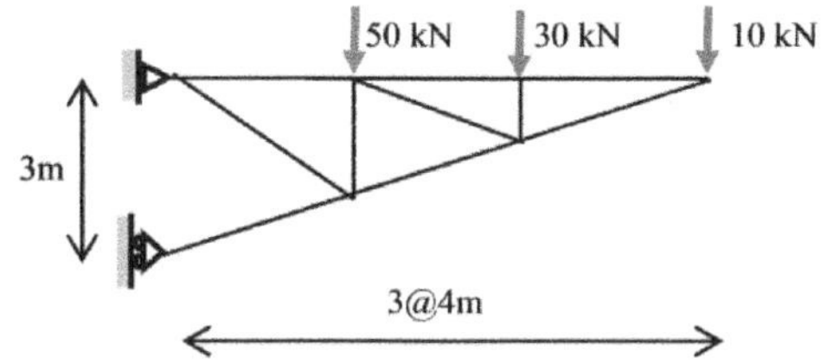

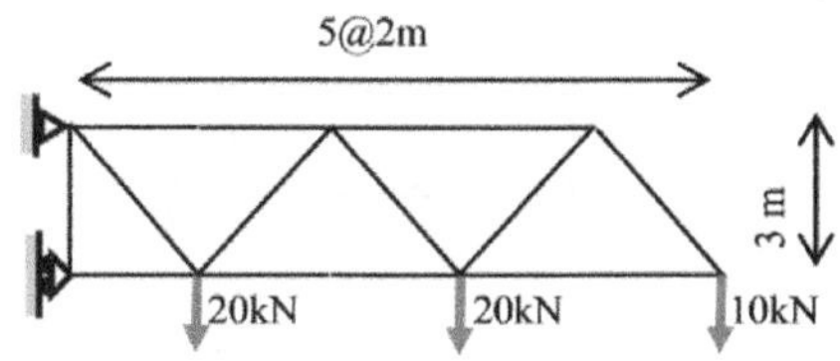

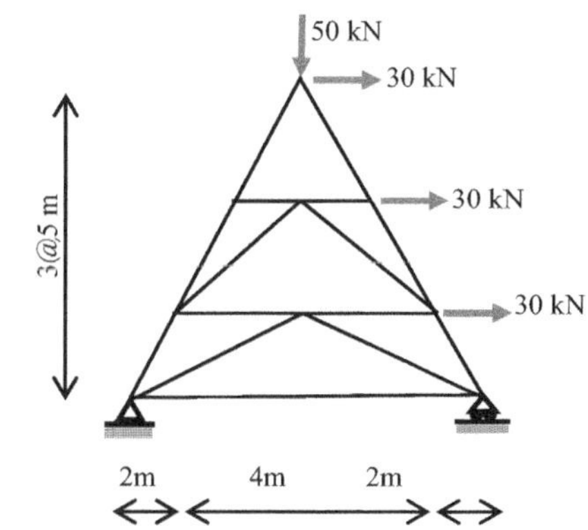

Exercise 5

Using the matrix method, determine the forces on the bars of the following trusses and draw a diagram of internal forces:

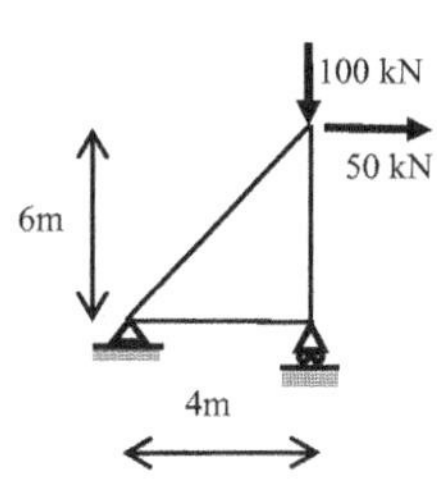

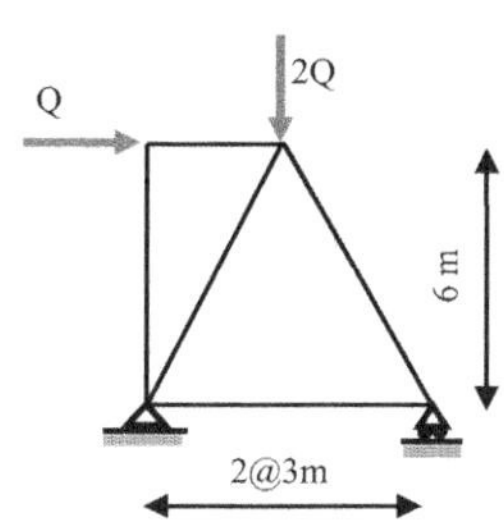

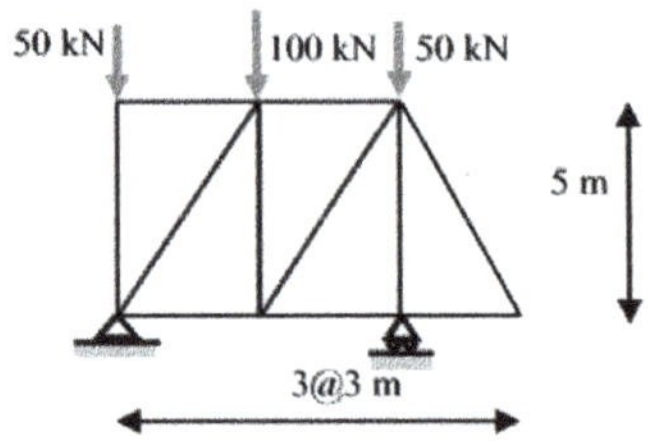

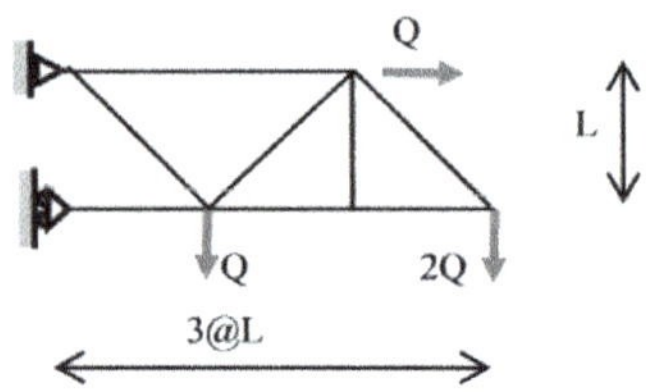

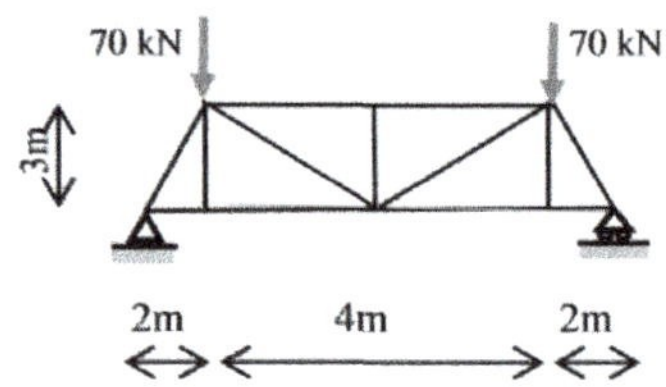

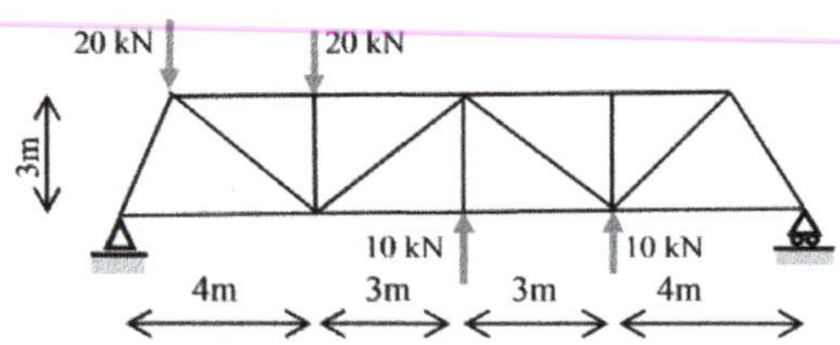

Exercise 6

Determine the forces on the bars of the following trusses and draw their diagrams:

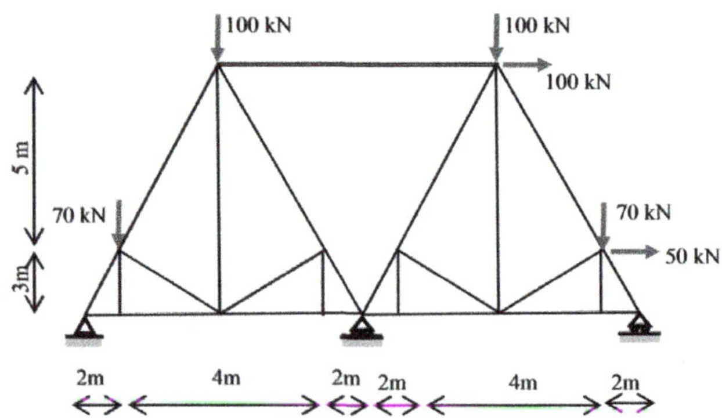

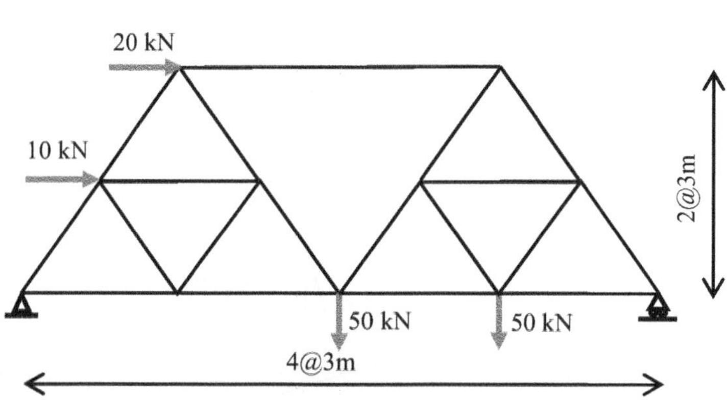

20 kN
10 kN
50 kN
50 kN
4@3m
2@3m

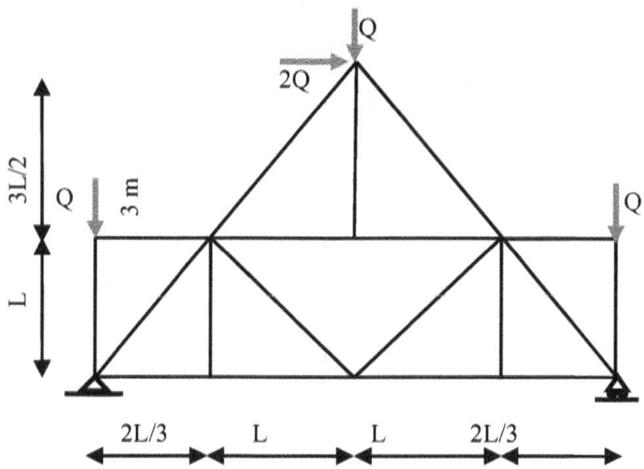

Q
2Q
3L/2
Q
3 m
Q
L
2L/3
L
L
2L/3

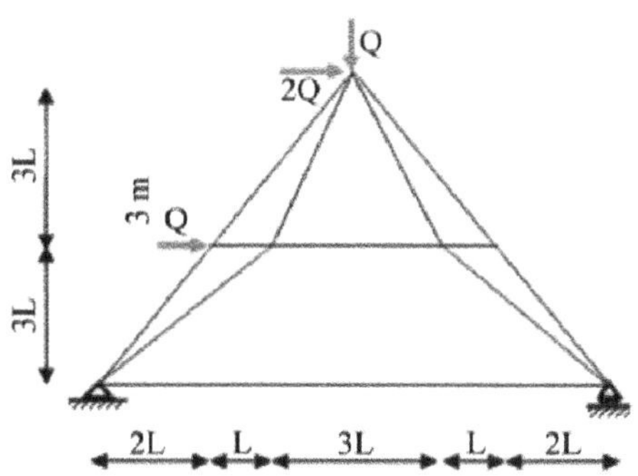

Q
2Q
3L
3 m
Q
3L
2L
L
3L
L
2L

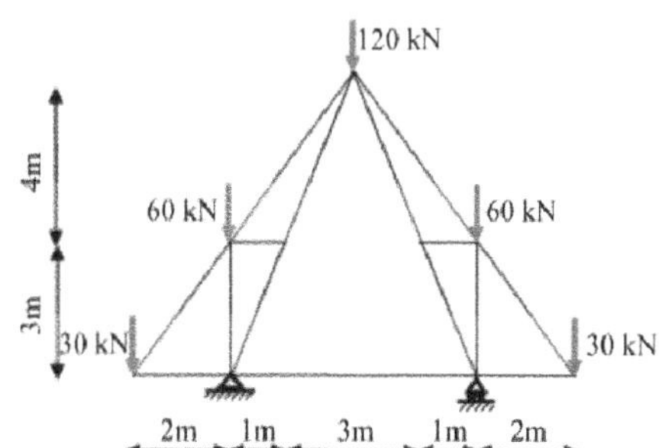

120 kN
4m
60 kN
60 kN
3m
30 kN
30 kN
2m
1m
3m
1m
2m

Exercise 7

Determine the forces on the bars of the following trusses and draw their diagrams:

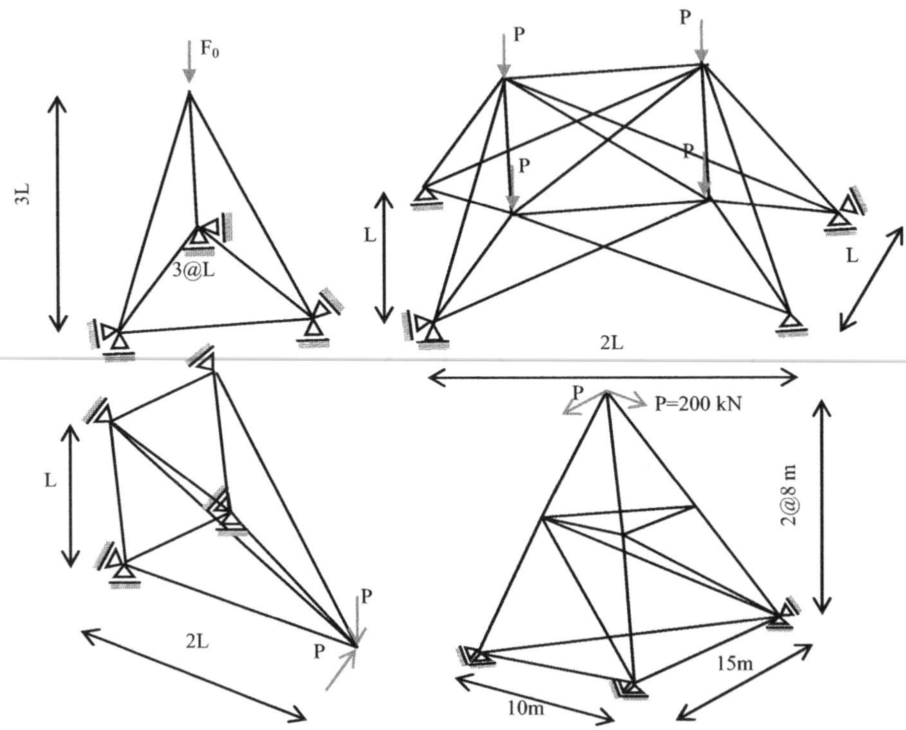

5

Internal Analysis of Beams and Frames

The teaching objectives for this chapter are as follows:

– determining the variations of internal actions along a beam or frame;

– identifying the relationship between the applied loading, the shear force and the bending moment of flexed structures;

– distinguishing between statically determinate, statically indeterminate and unstable beams and frames;

– establishing diagrams of internal actions;

– introduction to the analysis of statically indeterminate beams and frames.

We will first define the internal actions at any section and then generalize their calculation along plane beams and frames. In addition, we will formulate the relationships between the applied loading, the shear force and the bending moment. Finally, the analysis procedure for statically determinate beams and frames will be presented.

5.1. Normal force, shear force and bending moment

Consider a simply supported beam acted by any load q (x) (Figure 5.1).

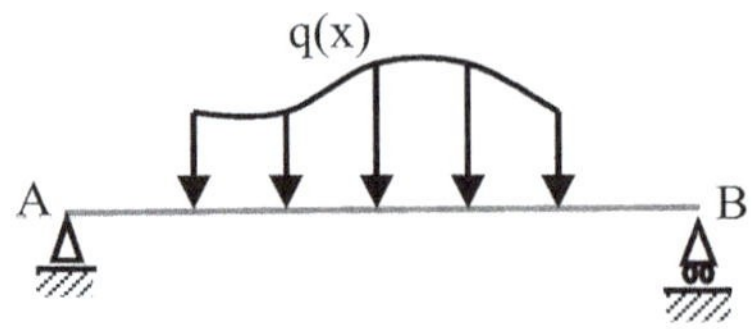

Figure 5.1. *Studied beam*[1]

We apply static equations to the free-body diagram (Figure 5.2). To obtain it, we substitute support reactions with a system of equivalent forces.

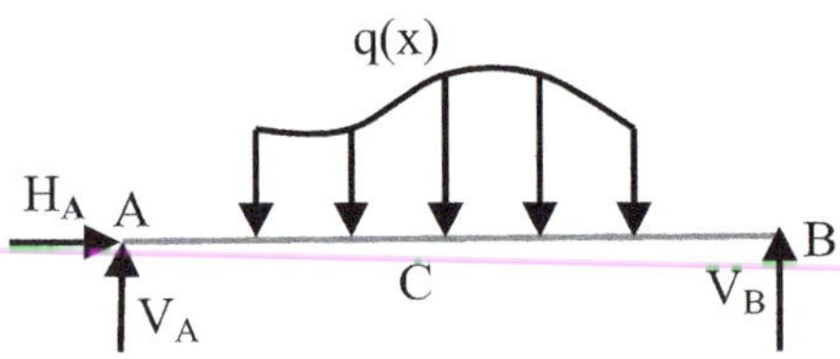

Figure 5.2. *Free-body beam*

We generally use the method of sections to determine internal actions. In this example, it is necessary to calculate the internal actions at point C. Thus, the beam (AB) is cut by a section perpendicular to the central axis, which divides it into two portions (AC) and (BC) (Figure 5.3).

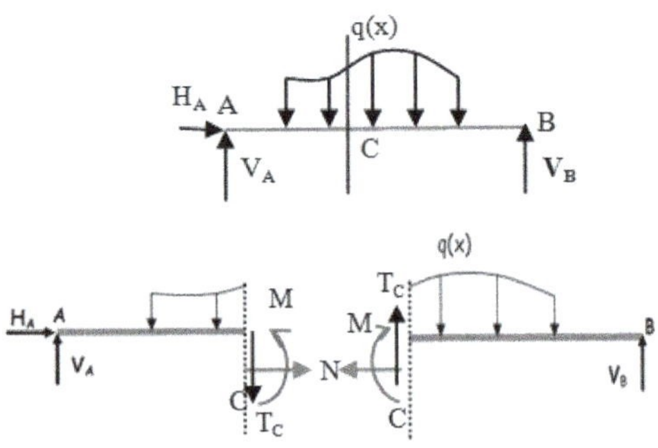

Figure 5.3. *Internal actions at any section*

1 All of the figures in this chapter are available to view in full color at www.iste.co.uk/khalfallah/analysis1.zip.

The internal actions at point C for both portions are N_C, T_C and M_C. The axial force N_C on the central axis of the beam is called the normal force while the force perpendicular to the central axis is the tangential or shear force. Finally, the internal moment is called the bending moment.

The portion (AC) of the beam (AB) (Figure 5.3) shows the internal actions related to the section passing through point C. The internal actions of portion (CB) are equal to those of portion (AC) as well as in the opposite direction.

Internal actions related to a section can be calculated using static equations at one of the two portions of the beam.

$$\sum F_x = 0 \quad [5.1]$$

$$\sum F_y = 0 \quad [5.2]$$

$$\sum M_i = 0 \quad [5.3]$$

The relationship [5.1] allows us to deduce the normal force, which is equal to the algebraic sum of all the components of external forces acting on the portion, in the direction of the central axis. The relationship [5.2] allows us to determine the shear force that is equal to the algebraic sum of all the components of applied forces on this portion, perpendicular to the central axis. Finally, the relationship [5.3] allows us to calculate the bending moment M_C, which is assumed to be equal to the algebraic sum of all the moments caused by applied external actions on the portion considered.

5.2. Sign convention

The sign convention given to the reduction elements is projected in Figure 5.4. The normal force is assumed to be positive if it generates tensile stresses. The shearing force is considered to be positive if it is directed down toward the left portion. Finally, the bending moment is positive if it causes compressive stresses in the upper part of the beam.

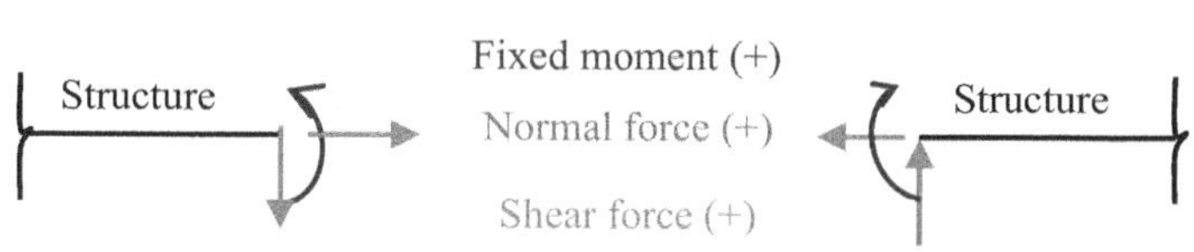

Figure 5.4. *Sign convention*

5.3. Beam analysis procedure

We summarize the steps for analyzing a statically determinate beam:

1) support reactions of the statically determinate beam are calculated by applying equilibrium equations and additional conditions if the beam is statically indeterminate;

2) the beam is divided by a section into two portions where we want to compute the internal actions;

3) we consider one of the two portions. Applying static equations allows us to calculate the internal actions relative to this section, respecting the sign convention;

4) careful application of equilibrium equations allows us to determine the internal actions, which are the normal force, the shear force and the bending moment according to the position of the section;

5) it is possible to check the calculation by taking the second portion of the beam. Internal actions must be identical for both portions;

6) finally, after determining the expressions of internal actions, we can draw their diagrams. Internal action diagrams are used to select critical sections and their corresponding values.

EXAMPLE 5.1.–

Determine the internal actions at point B of the beam (Figure 5.5).

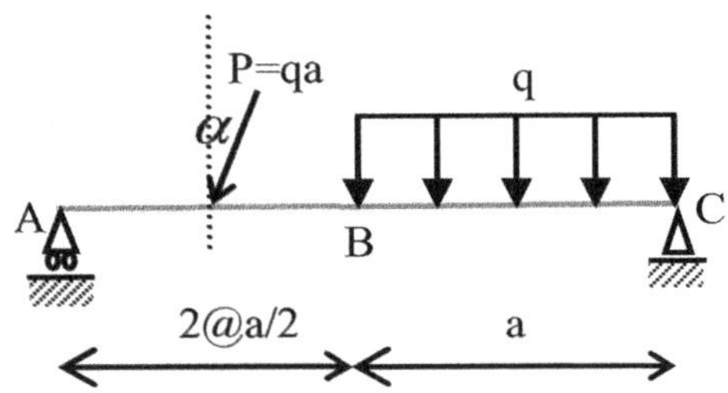

Figure 5.5. *Given beam*

Support reactions

We consider the free-body beam (Figure 5.6).

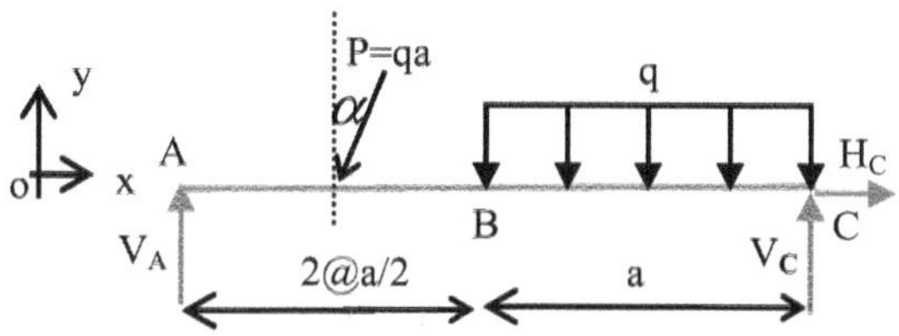

Figure 5.6. *Free-body beam*

We apply static equations.

$$\sum M_C = 0: \ V_A 2a - P\cos\alpha(a + \frac{a}{2}) - qa\frac{a}{2} = 0$$

$$V_A = \frac{qa}{4}(1 + 3\cos\alpha)$$

$$\sum F_x = 0: \ H_c - P\sin\alpha = 0$$

$$H_c = P\sin\alpha$$

$$\sum F_y = 0: \ V_A + V_C - P\cos\alpha - qa = 0$$

$$V_C = \frac{qa}{4}(3 + \cos\alpha)$$

Internal actions

We consider a section of the beam (AC) passing through point B (Figure 5.7).

Portion (I)

The equilibrium of the portion (I) allows writing

$$\sum F_x = 0: \ N_B - P\sin\alpha = 0$$

$$N_B = P\sin\alpha$$

$$\sum F_y = 0: \ -T_B + V_A - P\cos\alpha = 0$$

$$T_B = \frac{qa}{4}(1-\cos\alpha)$$

$$\sum M_A = 0: \; M_B - P\cos\alpha.\frac{a}{2} - T_B a = 0$$

$$M_B = \frac{qa^2}{4}(1+\cos\alpha)$$

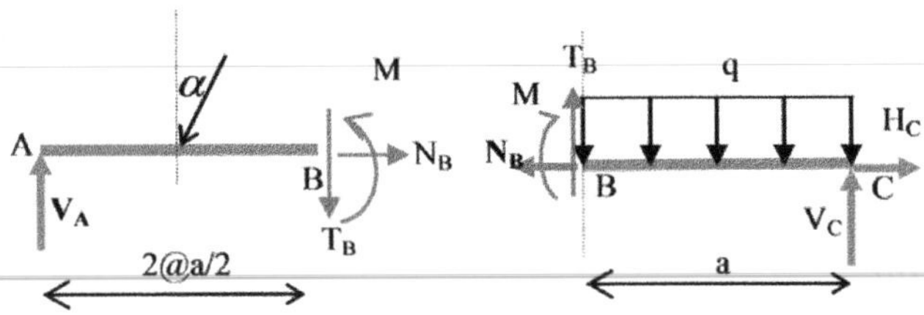

Figure 5.7. *Reduction elements of section B*

We consider part (II) (Figure 5.7).

$$\sum F_x = 0: \; H_C - N_B = 0$$

$$N_B = P\sin\alpha$$

$$\sum F_y = 0: \; T_B + V_C - qa = 0$$

$$T_B = \frac{qa}{4}(1-\cos\alpha)$$

$$\sum M_C = 0: \; M_B - qa.\frac{a}{2} + T_B a = 0$$

$$M_B = \frac{qa^2}{4}(1+\cos\alpha)$$

Internal actions at point B are:

– normal force: $N_B = P\sin\alpha$;

– tangential force: $T_B = \dfrac{qa}{4}(1-\cos\alpha)$;

– bending moment: $M_B = \dfrac{qa^2}{4}(1+\cos\alpha)$.

EXAMPLE 5.2.–

Determine the internal actions at point C of the beam (Figure 5.8).

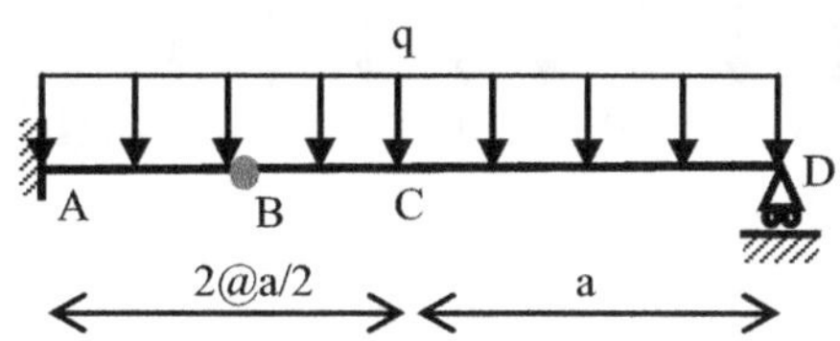

Figure 5.8. *Given beam*

The number of unknowns is equal to 4.

The number of static equations is equal to 3. Adding an additional equation $M_B = 0$ (a hinge). Finally, the number of equations becomes equal to the number of unknowns.

Support reactions

$$\sum M_B = 0:\ V_D.\frac{3}{2}a - \frac{1}{2}q(\frac{3a}{2})^2 = 0$$

$$V_D = \frac{3qa}{4}$$

$$\sum F_x = 0:\ H_A = 0$$

$$\sum F_y = 0:\ V_A + V_D - 2qa = 0$$

$$V_A = \frac{5qa}{4}$$

$$\sum M_B = 0 : V_A . \frac{1}{2} a - M_A - q \frac{a}{2} . \frac{a}{4} = 0$$

$$M_A = \frac{1}{2} q a^2$$

Figure 5.9 plots the support reactions of the beam.

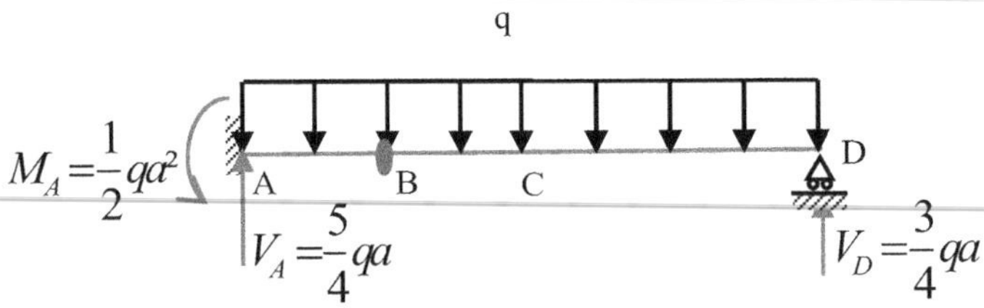

Figure 5.9. *Support reactions*

Internal actions

The beam (AD) is divided by a section passing through point C (Figure 5.10).

The equilibrium of the portion allows us to write

$$\sum F_x = 0 : N_C = 0$$

$$\sum F_y = 0 : T_C + V_D - qa = 0$$

$$T_C = \frac{1}{4} qa$$

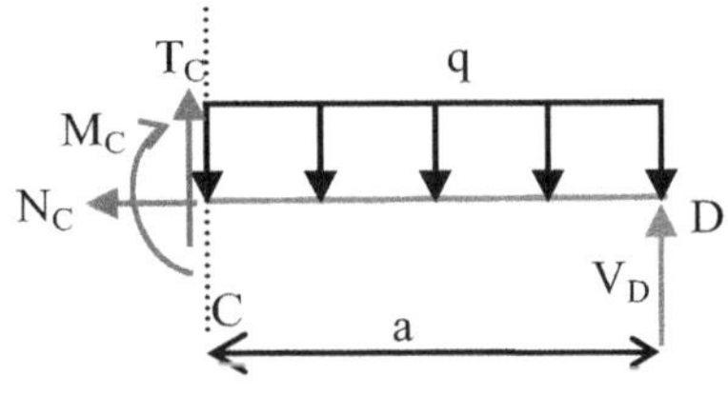

Figure 5.10. *Right portion*

$$\sum M_C = 0 : \ M_C + qa.\frac{a}{2} - V_D a = 0$$

$$M_B = \frac{qa^2}{4}$$

Thus, internal actions at point C are:

– normal force: $N_C = 0$;

– tangential force: $T_C = \frac{1}{4} qa$;

– bending moment: $M_B = \frac{qa^2}{4}$.

5.4. Diagrams of internal actions

Calculations of internal actions are presented in section 5.3. The internal actions vary along the beam and this variation allows us to work out the magnitude and direction of each internal action from their diagrams. The goal is therefore to determine the formulas of internal actions according to the abscissa of the section. The successive positions of the section depend on changing external actions. The magnitude of the evaluated internal actions plotted on the *y*-axis according to the abscissa of the section result in the diagram of this action.

EXAMPLE 5.3.–

Give the variation of internal actions of the beam (Figure 5.5) and draw their diagrams.

$$0 \le x \le \frac{a}{2}$$

The equilibrium of the element allows us to write

$$N(x) = 0$$

$$T(x) = V_A = \frac{qa}{4}(1 + 3\cos\alpha)$$

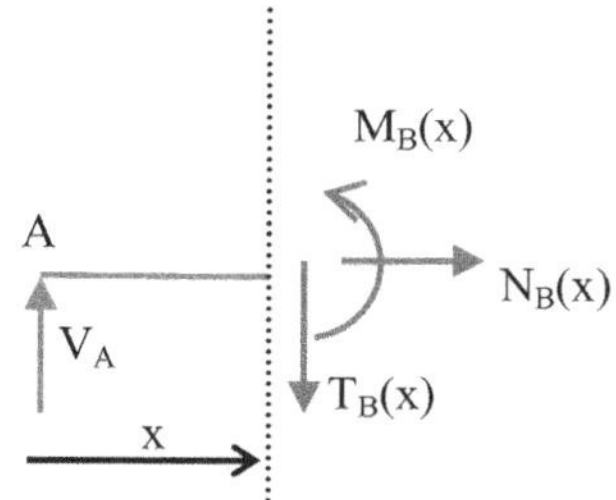

$$M(x) = V_A x = \frac{qa}{4}(1+3\cos\alpha)x$$

$$M(0) = 0$$

$$M(a) = \frac{qa^2}{4}(1+3\cos\alpha)$$

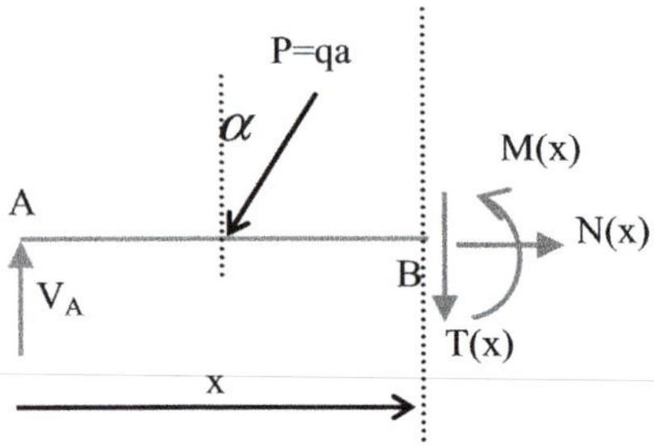

$$\frac{a}{2} \le x \le a$$

$$N(x) = P\sin\alpha$$

$$T(x) = V_A - P\cos\alpha = \frac{qa}{4}(1-\cos\alpha)$$

$$M(x) = V_A x - P\cos\alpha(x-\frac{a}{2}) = \frac{qa}{4}(1-\cos\alpha)x + \frac{qa^2}{2}\cos\alpha$$

$$M(\frac{a}{2}) = \frac{qa^2}{8}(3+\cos\alpha)$$

$$M(a) = \frac{qa^2}{4}(1+\cos\alpha)$$

$$0 \le Z \le a$$

$$N(z) = P\sin\alpha$$

$$T(z) = qz - V_C = qz - \frac{qa}{4}(3+\cos\alpha)$$

$$T(0) = -\frac{qa}{4}(3+\cos\alpha)$$

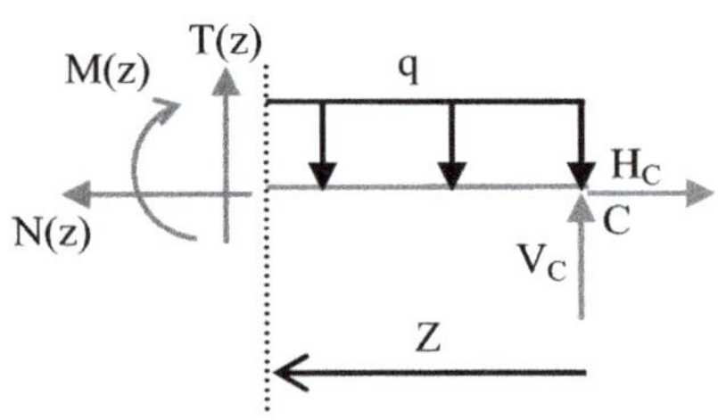

$$T(a) = \frac{qa}{4}(1-\cos\alpha)$$

$$M(z) = V_C z - \frac{1}{2}qz^2 = \frac{qa}{4}(3+\cos\alpha)z - \frac{1}{2}qz^2$$

$$M(0) = 0$$

$$M(a) = \frac{qa^2}{4}(1 + \cos\alpha)$$

The diagrams of internal actions are given by Figure 5.11.

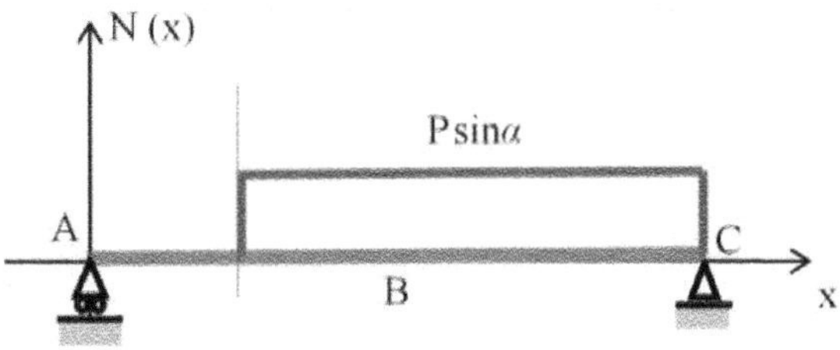

Normal force diagram

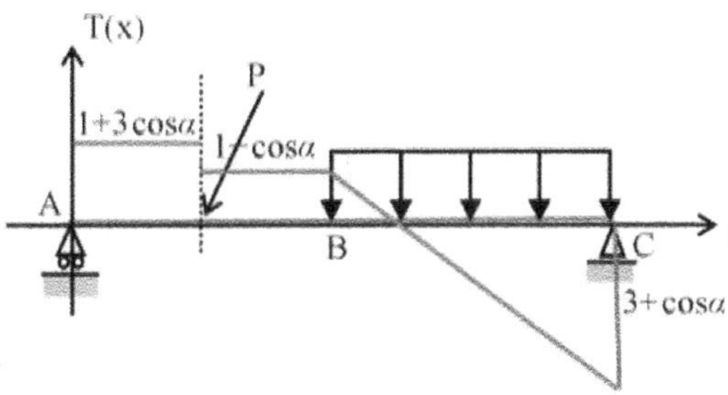

Shear force diagram $*(\frac{qa}{4})$

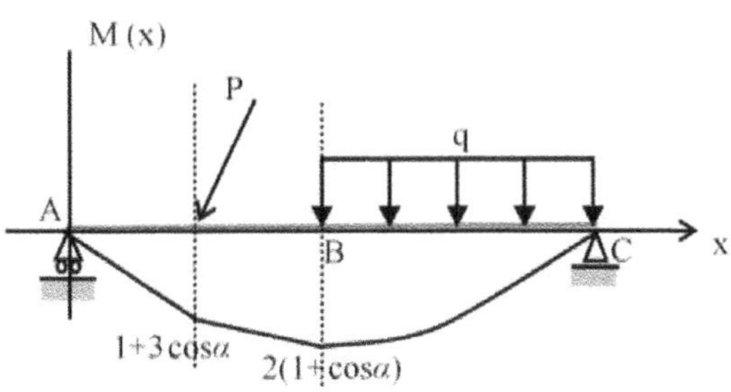

Bending moment diagram $*(\frac{qa^2}{8})$

Figure 5.11. *Diagram of internal actions*

5.5. Relationship between loading, shear force and bending moment

A beam element having the length Δx with applied loading is shown in Figure 5.12.

The equilibrium equations of the beam element are given as

$$\sum F_y = 0 \ : \ T(x) - \mathrm{q}(x).\Delta x - (T(x) + \Delta T(x)) = 0 \qquad [5.4]$$

$$\frac{\Delta T(x)}{\Delta x} = -q(x) \qquad [5.5]$$

$$\Delta x \to 0, \ \frac{dT(x)}{dx} = -q(x) \qquad [5.6]$$

$$\sum M_O = 0: \ \mathrm{M}(x) + T(x).\Delta x - (\mathrm{M}(x) + \Delta M(x)) - \mathrm{q}(x).\Delta x.\frac{\Delta x}{2} = 0 \qquad [5.7]$$

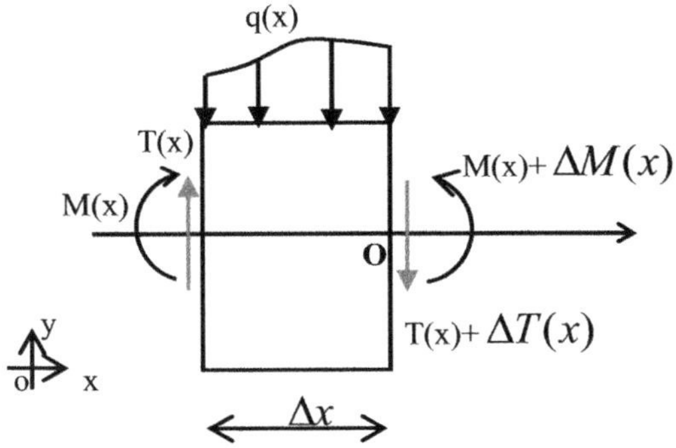

Figure 5.12. *Study of an element* Δx

The term $\frac{1}{2}\mathrm{q}(x).\Delta x^2$ can be neglected (higher order in Δx).

The relationship [5.5] is written as

$$\mathrm{M}(x) + T(x).\Delta x - \mathrm{M}(x) - \Delta M(x) = 0 \qquad [5.8]$$

Or

$$\frac{\Delta \mathrm{M}(x)}{\Delta x} = T(x) \qquad [5.9]$$

$$\Delta x \to 0,\ \frac{dM(x)}{dx} = T(x) \qquad [5.10]$$

Deriving the relationship [5.10], we obtain

$$\frac{d^2M(x)}{dx^2} = \frac{dT(x)}{dx} \qquad [5.11]$$

Substituting the relationship [5.6] with [5.11], we get

$$\frac{d^2M(x)}{dx^2} = -q(x) \qquad [5.12]$$

Equations [5.10] and [5.12] describe the relationship between applied loading, bending moment and shear force.

5.6. Static determinacy, static indeterminacy and instability of beams and frames

In general, the static determinacy, static indeterminacy and instability criteria of a plane structure containing *b* bars, *n* joints, *r* unknowns and *k* additional conditions is deduced by the following formula:

$$\text{f} = (3\text{b} + \text{r}) - (3\text{n} + \text{k}) \qquad [5.13]$$

The term (3b + r) represents the number of unknowns of the problem and (3n + k) represents the independent equations. We can distinguish:

– if $f \succ 0$: ($(3b+r) \succ (3n+k)$), the beam or frame is statically indeterminate of the order f;

– if $f = 0$: ($(3b+r) = (3n+k)$), the structure is statically determinate;

– finally if $f \prec 0$, the structure is unstable.

EXAMPLE 5.4.–

Determine the degrees of static indeterminacy of the structures (Figure 5.13).

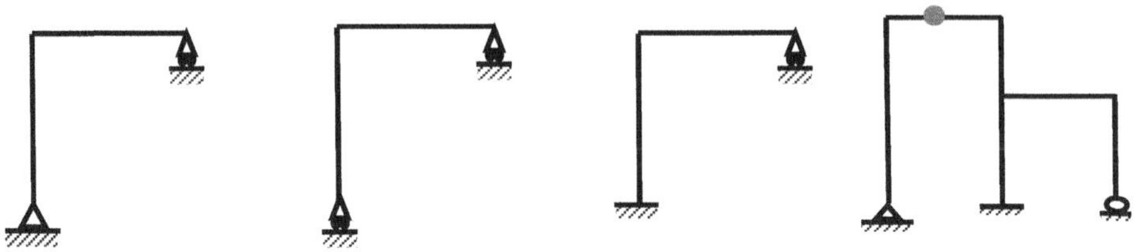

Figure 5.13. *Given beams and frames*

The static determinacy, static indeterminacy and instability of beams and frames (Figure 5.13) are grouped in Table 5.1.

Beam	b	r	n	k	f	Nature of the beam
P1	1	3	2	0	0	Statically determinate
P2	1	2	2	0	-1	Unstable
P3	1	4	2	0	1	Statically indeterminate
P4	1	5	2	1	1	Indeterminate

Frame	b	r	n	k	f	Nature of the frame
Por1	2	3	3	0	0	Statically determinate
Por2	2	2	3	0	-1	Unstable
Por3	2	4	3	0	1	Indeterminate
Por4	6	6	7	1	2	Indeterminate

Table 5.1. *Degree of static indeterminacy beams and frames*

5.7. Plane frame analysis procedure

The following steps are used to analyze plane frames:

1) Verify the static determinacy of a plane frame. If the degree of static indeterminacy is strictly greater than zero, it is possible to analyze the frame in a

particular way by adding hinges. This analysis depends on the number of hinges and where they are on the frame.

2) Determining support reactions is feasible using static equations and additional conditions, if they exist. This is valid if the number of equations is equal to the number of unknowns of the problem.

3) It is necessary to formulate the expressions of internal actions in each region of each bar of the frame. After presenting the expressions of internal actions, we can calculate the magnitudes at the ends of each bar.

4) We construct the diagrams of internal actions for each bar of the frame.

EXAMPLE 5.5.–

Draw the diagrams of internal actions (Figure 5.14).

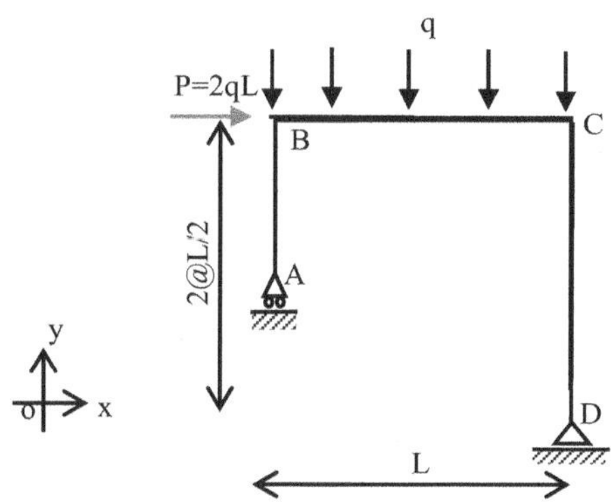

Figure 5.14. *Given frame*

Frame (Figure 5.14) is statically determinate.

Support reactions

$$\sum M_A = 0 \; : \; V_D . L - \frac{1}{2} q L^2 - P \frac{L}{2} = 0$$

$$V_D = \frac{3}{2} q L$$

$$\sum F_x = 0 : \; H_D + 2qL = 0$$

$$H_D = -2qL$$

$$\sum F_y = 0 : V_A + V_D - qL = 0$$

$$V_A = -\frac{1}{2}qL$$

Figure 5.15 shows support reactions.

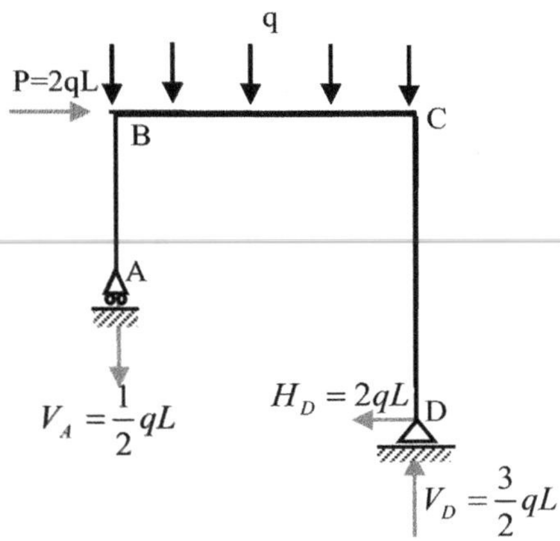

Figure 5.15. *Support reactions*

Internal actions

We divide element (AB) by a section on the *y*-axis (Figure 5.16).

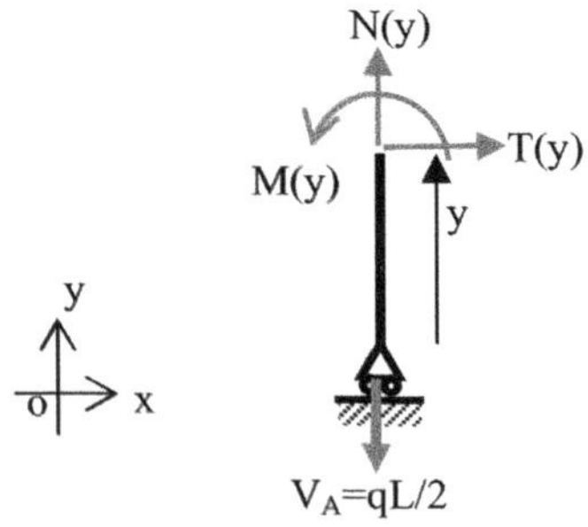

Figure 5.16. *Bar (AB)*

The equilibrium of the portion (Figure 5.16) allows writing

$$0 \le y \le \frac{L}{2}$$

$$\sum F_x = 0: \; T(y) = 0$$

$$\sum F_y = 0: \; N(y) - V_A = 0$$

$$N(y) = \frac{1}{2} qL$$

$$\sum M_A = 0: \; M(y) = 0$$

We divide bar (AB) by a section on the x-axis (Figure 5.16).

$$0 \le x \le L$$

$$\sum F_x = 0: \; N(x) + 2qL = 0$$

$$N(x) = -2qL$$

$$\sum F_y = 0: \; -T(x) - \frac{1}{2} qL - qx = 0$$

$$T(0) = -\frac{1}{2} qL$$

$$T(\mathrm{L}) = -\frac{3}{2} qL$$

$$\sum M_Q = 0: \; M(x) + \frac{1}{2} qx^2 + \frac{1}{2} qLx = 0$$

$$M(x) = -\frac{1}{2} qx^2 - \frac{1}{2} qLx$$

$$M(0) = 0$$

$$M(\mathrm{L}) = -qL^2$$

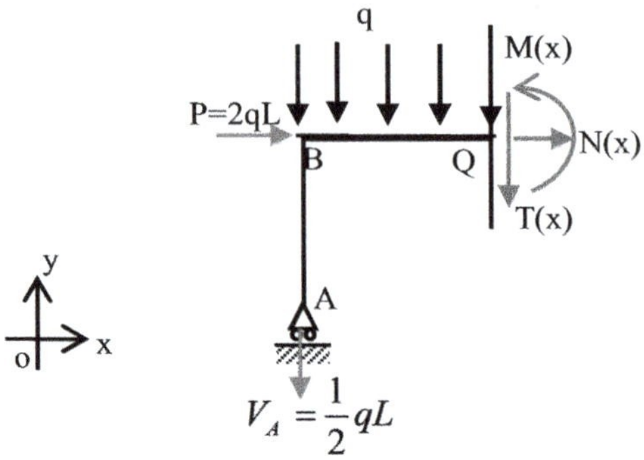

Figure 5.17. *Bar (BC)*

Finally, we divide bar (DC) by one on the y-axis (Figure 5.18).

$$0 \le y \le L$$

$$\sum F_x = 0: \ T(x) + H_D = 0$$

$$T(y) = -2qL$$

$$\sum F_y = 0: \ \mathrm{N(y)} + \mathrm{V}_D = 0$$

$$\mathrm{N(y)} = -\frac{3}{2}qL$$

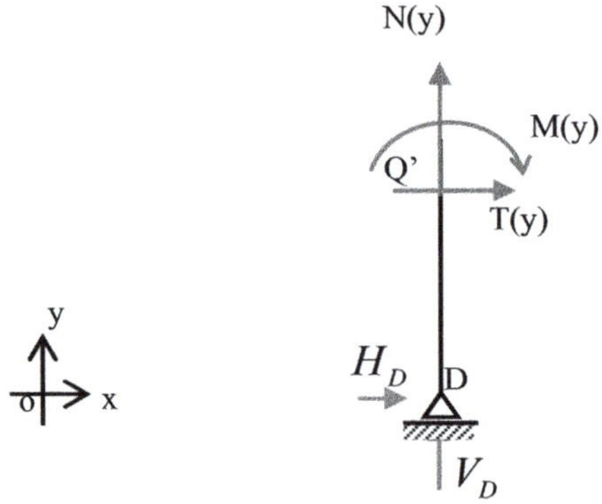

Figure 5.18. *Bar (CD)*

$$\sum M_{Q'} = 0 : M(\mathrm{y}) - \mathrm{H}_D\, y = 0$$

$$M(\mathrm{y}) = -2qLy$$

$$M(0) = 0$$

$$M(\mathrm{L}) = -2qL^2$$

Figure 5.19 plots the diagrams of internal actions.

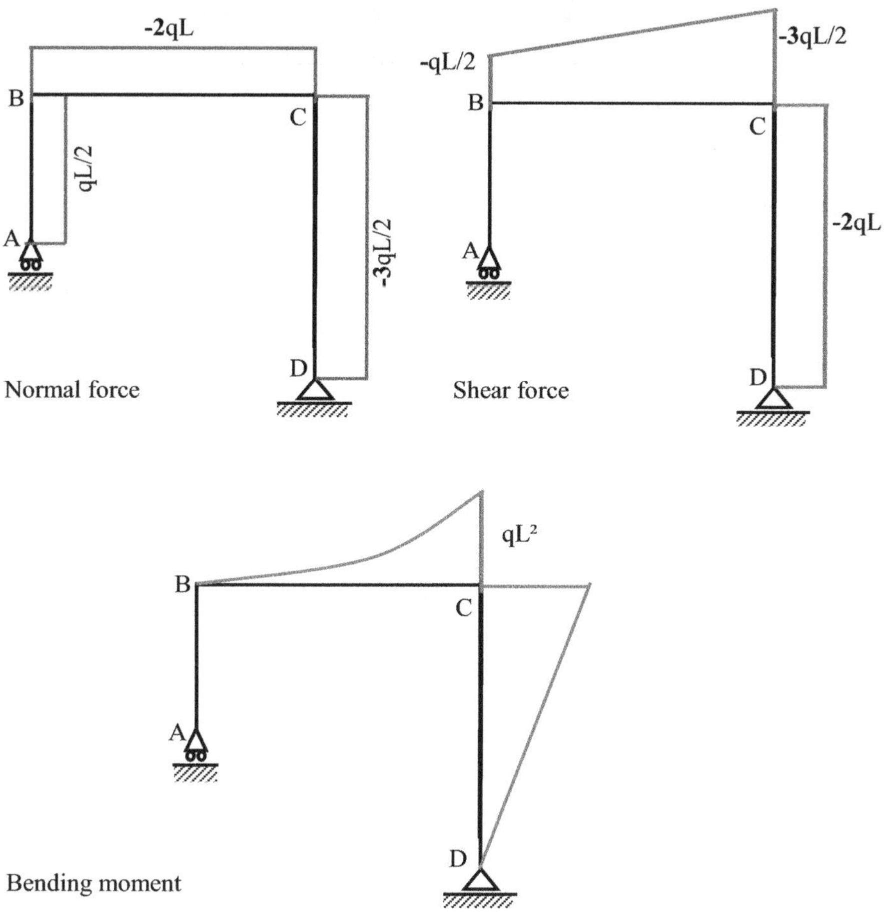

Figure 5.19. *Diagram of internal actions*

EXAMPLE 5.6.–

Draw the diagrams of internal actions of the frame (Figure 5.20).

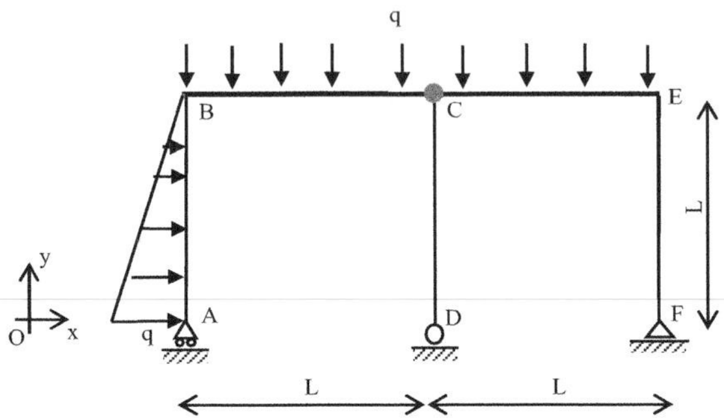

Figure 5.20. *Given frame*

Number of equations = 4.

Number of unknowns = 4.

Support reactions

$$\sum F_x = 0 : H_F + \frac{1}{2} qL = 0$$

$$H_F = -\frac{1}{2} qL$$

$$\sum M_C = 0 : V_A.L - \frac{1}{2} qL^2 - \frac{1}{2} qL(\frac{2}{3} L) = 0$$

$$V_A = \frac{5}{6} qL$$

$$\sum M_C = 0 : V_F.L + H_F L - \frac{1}{2} qL^2 = 0$$

$$V_F = qL$$

$$\sum F_y = 0 : V_A + V_D + V_F - 2qL = 0$$

$$V_D = \frac{1}{6} qL$$

Figure 5.21 presents the support reactions of the frame (Figure 5.20).

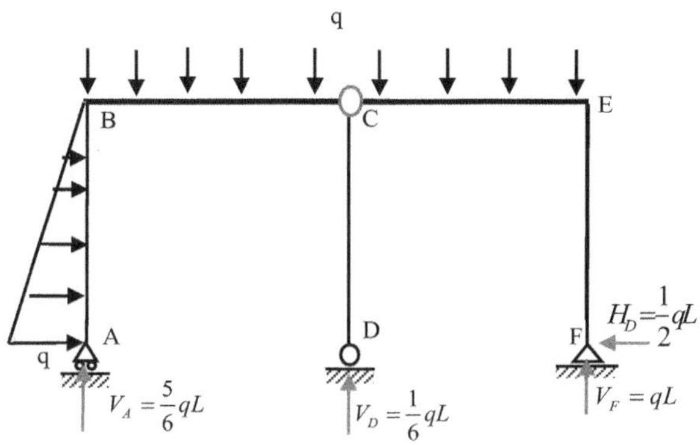

Figure 5.21. *Support reactions*

Internal actions

Bar (FE)

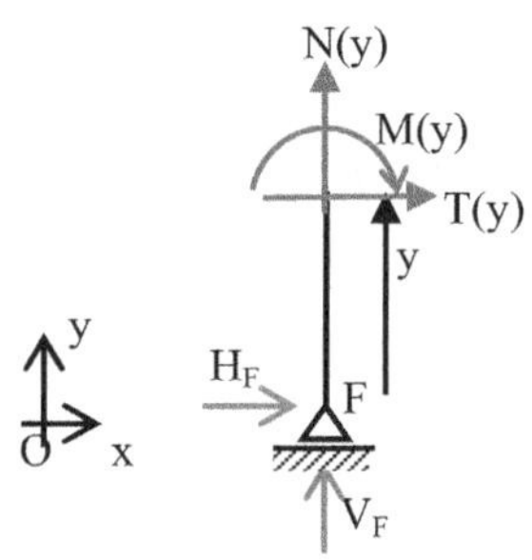

$$0 \le y \le L$$

$$\sum F_x = 0 : T(y) + H_F = 0$$

$$T(y) = \frac{1}{2} qL$$

$$\sum F_y = 0 : N(y) - V_F = 0$$

$$N(y) = -qL$$

$$\sum M_A = 0 : M(y) - H_F \, y = 0$$

$$M(y) = -\frac{1}{2} qLy$$

$$M(0) = 0$$

$$M(L) = -\frac{1}{2}qL^2$$

Bar (EC)

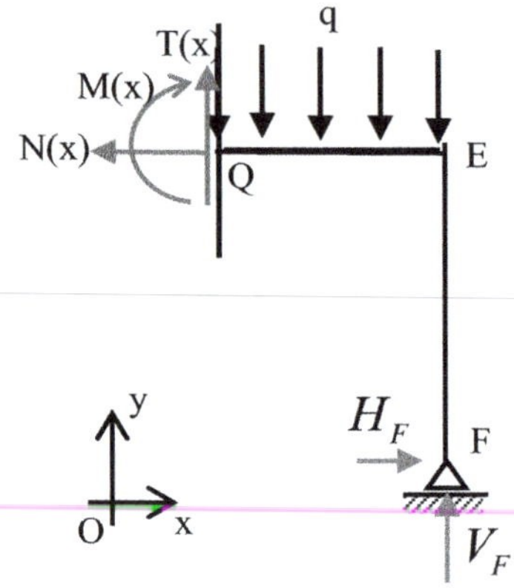

$$0 \le x \le L$$

$$\sum F_x = 0: \ -N(x) + H_F = 0$$

$$N(x) = -\frac{1}{2}qL$$

$$\sum F_y = 0: \ T(x) + V_F - qx = 0$$

$$T(x) = qx - qL$$

$$T(0) = -qL$$

$$T(L) = 0$$

$$\sum M_Q = 0: \ M(x) + \frac{1}{2}qx^2 - V_F x - H_F L = 0$$

$$M(x) = qLx - \frac{1}{2}qx^2 + \frac{1}{2}qL^2$$

$$M(0) = \frac{1}{2}qL^2$$

$$M(L) = qL^2$$

Bar (DC)

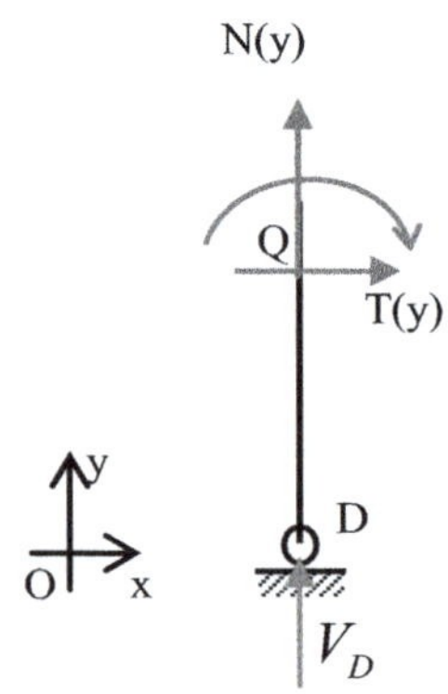

$$0 \le y \le L$$

$$\sum F_x = 0: \ T(y) = 0$$

$$\sum F_y = 0: \ N(y) + V_D = 0$$

$$N(y) = -\frac{1}{6}qL$$

$$\sum M_Q = 0 : M(\text{y}) = 0$$

Bar (AB)

Similarly, we divide the bar (AB) by a section on the y-axis.

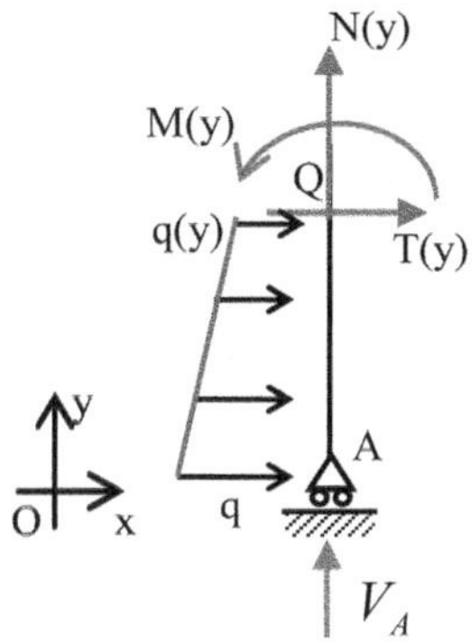

$$0 \le y \le L$$

$$q(y) = q(1 - \frac{y}{L})$$

$$\sum F_x = 0 : T(x) + \frac{q + q(y)}{2} y = 0$$

$$T(y) = \frac{1}{2} q \frac{y^2}{L} - qy$$

$$T(0) = 0$$

$$T(L) = -\frac{1}{2} qL$$

$$\sum F_y = 0 : \text{N(y)} + \text{V}_A = 0$$

$$\text{N(y)} = -\frac{5}{6} qL$$

$$\sum M_Q = 0 : M(\text{y}) + \text{q}(y) y \frac{y}{2} + \frac{1}{2}(q - q(y)).y.\frac{2}{3} y = 0$$

$$M(y) = -\frac{1}{3} q \frac{y^3}{L} - qy^2$$

$$M(0) = 0$$

$$M(\text{L}) = -\frac{1}{3} qL^2$$

Bar (BC)

Finally, we divide the bar (BC) by one on the x-axis.

$$0 \leq x \leq L$$

$$\sum F_y = 0: \ -T(x) - q.x + V_A = 0$$

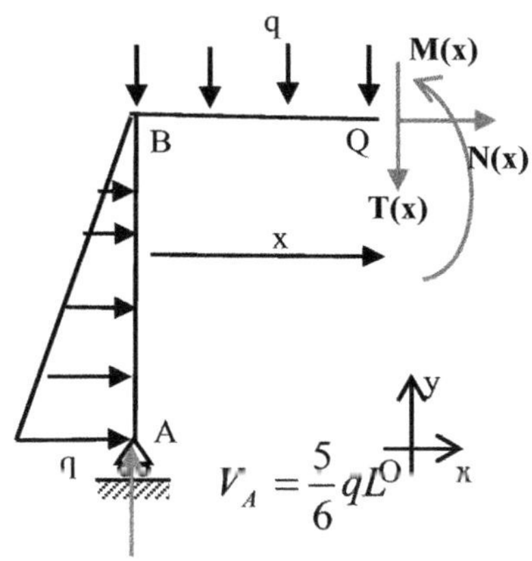

$$T(x) = \frac{5}{6}qL - qx$$

$$\mathrm{T}(0) = \frac{5}{6}qL$$

$$\mathrm{T}(\mathrm{L}) = -\frac{1}{6}qL^2$$

$$\sum F_x = 0: \ \mathrm{N}(x) + \frac{1}{2}qL = 0$$

$$\mathrm{N}(x) = -\frac{1}{2}qL$$

$$\sum M_Q = 0: \ M(x) - \mathrm{V}_A x + \frac{1}{2}qL\left(\frac{2L}{3}\right) + \frac{1}{2}qx^2 = 0$$

$$M(x) = -\frac{1}{2}qx^2 + \frac{5}{6}qLx - \frac{1}{3}qL^2$$

$$M(0) = -\frac{1}{3}qL^2$$

$$M(L) = 0$$

Figure 5.22 shows the diagrams of internal actions.

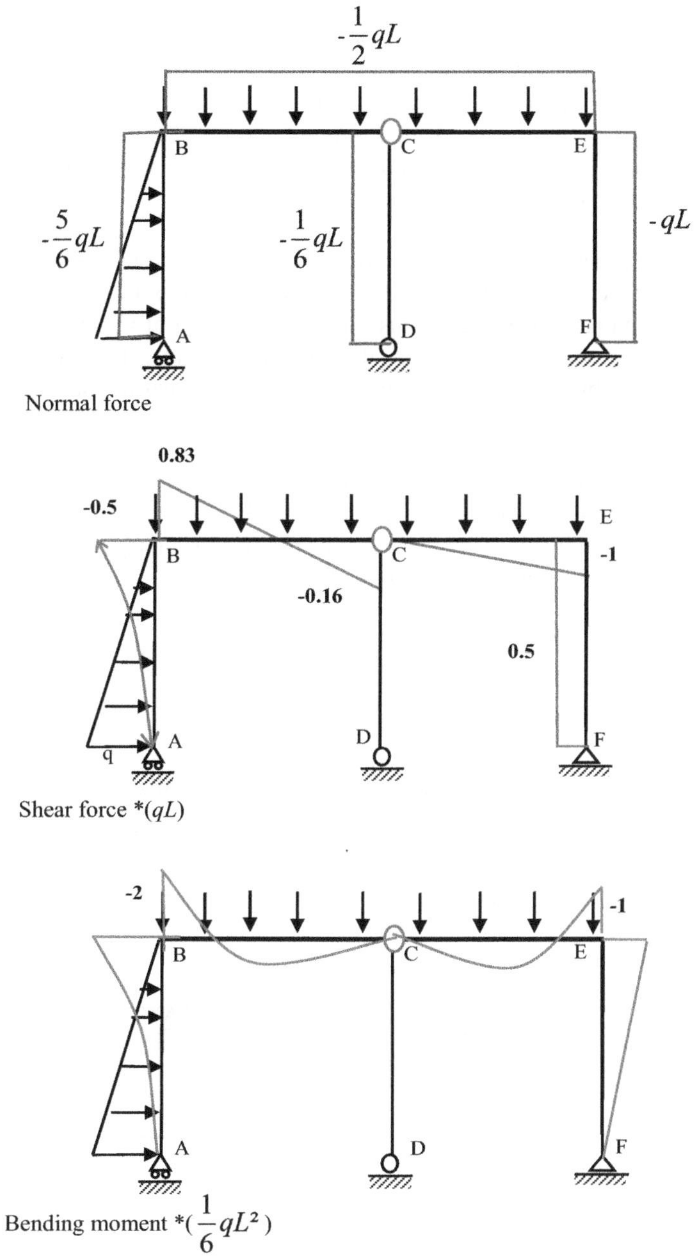

Figure 5.22. *Diagrams of internal actions*

5.8. Conclusion

Throughout this chapter, we have shown the procedure for determining internal actions at the section level of a statically determinate beam or frame. Applying static equations means that we can determine internal actions. Their variations according to the position of the section allow us to establish diagrams of the normal force, the tangential force and the bending moment.

In general, the axial force relative to a section is the algebraic sum of all the components of external actions, including support reactions in the direction of the axial force. The tangential force is thus the algebraic sum of all the components of external actions along the axis with the tangential force. Finally, the bending moment is the algebraic sum of all the moments of external actions applied to the studied portion.

A frame is considered statically determinate if internal actions can be determined using static equations. So, for a frame containing b bars, n joints, r support reactions and k additional relationships, the degree of static indeterminacy of the frame can be calculated using the formula $f = (3b + r) - (3n + k)$.

5.9. Problems

Exercise 1

Determine the degree of static indeterminacy and distinguish between the statically determinate, statically indeterminate and unstable beams among the following beams:

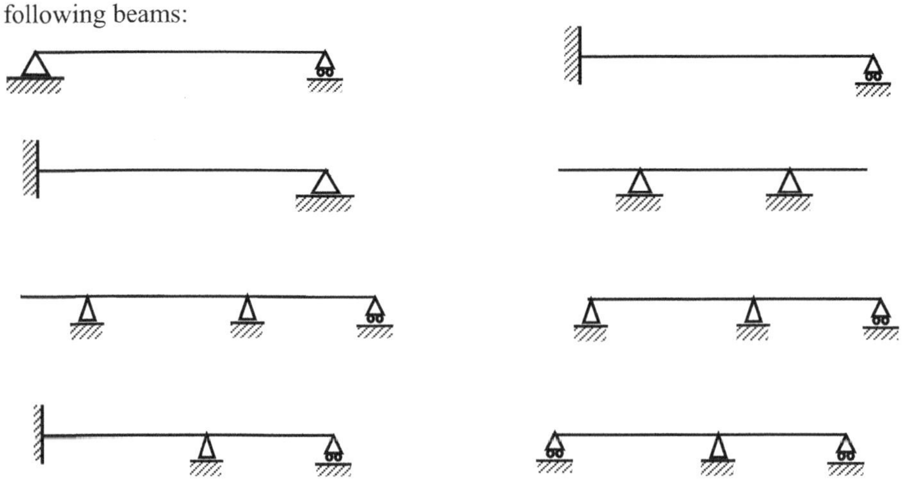

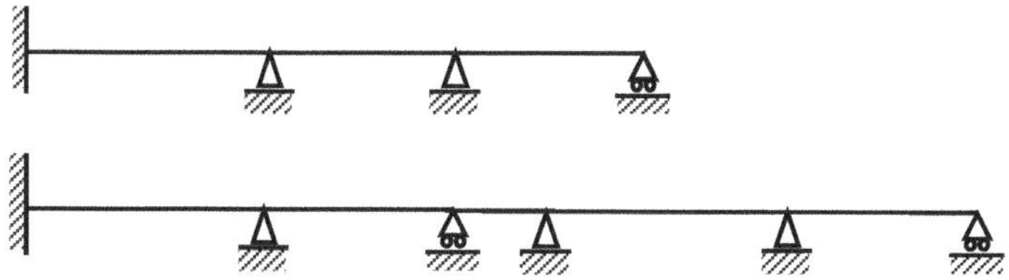

Exercise 2

Determine the degree of static indeterminacy and distinguish among the following frames: the statically determinate, statically indeterminate and unstable frames:

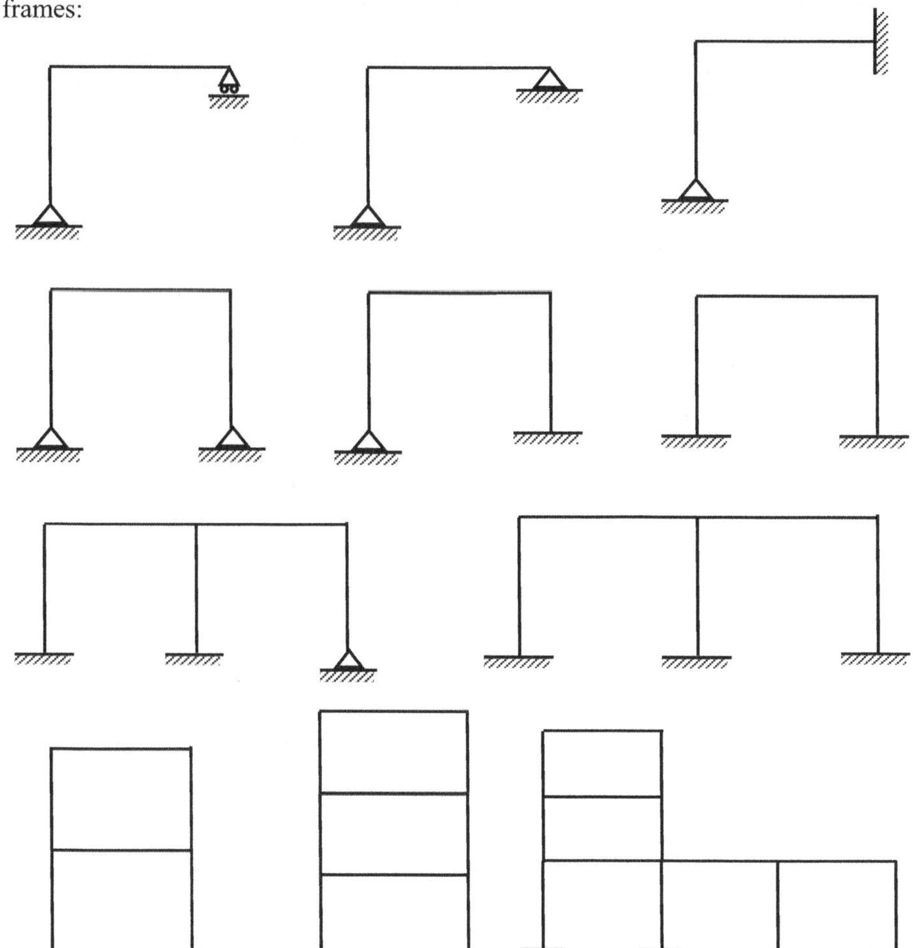

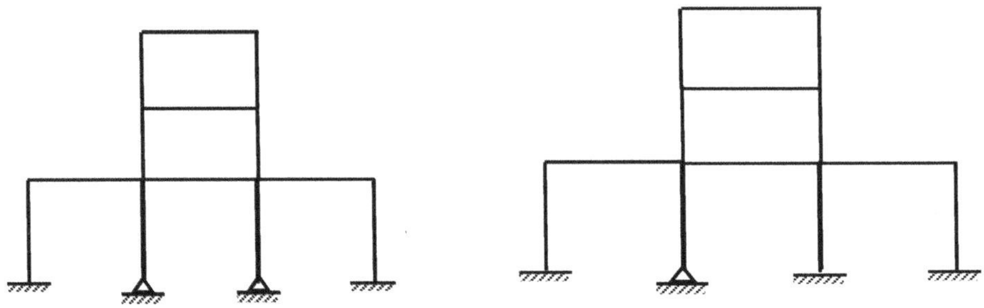

Exercise 3

Determine internal actions at points A and B for the following beams:

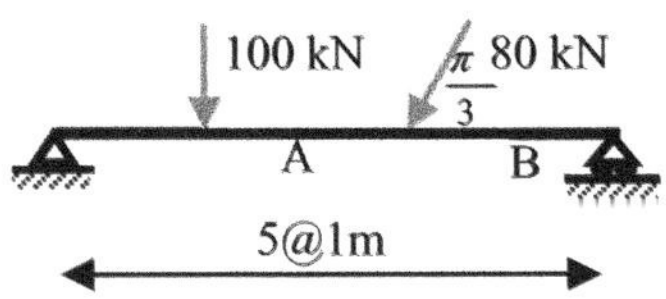

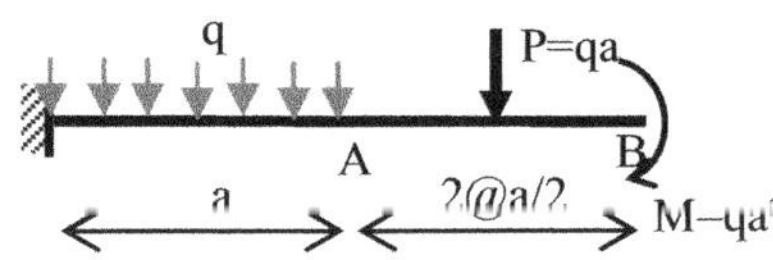

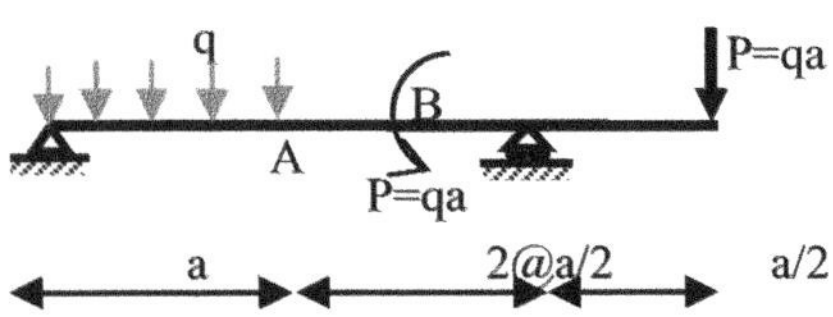

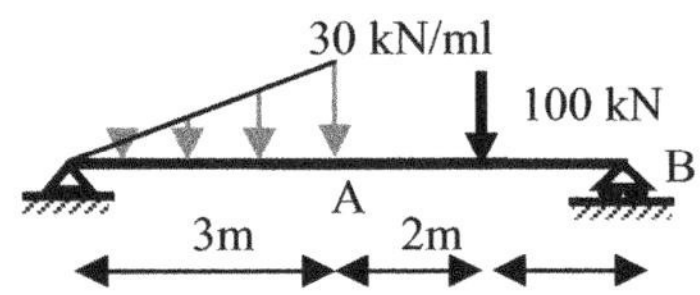

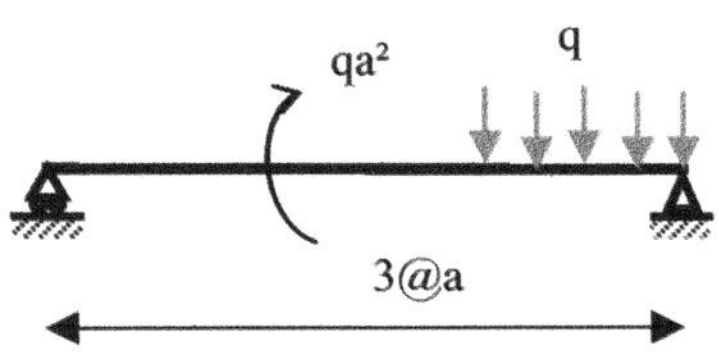

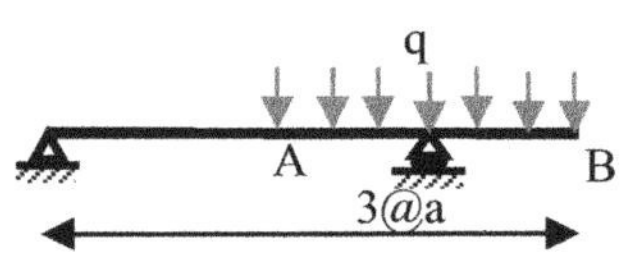

Exercise 4

Give the expressions of internal actions and draw the diagrams of the following beams:

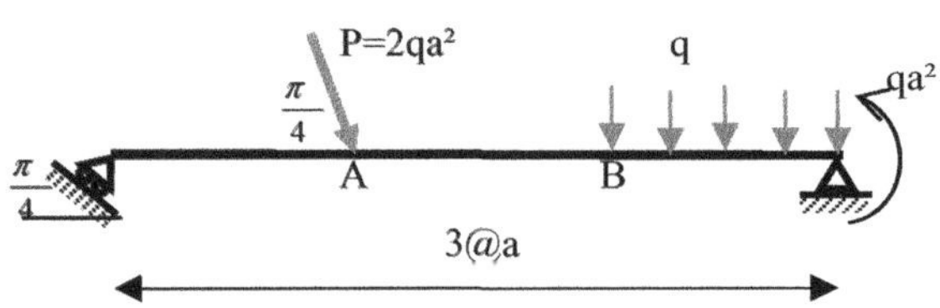

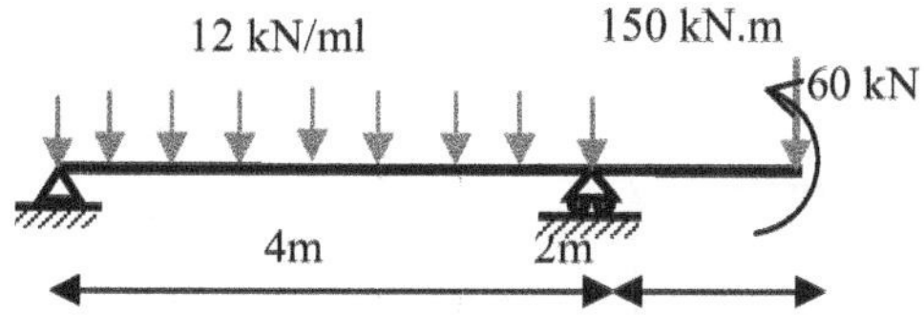

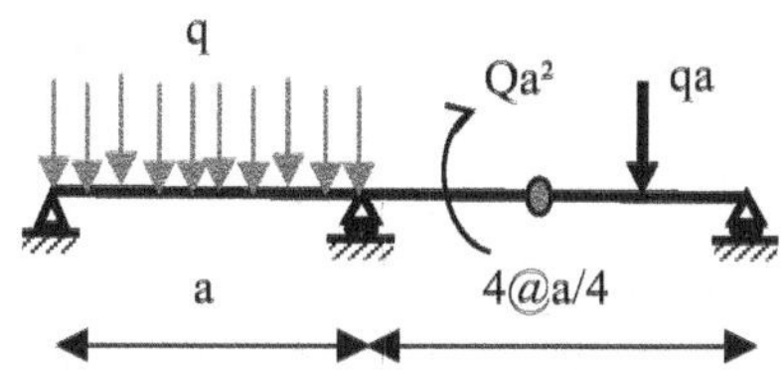

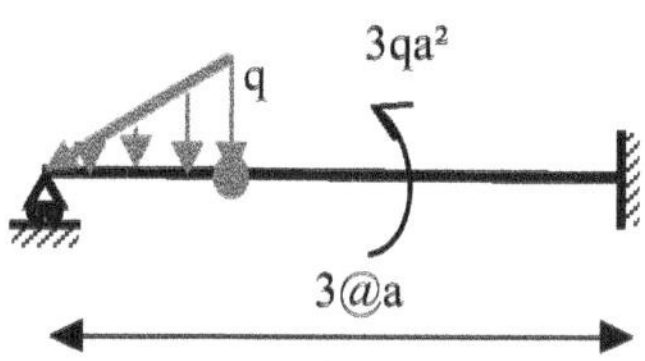

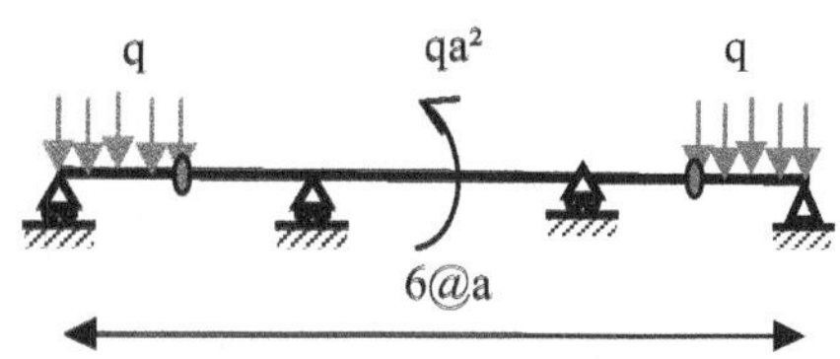

Exercise 5

Give the expressions of internal actions and draw the diagrams of the following frames:

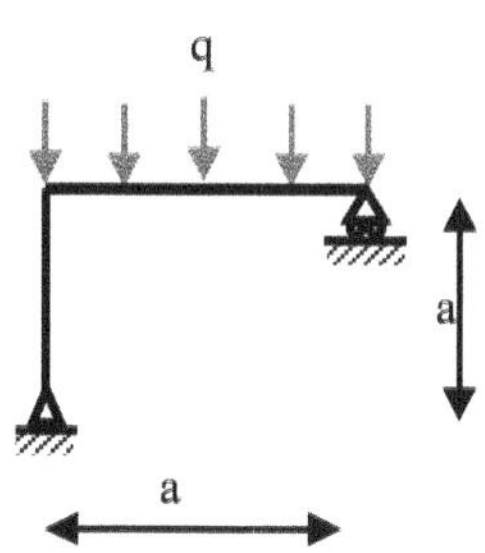

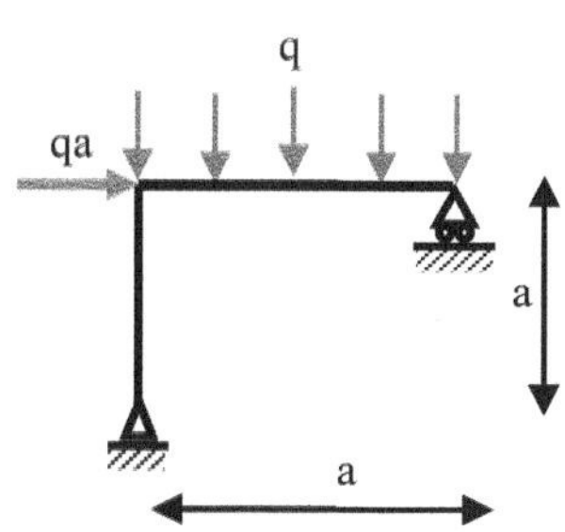

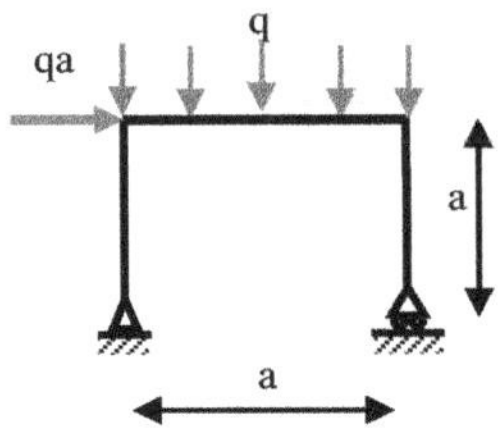
qa
q
a
a

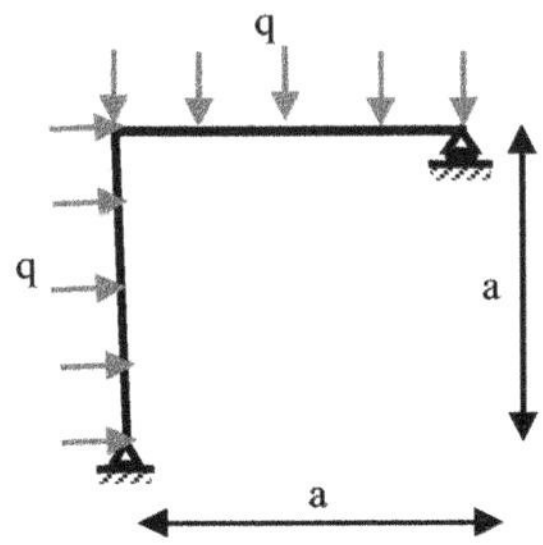
q
q
a
a

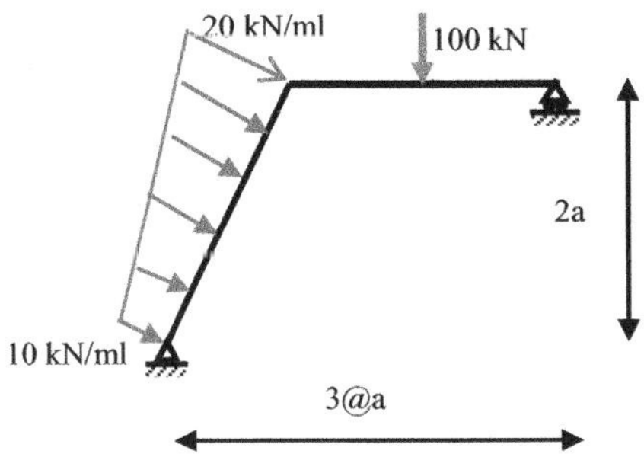
20 kN/ml
100 kN
2a
10 kN/ml
3@a

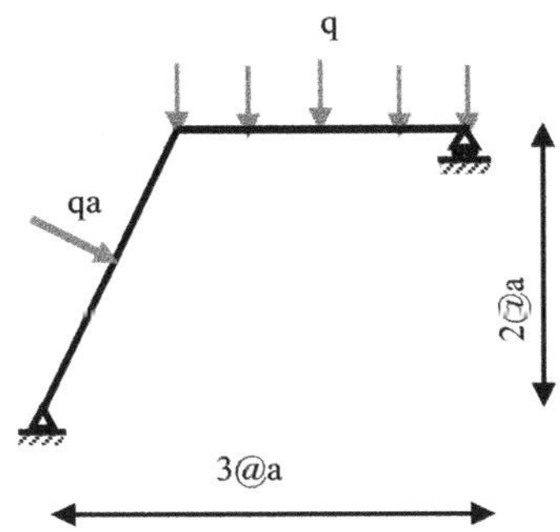
q
qa
2@a
3@a

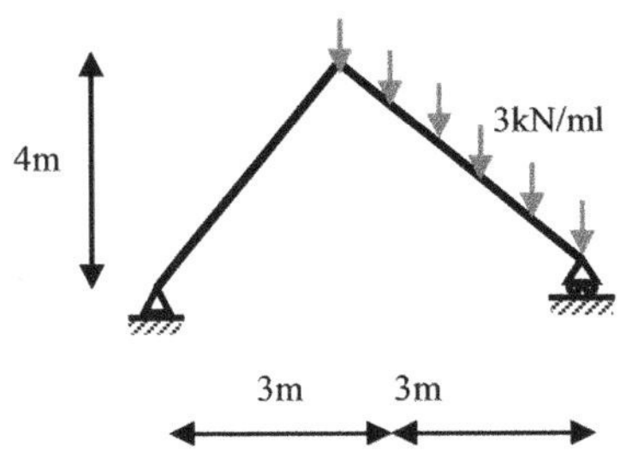
4m
3kN/ml
3m
3m

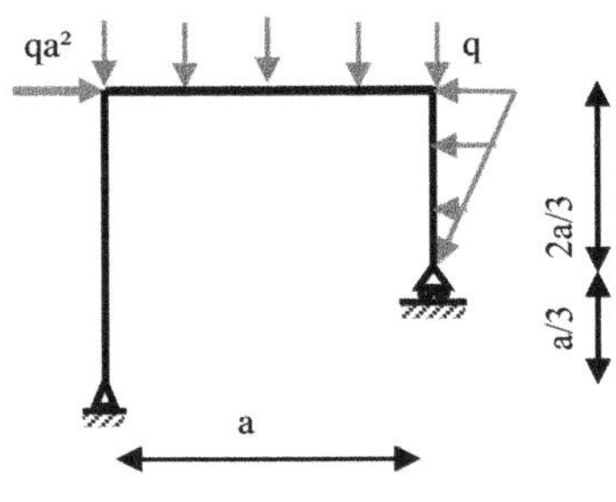
qa²
q
2a/3
a/3
a

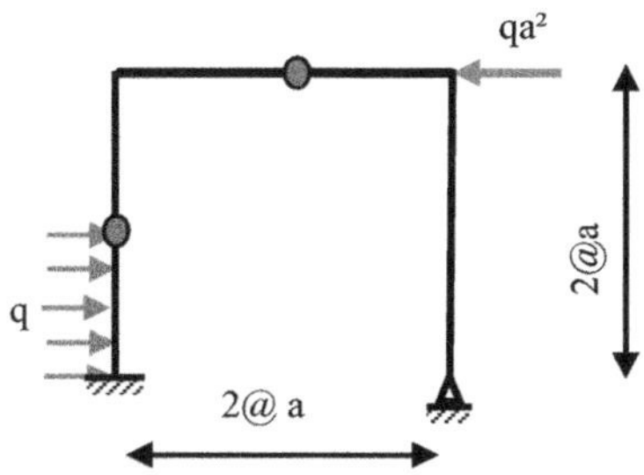
qa²
2@a
q
2@ a

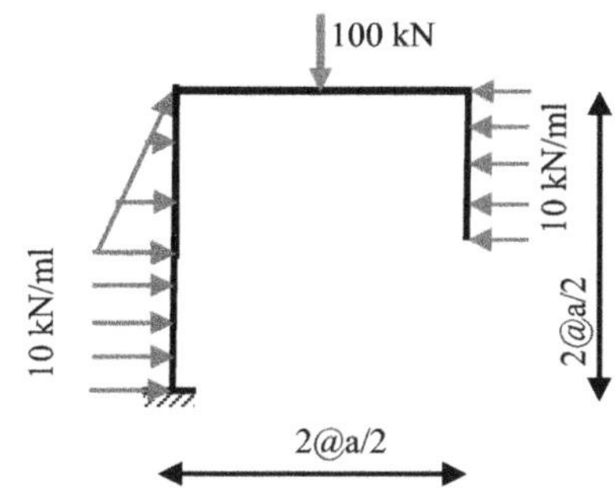
100 kN
10 kN/ml
10 kN/ml
2@a/2
2@a/2

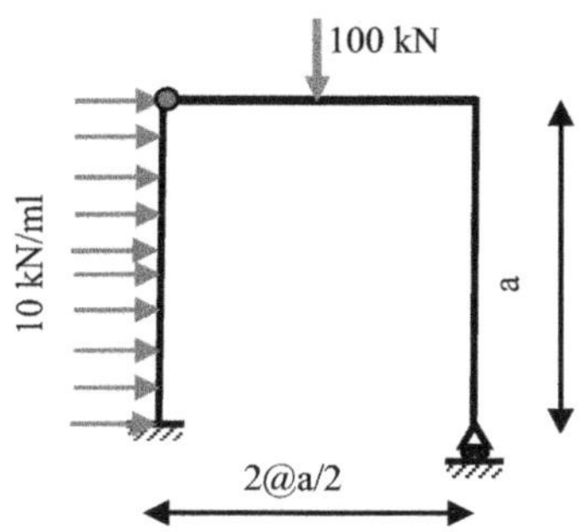
100 kN
10 kN/ml
a
2@a/2

6

Deflections of Elastic Beams: Energy Methods

The teaching objectives for this chapter are as follows:

– deriving the expressions of slopes and deflections at any point by using one of the deflection calculation methods;

– distinguishing between deflection calculation methods;

– implementing each calculation method.

This chapter focuses on the presentation and formulation of geometric methods for calculating the deflections of elastic structures. The deflection calculation methods presented in this chapter are as follows: (1) differential equation for beam deflection (elastic line) method, (2) direct integration method, (3) moment-area method and (4) fictitious beam method.

6.1. Elastic deflection of beams

An external action may be a mechanical force, a temperature variation, a settlement of support or a mechanical manufacturing error. Applying one or more actions to a structure generates internal actions within it and consequently deflections.

In structural mechanics, the deflection must be limited to meet resistance criterion, shape stability, cracking, etc. In addition, the deflection calculation plays a major role in the design of structures and pieces. It can be used as a factor to verify the resistance criterion or the structural response.

In this chapter, we are interested in calculating linear elastic deflections of statically determinate beams. For this reason, structural behavior can be modeled by a spring whose stiffness is equal to the rigidity of the studied structure. Deflections inside a structure are due to the effect of the internal actions, which originate at the level of each section. Deflections in successive sections constitute elastic deflection. We define the displacement of a point by the distance between the deformed position and its initial configuration. In the same way, the slope of a section is measured as being the tangent to the deflection at a given point.

Consider beam (ABC) acted on by load q (x) (Figure 6.1).

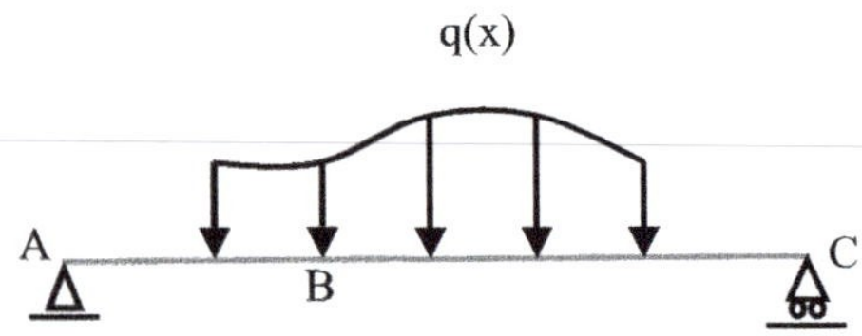

Figure 6.1. *Studied beam*[1]

Before applying load q (x), the beam is horizontal and non-deformed (the dead weight of the beam is neglected). After applying load q (x), the beam begins to deform and continues to do so until the internal energy is equal to the work of the external actions. In this case, the beam keeps its final deformed state (Figure 6.2).

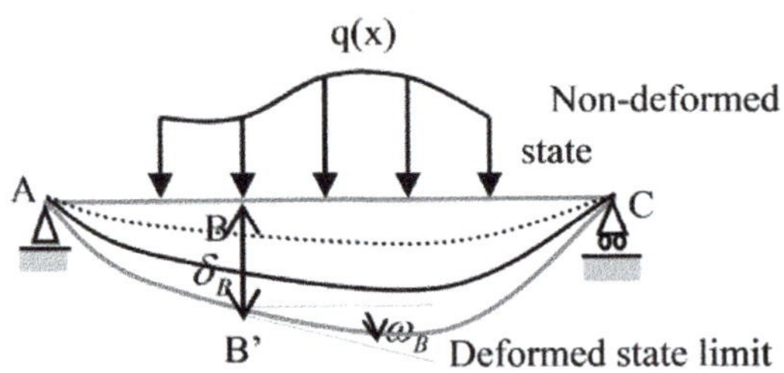

Figure 6.2. *Deformed beam*

In Figure 6.2, δ_B and ω_B represent the displacement and the slope of the section passing through point B, respectively. The locus of the displacements along the beam constitutes elastic line deflection (AC) and the tangent at each point forms the slope curve of the line (AC).

1 All of the figures in this chapter are available to view in full color at www.iste.co.uk/khalfallah/analysis1.zip.

6.2. Calculation of deflections

The deflection of a beam can be deduced by the following methods.

6.2.1. *Method of the differential equation of the elastic line*

Consider a statically determinate beam (Figure 6.1) whose deflection is shown in (Figure 6.2). To derive the differential equation of the elastic line, we extract the length Δx and the curvature $d\theta$. The element is stressed by a bending moment M(x) (Figure 6.3).

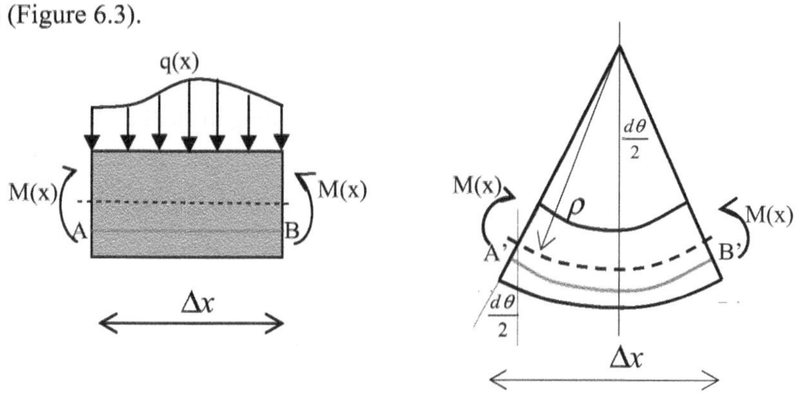

Figure 6.3. *Deflection of element Δx*

The longitudinal fiber (AB) before deflection becomes (A'B') after deflection, which is located in the stretched part of the beam, so it undergoes lengthening:

$$\Delta = A'B' - AB \qquad [6.1]$$

$$tg\frac{d\theta}{2} \cong \frac{d\theta}{2} = \frac{\Delta/2}{y} \qquad [6.2]$$

y is the ordinate of fiber (AB).

$$\Delta = y.d\theta \qquad [6.3]$$

The deflection of fiber (AB) can be deduced by

$$\varepsilon(x) = \frac{\Delta}{dx} = \frac{y.d\theta}{dx} \qquad [6.4]$$

Similarly, we have

$$\sin\frac{d\theta}{2} \simeq \frac{d\theta}{2} = \frac{dx}{2\rho} \qquad [6.5]$$

with

$$d\theta = \frac{dx}{\rho} \qquad [6.6]$$

$$\varepsilon(x) = \frac{y}{\rho} \qquad [6.7]$$

The normal stress in fiber (AB) is calculated according to Hooke's law by:

$$\sigma(y) = E\varepsilon(x) = E\frac{y}{\rho} \qquad [6.8]$$

The bending moment due to stress distribution $\sigma(y)$ is:

$$M(x) = \int_{\Omega} \sigma(y) y d\Omega \qquad [6.9]$$

Calculation of integral [6.9] by substituting the relationship [6.8] leads to:

$$\frac{1}{\rho} = \frac{M(x)}{E.I_Z} \qquad [6.10]$$

In Cartesian coordinates, the relationship [6.10] is written mathematically as:

$$\frac{1}{\rho} = \frac{\frac{d^2y(x)}{dx^2}}{\left[1+(\frac{dy(x)}{dx})^2\right]^{3/2}} \qquad [6.11]$$

where y (x) is the deflection relationship of the central fiber. In the case of small slopes, the higher order quantity $(\frac{dy}{dx})^2$ is neglected. The relationship [6.11] becomes:

$$\frac{1}{\rho} = \frac{d^2y}{dx^2} \qquad [6.12]$$

Taking into account the relationships [6.10] and [6.12], we can write:

$$\frac{d^2y}{dx^2} = \frac{M(x)}{E.I_Z} \qquad [6.13]$$

The relationship [6.13] is the differential equation for beam deflection (elastic line), which was established by Bernoulli–Euler.

The relationship [6.13] can also be written as:

$$\frac{d\omega(x)}{dx} = \frac{M(x)}{E.I_Z} \qquad [6.14]$$

EXAMPLE 6.1.–

Determine the slopes at ends A and B and the deflection at the middle of a beam stressed by a uniform load of magnitude q and flexural rigidity EI (Figure 6.4).

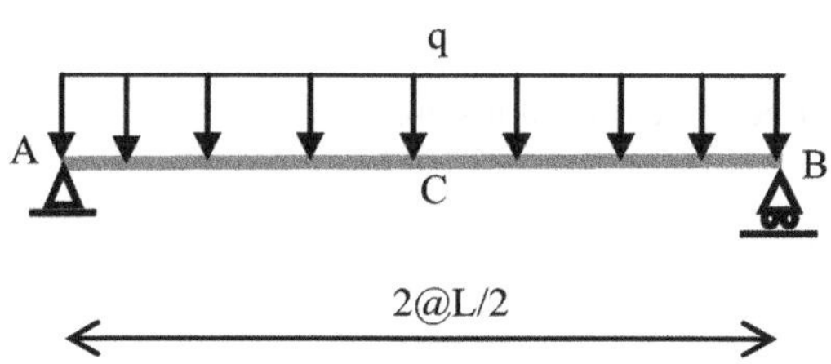

Figure 6.4. *Given beam*

The expression of the bending moment is given by:

$$M(x) = \frac{qL}{2}x - \frac{1}{2}qx^2$$

The relationship [6.12] is written as:

$$y''(x) = \frac{q}{2E.I_Z}(Lx - x^2)$$

Integrating this equation leads to formulating the slope expression:

$$\omega(x) = \frac{dy}{dx} = \frac{q}{2E.I}\left(\frac{Lx^2}{2} - \frac{x^3}{3} + \alpha\right)$$

The deflection equation of the beam can be deduced by:

$$\delta(x) = y(x) = \frac{q}{2E.I}\left(\frac{Lx^3}{6} - \frac{x^4}{12} + \alpha x + \beta\right)$$

α and β are the integration constants, which can be determined by using limit conditions.

$$y(0) = 0 \ : \ \beta = 0$$

$$y(x = L) = 0 \ : \ \alpha = -\frac{L^3}{12}$$

The slope and deflection expressions are

$$\omega(x) = \frac{q}{2E.I}\left(\frac{Lx^2}{2} - \frac{x^3}{3} - \frac{L^3}{12}\right)$$

$$y(x) = \frac{q}{2E.I}\left(\frac{Lx^3}{6} - \frac{x^4}{12} - \frac{L^3}{12}x\right)$$

These equations allow us to calculate the deflection in the middle and the slope at the ends of the beam (Table 6.1).

x	0	L/2	L
$\omega(x)$	$-\frac{qL^3}{24E.I_Z}$	0	$\frac{qL^3}{24E.I_Z}$
$y(x)$	0	$-\frac{5qL^4}{384E.I_Z}$	0

Table 6.1. *Beam deflections*

6.2.2. *Direct integration method*

The successive integration of the relationship [6.13] leads first to the formulation of the slope expression and second to the formulation of deflections for all points on the beam.

The slope expression [6.14] is:

$$\omega(x) = \frac{dy}{dx} = \int \frac{M(x)}{E.I_z} dx \qquad [6.15]$$

The deflection expression can be deduced by:

$$y(x) = \int\int \frac{M(x)}{E.I_z} dx \qquad [6.16]$$

For each relationship [6.15] and [6.16], an integration constant must be introduced, which can be established by limit conditions.

EXAMPLE 6.2.–

We can formulate deflection and slope expressions of the beam (Figure 6.5). The flexural rigidity (EI) is assumed to be constant.

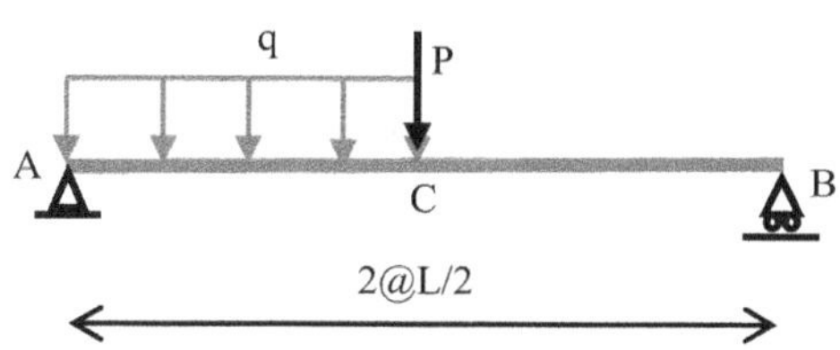

Figure 6.5. *Given beam*

The bending moments expressions in the two intervals are given as:

$$0 \le x \le \frac{L}{2}$$

$$M(x) = \frac{1}{2}Px + \frac{3}{8}qLx - \frac{1}{2}qx^2$$

$$\frac{L}{2} \leq x \leq L$$

$$M(x) = \frac{1}{2}PL - \frac{1}{2}Px - \frac{1}{8}qLx + \frac{1}{8}qL^2$$

The deflection and the slope expressions of each section are given as:

$$0 \leq x \leq \frac{L}{2}$$

$$\omega_1(x) = \frac{1}{EI}\left[\frac{1}{4}Px^2 + \frac{3}{16}qLx^2 - \frac{1}{6}qx^3 + C_1\right]$$

$$y_1(x) = \frac{1}{EI}\left[\frac{1}{12}Px^3 + \frac{1}{16}qLx^3 - \frac{1}{24}qx^4 + C_1 x + C_2\right]$$

$$\frac{L}{2} \leq x \leq L$$

$$\omega_2(x) = \frac{1}{EI}\left[-\frac{1}{4}Px^2 + \frac{1}{2}PLx - \frac{1}{16}qLx^2 + \frac{1}{8}qL^2x + C_3\right]$$

$$y_2(x) = \frac{1}{EI}\left[-\frac{1}{12}Px^3 + \frac{1}{4}PLx^2 - \frac{1}{48}qLx^3 + \frac{1}{16}qL^2x^2 + C_3 x + C_4\right]$$

Limit conditions and the continuity of the deflections are written as:

$$\omega_1\left(\frac{L}{2}\right) = \omega_2\left(\frac{L}{2}\right)$$

$$y_1\left(\frac{L}{2}\right) = y_2\left(\frac{L}{2}\right)$$

$$y_1(0) = y_2(L) = 0$$

To simplify the calculation, we put $P = qL$. The preceding formulae give:

$$C_1 + C_3 = \frac{1}{8}PL^2 + \frac{1}{48}qL^3$$

$$C_2 = 0$$

$$C_3 L + C_4 = -\frac{1}{6}PL^3 + \frac{1}{16}qL^4$$

$$C_1 \frac{L}{2} + C_2 - C_3 L - C_4 = \frac{1}{24}PL^3 - \frac{1}{196}qL^4$$

Integration constants can be deduced:

$$C_1 = -\frac{331}{3528}qL^3\,, \quad C_2 = 0\,, \quad C_3 = -\frac{809}{7056}qL^3\,, \quad C_4 = \frac{956}{7056}qL^4$$

The slope and deflection expressions are written as:

$$0 \le x \le \frac{L}{2}$$

$$\omega_1(x) = \frac{1}{EI}\left[\frac{1}{4}Px^2 + \frac{3}{16}qLx^2 - \frac{1}{6}qx^3 - \frac{331}{3528}qL^3\right]$$

$$y_1(x) = \frac{1}{EI}\left[\frac{1}{12}Px^3 + \frac{1}{16}qLx^3 - \frac{1}{24}qx^4 - \frac{331}{3528}PLx\right]$$

$$\frac{L}{2} \le x \le L$$

$$\omega_2(x) = \frac{1}{EI}\left[-\frac{1}{4}Px^2 + \frac{1}{2}PLx - \frac{1}{16}qLx^2 + \frac{1}{8}qL^2x - \frac{809}{7056}PL^2\right]$$

$$y_2(x) = \frac{1}{EI}\left[-\frac{1}{12}Px^3 + \frac{1}{4}PLx^2 - \frac{1}{48}qLx^3 + \frac{1}{16}qL^2x^2 - \frac{809}{7056}PL^3x + \frac{956}{7056}PL^4\right]$$

6.2.3. *Moment-area method*

The moment-area method was established by Charles E. Greene in 1873. Once the expression of bending moment variation is established, it is possible to derive the slope expression using the relationship [6.14].

$$\omega(x) = \frac{dy}{dx} = \int \frac{M(x)}{E.I_z} dx \qquad [6.17]$$

By integrating the relationship [6.17], the deflection expression is written as:

$$y(x) = \frac{d\omega(x)}{dx} = \int \left[\int \frac{M(x)}{E.I_z} dx \right] dx \qquad [6.18]$$

From the relationship [6.14], we can write:

$$d\omega(x) = \frac{M(x)}{E.I_z} dx \qquad [6.19]$$

The quantity $\frac{M(x)}{E.I_z} dx$ represents the infinitesimal area of the curve $\frac{M(x)}{E.I_z}$ along the length dx and the axis (ox) (Figure 6.6).

Using the relationship [6.19] makes it possible to calculate the difference in slopes between points A and B (Figure 6.7).

$$\int_A^B d\omega(x) = \int_A^B \frac{M(x)}{E.I_z} dx \qquad [6.20]$$

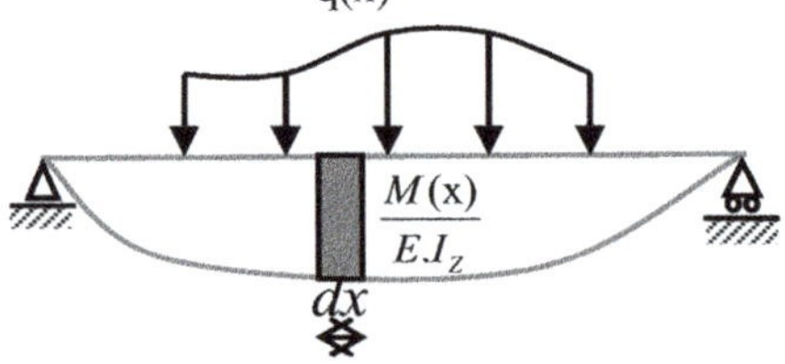

Figure 6.6. *Deformed beam*

The difference between slopes noted ω_{BA} is written as:

$$\omega_{BA} = \omega_B - \omega_A = \int_A^B \frac{M(x)}{E.I_Z} dx \qquad [6.21]$$

ω_{BA} is the relative angle between the tangents to the curve at points A and B, respectively, and $\int_A^B \frac{M(x)}{E.I_Z} dx$ is the area of the bending moment diagram per unit of flexural rigidity between points A and B (Figure 6.7).

THEOREM 6.1.– The slope variation between the tangents at two points of the elastic deflection is equal to the area of the diagram $\frac{M(x)}{E.I_Z}$ between the sections passing through these two points.

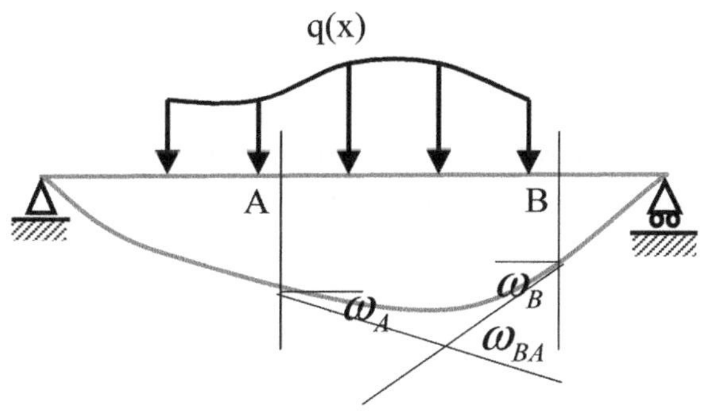

Figure 6.7. *Relationships of sections A and B*

We consider an infinitesimal element of length Δx between two very similar sections passing through points A and B (Figure 6.8). Relative slipping between the boundaries of this element is given as:

$$d\Delta = \bar{x}.\, d\omega \qquad [6.22]$$

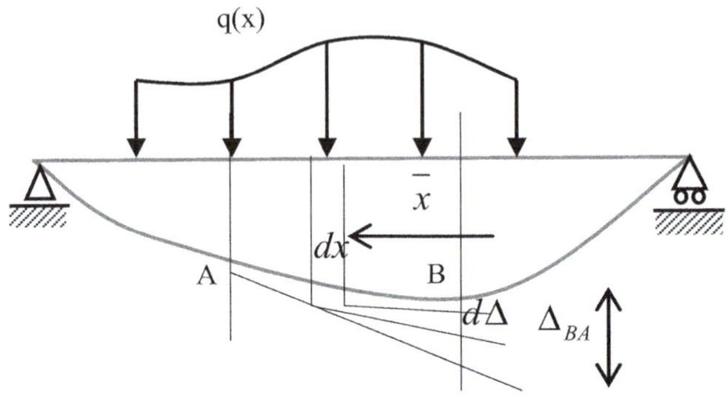

Figure 6.8. *Displacements of points A and B*

Substituting the relationship [6.19] with [6.22], we obtain:

$$d\Delta = \bar{x}.\frac{M(x)}{E.I_Z}dx\ , \qquad [6.23]$$

where $\bar{x}$ is the distance between point B and the element of length $d\Delta$. Integration of the relationship [6.23] between points A and B leads to:

$$\int_A^B d\Delta = \int_A^B \bar{x}.\frac{M(x)}{E.I_Z}dx \qquad [6.24]$$

The relationship [6.24] can also be written as:

$$\Delta_{BA} = \Delta_B - \Delta_A = \int_A^B \bar{x}.\frac{M(x)}{E.I_Z}dx\ , \qquad [6.25]$$

where Δ_{BA} represents the lateral slipping of point B relative to point A.

THEOREM 6.2.– The relative deflection between two points A and B is equal to the static moment of the bending area moment per unit of flexural rigidity ($\frac{M(x)}{E.I_Z}$) between points A and B [6.25].

EXAMPLE 6.3.–

A beam (ACB) of flexural rigidity EI is stressed by a concentrated load on point C (Figure 6.9). Calculate the slope and displacement at point C.

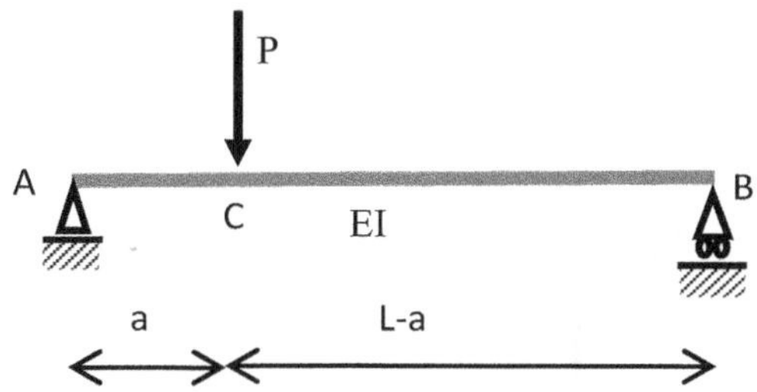

Figure 6.9. *Given beam*

$$0 \le x \le a$$

$$V_A = P(1-\frac{a}{L})$$

$$M(x) = P(1-\frac{a}{L})x$$

Applying equation [6.21] between points A and C allows us to write:

$$\omega_{CA} = \omega_C - \omega_A = \int_A^C \frac{M(x)}{E.I_Z} dx = \frac{Pa^2(1-\frac{a}{L})}{2E.I_Z}$$

The slope of the section passing through point C is given as:

$$\omega_C = \omega_A - \omega_{CA}$$

We apply equation [6.18] between points A and B:

$$\omega_{BA} = \omega_B - \omega_A = \int_A^C \frac{M(x)}{E.I_Z} dx$$

The slope of point A is calculated by:

$$\omega_A = \frac{\Delta_{BA}}{L}$$

with $\Delta_{AB} = \int_A^B \bar{x}\frac{M(x)}{E.I_Z}dx$

The bending moment diagram can be divided into two parts (Figure 6.10).

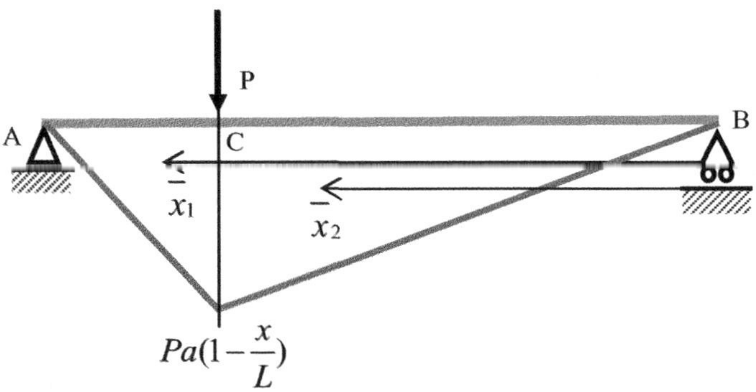

Figure 6.10. *Bending moment diagram*

$$\Delta_{AB} = \int_A^C \bar{x}_1 \frac{M_1(x)}{E.I_Z}dx + \int_C^B \bar{x}_2 \frac{M_2(x)}{E.I_Z}dx$$

$$\Delta_{AB} = \frac{1}{6}\frac{Pa^2}{E.I_Z}(1-\frac{a}{L})(2\frac{L^2}{a}-L)$$

The slopes at joints A and C can be deduced as:

$$\omega_A = \frac{1}{6}\frac{Pa^2}{E.I_Z}(1-\frac{a}{L})(2\frac{L}{a}-1)$$

$$\omega_C = \frac{1}{3}\frac{Pa^2}{E.I_Z}(1-\frac{a}{L})(\frac{L}{a}-2)$$

The lateral slipping between points A and C is calculated by:

$$\Delta_{CA} = \int_A^C \overline{x}_1 \frac{M(x)}{E.I_Z}dx = \frac{1}{6}\frac{Pa^3}{E.I_Z}(1-\frac{a}{L})$$

Therefore, the displacement at point C is written as:

$$\Delta_C = \omega_A.a - \Delta_{CA} = \frac{1}{3}\frac{PLa^2}{E.I_Z}(1-\frac{a}{L})^2$$

The slope at point C is evaluated as:

$$\omega_C = \frac{1}{3}\frac{Pa^2}{E.I_Z}(1-\frac{a}{L})(\frac{L}{a}-2)$$

$$\Delta_C = \frac{1}{3}\frac{PLa^2}{E.I_Z}(1-\frac{a}{L})^2$$

Numerical application

For P = 100 kN, a = 4 m, L = 10 m and $E \cdot I_Z = 10^5$ kN·m², we find:

$$\Delta_C = 19.2\,\text{mm} \text{ and } \omega_C = 1.6\times10^{-3}\,\text{rad.}$$

6.2.4. *Conjugate beam method*

The conjugate beam method was developed by H. Muller-Breslau in 1865. This method is applied in a similar way to the moment-area method. It uses the principles of static structures and the method is widely used in practice.

On the one hand, integration of relationships [5.4] and [6.14] means that we can express the shear force and slope in any section x.

$$T(x) = -\int q(x)dx \qquad [6.26]$$

$$\omega(x) = \int \frac{M(x)}{E.I} dx \qquad [6.27]$$

On the other hand, integration of relationships [5.9] and [6.13] means that we can formulate the bending moment and deflection expressions.

$$M(x) = -\int \left[\int q(x)dx \right] .dx \qquad [6.28]$$

$$y(x) = \int \left[\int \frac{M(x)}{E.I} dx \right] .dx \qquad [6.29]$$

Starting from the results obtained in [6.27] and [6.29], we can state the theorems of the conjugate beam method.

THEOREM 6.3.– The slope at one point of the real beam is equal to the shear force by the flexural rigidity at the same point on the conjugate beam.

THEOREM 6.4.– The deflection at one point of the real beam is equal to the bending moment by the flexural rigidity at the same point on the conjugate beam.

6.2.4.1. *Characteristics of the conjugate beam*

For the rigorous application of the conjugate beam method, the following conditions must be respected:

– the fictitious beam is the same length as the real beam;

– the loading on the conjugate beam is the variation of the bending moment by flexural rigidity (EI);

– the fictitious beam supports can be deduced from those of the real beam (Table 6.2).

Real beam	Kinematic characteristic	Fictitious beam	Corresponding stresses
	$\delta_A = 0$ $\omega_A \neq 0$	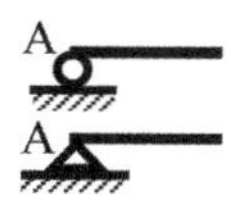	$M_A = 0$ $T_A \neq 0$
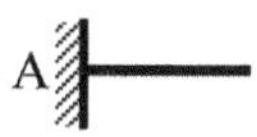	$\delta_A = 0$ $\omega_A = 0$		$M_A = 0$ $T_A = 0$
	$\delta_A \neq 0$ $\omega_A \neq 0$	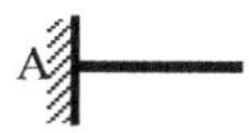	$M_A \neq 0$ $T_A \neq 0$
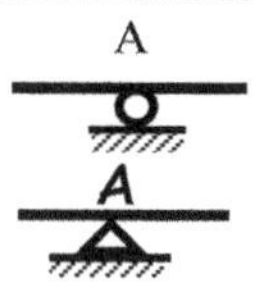	$\delta_A = 0$ $\omega_A \neq 0$	hinge 	$M_A = 0$ $T_A \neq 0$
joint 	$\delta_A \neq 0$ $\omega_A \neq 0$	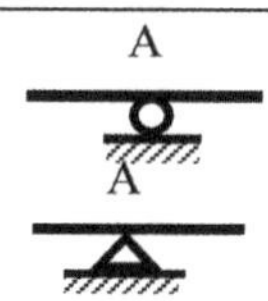	$M_A \neq 0$ $T_A \neq 0$

Table 6.2. *Supports of real and fictitious beams*

EXAMPLE 6.4.–

Calculate the deflection and slope at point A of the cantilevered beam (Figure 6.11).

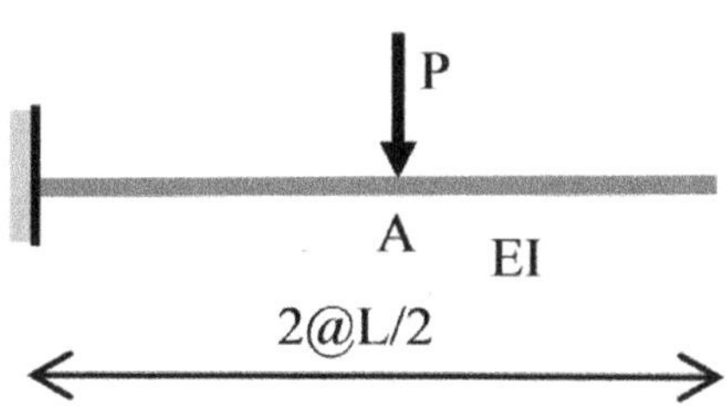

Figure 6.11. *Given cantilever*

The bending moment diagram is given by Figure 6.12.

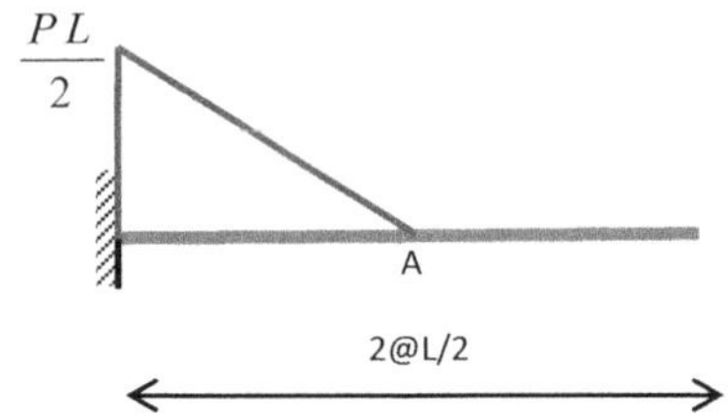

Figure 6.12. *Bending moment diagram*

From Figure 6.12 and respecting the hypotheses of the conjugate beam (Table 6.2), it is possible to construct the conjugate beam (Figure 6.12).

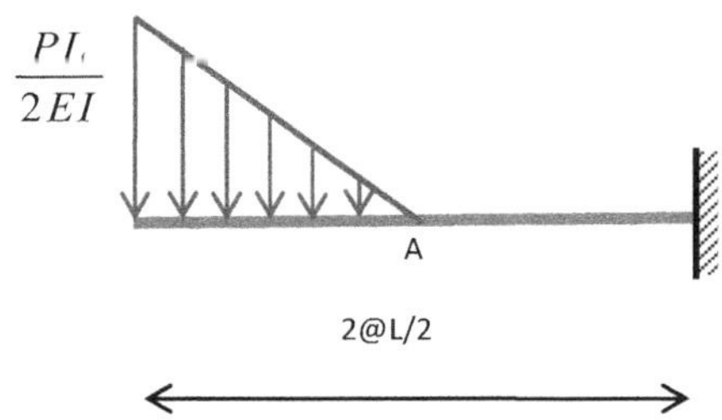

Figure 6.13. *Conjugate beam loading*

The shear force and the bending moment at point A are given as:

$$T_A = -\frac{1}{2}\left(\frac{PL}{2EI}\right)\frac{L}{2} = -\frac{PL^2}{8EI}$$

$$M_A = -\frac{1}{2}\frac{PL}{2EI}\frac{L}{2}.\left(\frac{L}{3}\right) = -\frac{PL^3}{24EI}$$

Hence, the vertical displacement and slope at point A are given as:

$$\delta_A = -\frac{PL^3}{24EI}$$

$$\omega_A = -\frac{PL^2}{8EI}$$

The deflection of beam (Figure 6.14) can be established from the results obtained.

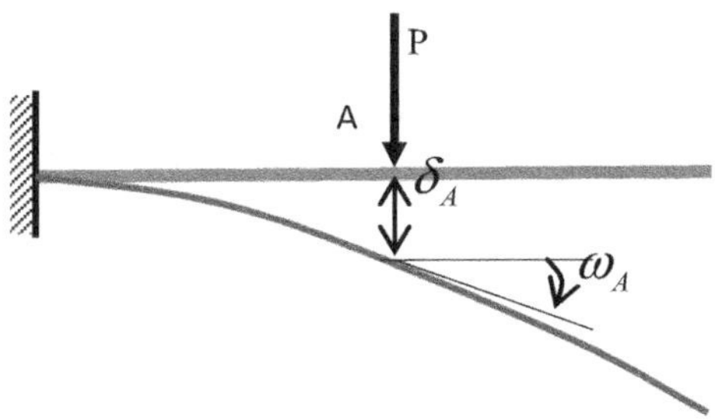

Figure 6.14. *Deformed cantilever*

6.3. Superposition principle

The superposition of effects principle is used when the beam is acted on by a set of external actions. The deflection (slope) at a point can be obtained by the algebraic sum of the deflections (slopes) generated by each load applied to the same point.

The slope and the deflection at any point of the beam are calculated by:

$$\omega = \sum_{i=1} \omega_i \qquad [6.30]$$

$$\delta = \sum_{i=1} \delta_i \qquad [6.31]$$

The deflection or slope at one point can be evaluated by one of the deflection calculation methods: the differential equation method, the direct integration method, the moment-area method and the conjugate beam method.

EXAMPLE 6.5.–

A beam is stressed by concentrated load P and by a uniformly distributed load of magnitude q. Evaluate the slope at point A and the deflection at point B (Figure 6.15). EI is assumed to be constant.

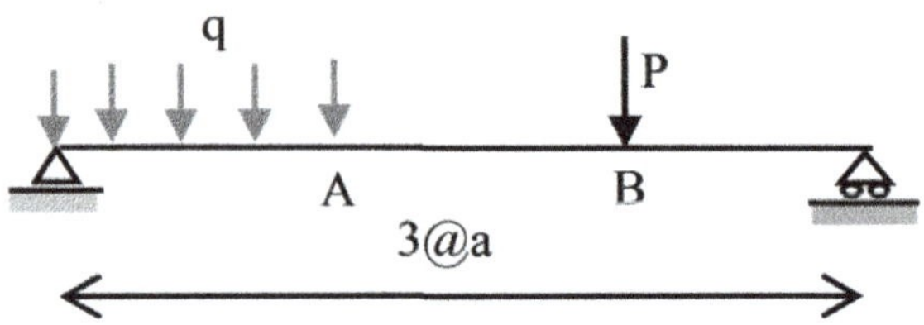

Figure 6.15. *Given beam*

Using the superposition effect leads to the decomposition of the effect of P and that of q.

– Case 1: uniformly distributed load (Figure 6.16).

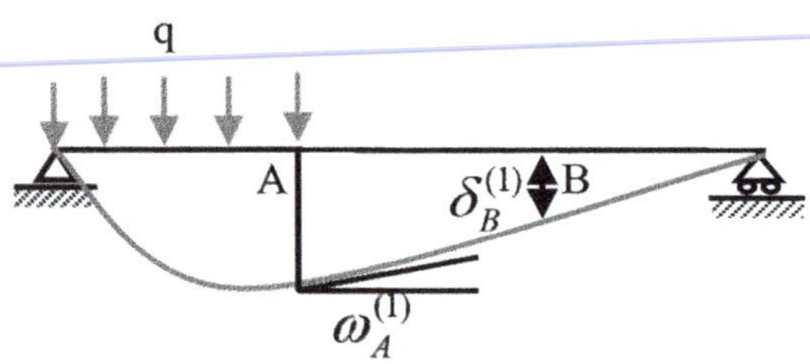

Figure 6.16. *Load force q*

– Case 2: concentrated load (Figure 6.17).

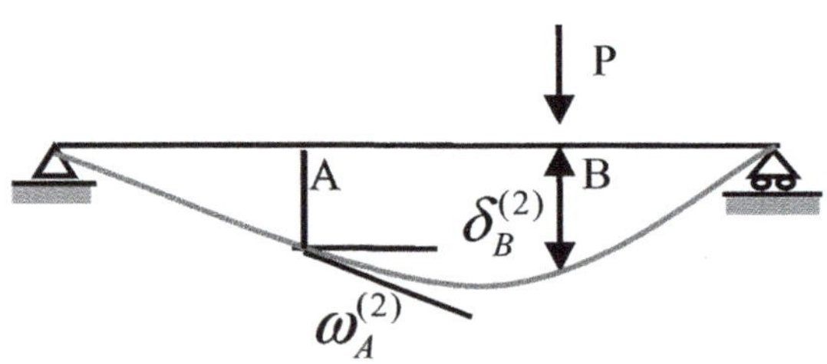

Figure 6.17. *Load force P*

Using the superposition principle allows us to write the following:

– the slope at point A:

$$\omega_A = \omega_A^{(P)} + \omega_A^{(q)}$$

– the displacement at point B:

$$\delta_A = \delta_A^{(P)} + \delta_A^{(q)}$$

The slopes $\omega_A^{(P)}$, $\omega_A^{(q)}$ and deflections $\delta_A^{(P)}$, $\delta_A^{(q)}$ at point A or B, respectively, can be calculated by a deflection calculation method.

EXAMPLE 6.6.–

Calculate the displacement at point C and the slope at point B of the beam (Figure 6.18).

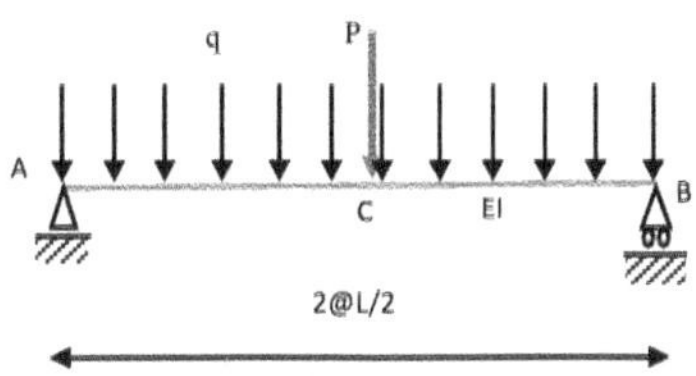

Figure 6.18. *Given beam*

– Uniform load effect:

$$\delta_C^{(q)} = -\frac{5}{384}\frac{qL^4}{EI}$$

$$\omega_B^{(q)} = +\frac{1}{24}\frac{qL^3}{EI}$$

– Concentrated force effect:

$$\delta_C^{(P)} = -\frac{1}{48}\frac{PL^3}{EI}$$

$$\omega_B^{(P)} = +\frac{1}{16}\frac{PL^2}{EI}$$

The superposition principle is used to obtain the vertical displacement at point C and the slope at point B:

$$\delta_C = -\frac{1}{48}\frac{PL^3}{EI} - \frac{5}{384}\frac{qL^4}{EI}$$

$$\omega_B = \frac{1}{16}\frac{PL^2}{EI} + \frac{1}{24}\frac{qL^3}{EI}$$

In the particular case of $P = qL$, we obtain:

$$\delta_C = -\frac{13}{384}\frac{qL^4}{EI}$$

$$\omega_B = \frac{5}{48}\frac{qL^3}{EI}$$

6.4. Conclusion

In this chapter, we have presented the geometric methods for calculating the deflections of elastic behavior beams. Using these methods makes it possible to calculate displacements and/or slopes at any point of a statically determinate beam.

The first integration of the differential equation method of deflection means that we can describe the slope formula. The second integration allows us to formulate the expression of deflection along a statically determinate beam.

On the same subject, the direct integration method provides the expressions of slope and deflection of a statically determinate beam. Integration constants can be evaluated by using the limit conditions of the beam with the external environment.

The graphical moment-area method produces two main theorems.

The slope variation between the tangents at two points of the elastic deflection is equal to the area of the diagram of $\frac{M(x)}{E.I_Z}$ between these two points.

The relative deflection between two points is equal to the static moment of the bending area moment $\frac{M(x)}{E.I_Z}$ between these two points.

The conjugate beam method is a graphical tool for calculating deflections. The fictitious beam is the same length as the real beam, and it is stressed by a load $\frac{M(x)}{E.I_Z}$. The shear force and the bending moment at a point of the fictitious beam are, respectively, the slope and the deflection of the real beam.

Finally, in the case where the beam is simultaneously stressed by a set of actions, it is acceptable to use the superposition principle to simplify the formulations for calculating slopes and deflections.

6.5. Problems

Exercise 1

Using the differential equation for beam deflection method, determine the expressions of deflection and slope. The flexural rigidity (EI) is assumed to be constant.

Deduce the slope at point A and the vertical displacement at point B.

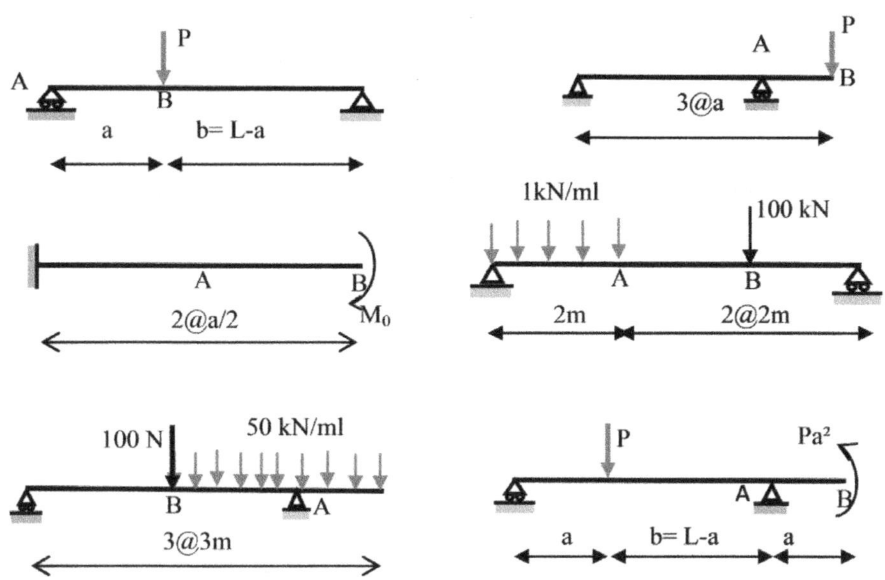

Exercise 2

Using the direct integration method, give the expressions of beam deflection and slope from Exercise 1.

Deduce the slope at point A and the deflection at point B.

Exercise 3

Using the moment-area method, calculate the deflection at point B and the slope at point A in the following beams. The flexural rigidity (EI) is assumed to be constant.

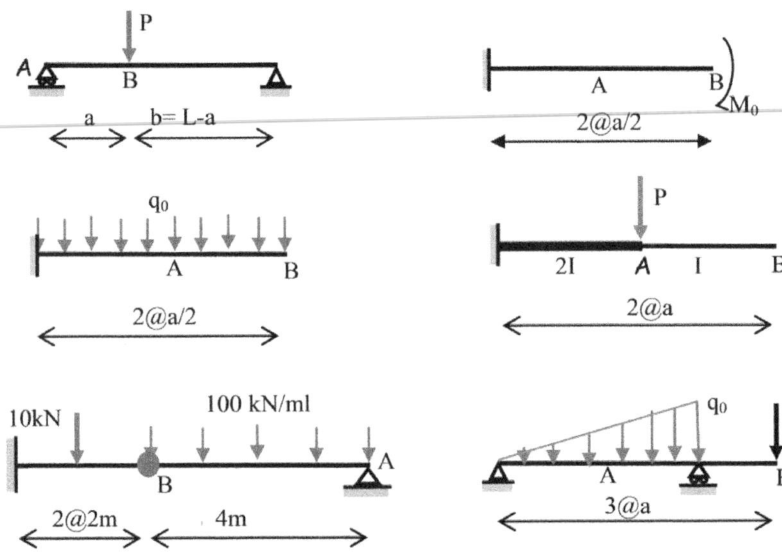

Exercise 4

Using the conjugate beam method, calculate the deflection at point B and the slope at point A in the following beams. The flexural rigidity (EI) is assumed to be constant.

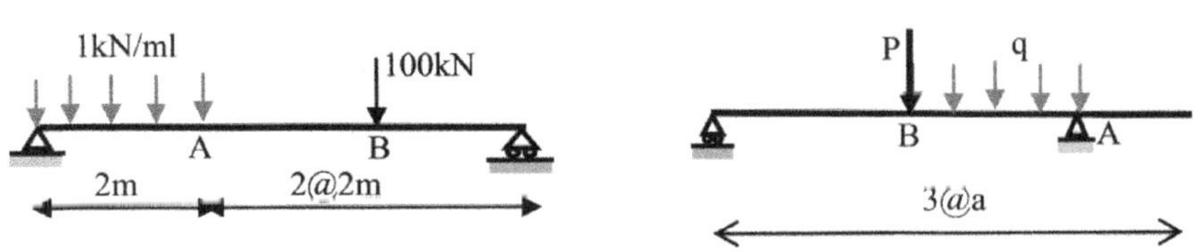

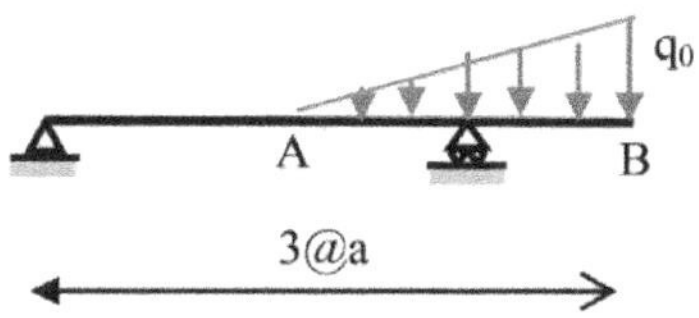

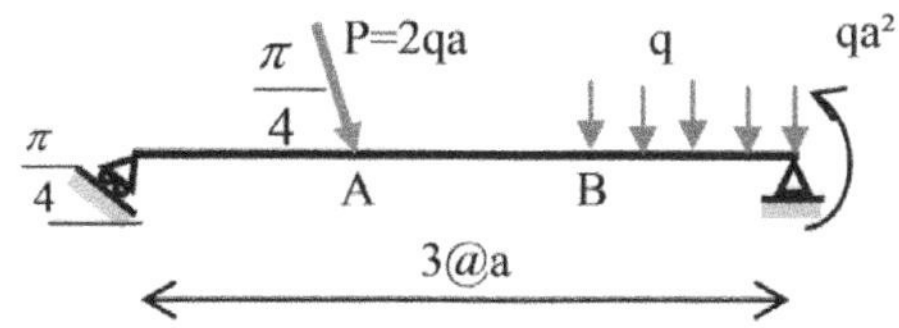

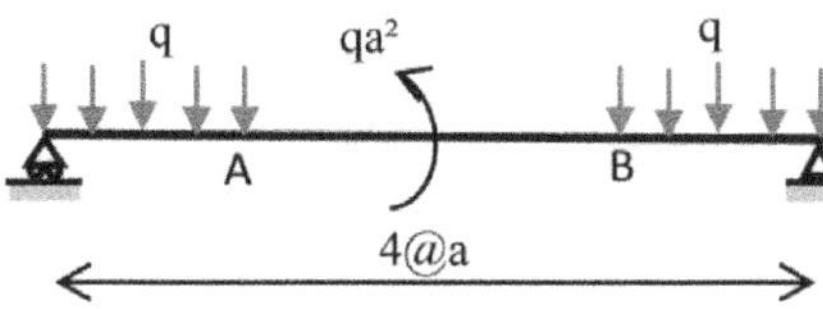

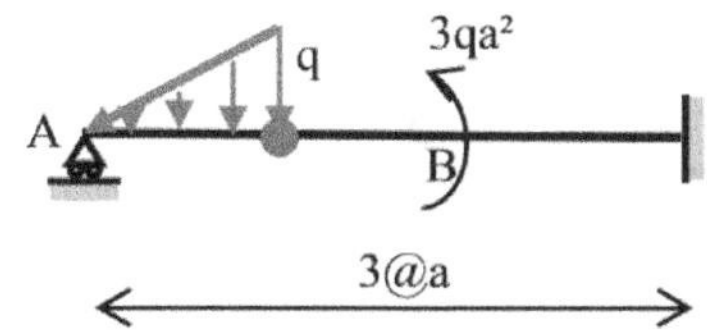

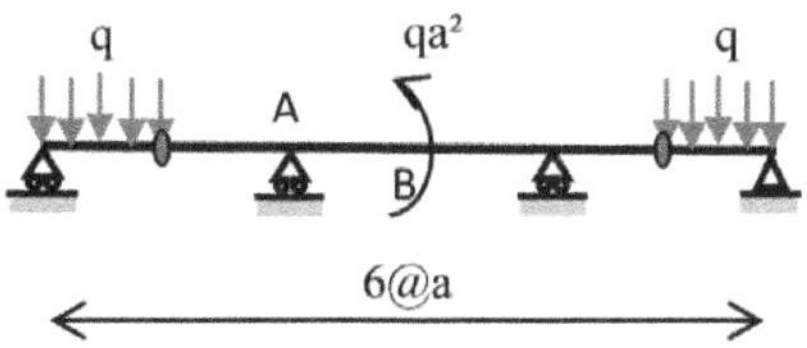

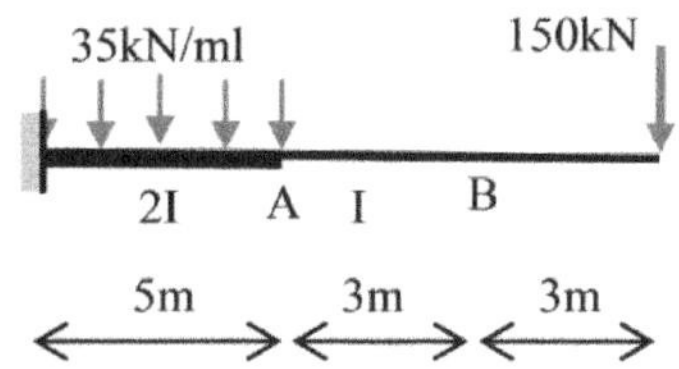

Exercise 5

Using the superposition of effects method, calculate the deflection at point B and the slope at point A in the following beams. The flexural rigidity (EI) is assumed to be constant.

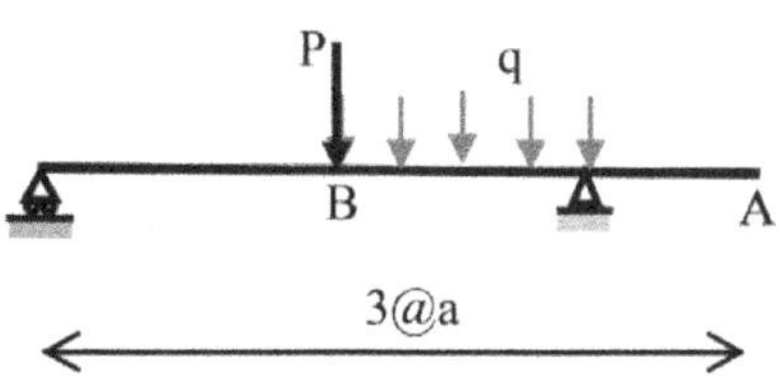

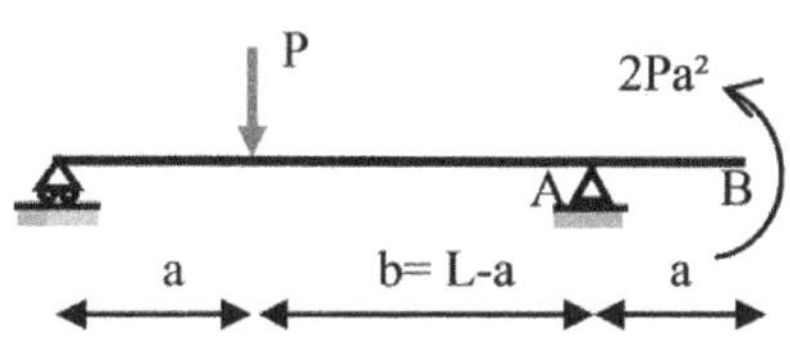

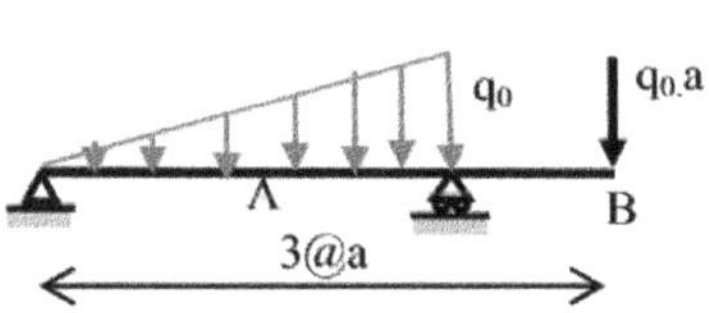

q0
q0.a
A
B
3@a

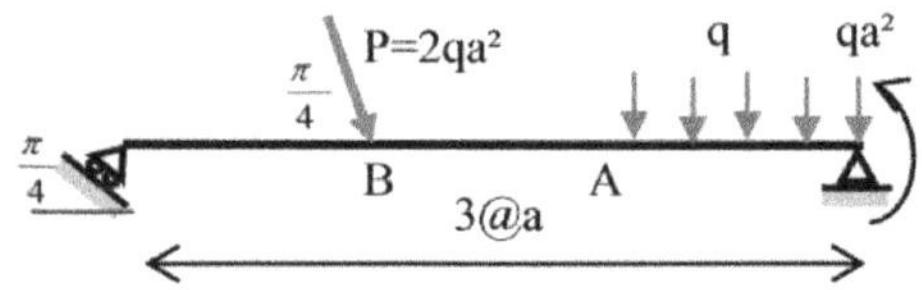

P=2qa²
π/4
π/4
q
qa²
B
A
3@a

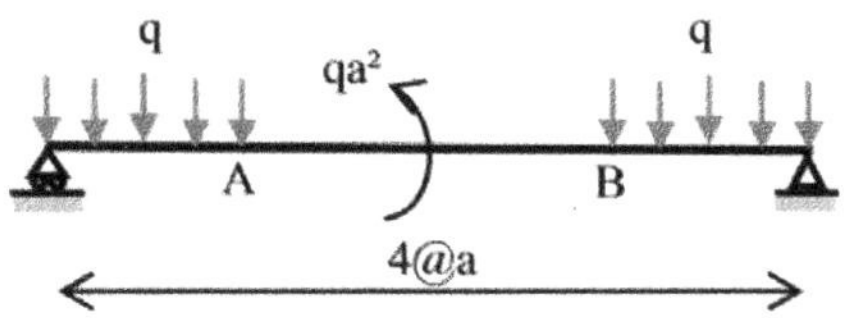

q
qa²
q
A
B
4@a

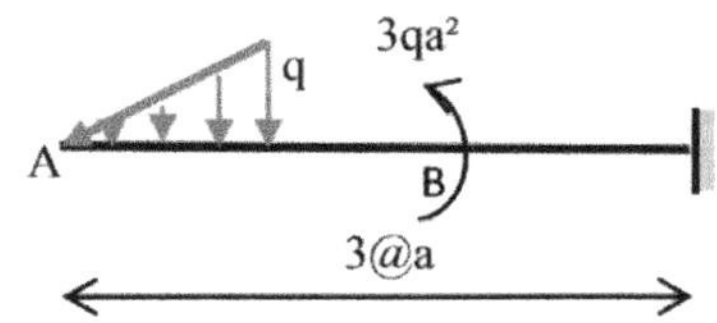

3qa²
q
A
B
3@a

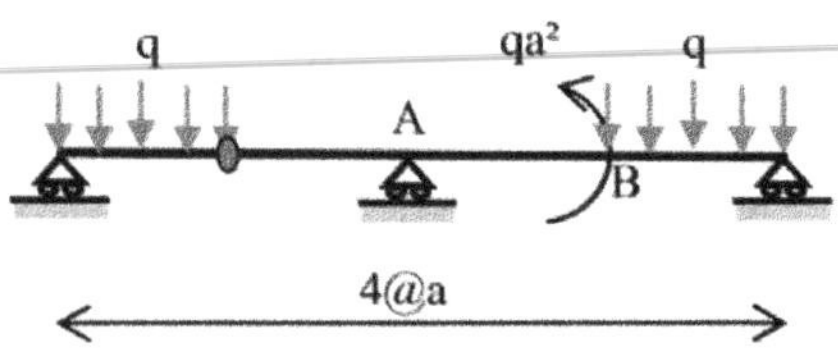

q
qa²
q
A
B
4@a

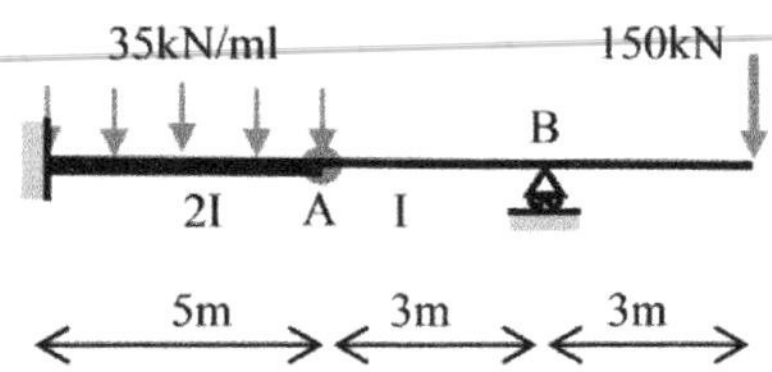

35kN/ml
150kN
B
2I
A
I
5m
3m
3m

7

Structural Deflections: Energy Methods

The teaching objectives for this chapter are as follows:

– understanding the notion of strain energy;

– formulating the internal energy of a structure;

– using the energy principle to calculate deflections;

– applying the principle of virtual work, Castigliano's theorem and the Maxwell Betti law.

This chapter is devoted to the presentation of energy methods to calculate elastic structure deflections. First, we present the concepts of the work of external actions, internal energy and the principle of virtual work. The latter is widely used to calculate deflections in trusses, beam and frames. Finally, we describe Castigliano's theorem and the Maxwell–Betti law.

7.1. Work of external actions

The work of an external action is defined by the scalar product of the forces vector (moments) and the displacements vector (slopes), which are generated. The expression of the elemental work of a force is given by

$$dW = \vec{F}.d\vec{\Delta} \qquad [7.1]$$

The vectors $\vec{F}$ and $d\vec{\Delta}$ form an angle α. The work expression is

$$dW = F.d\Delta.\cos\alpha \qquad [7.2]$$

The work of a force is considered to be positive if $-\frac{\pi}{2} \prec \alpha \prec \frac{\pi}{2}$, negative if $\frac{\pi}{2} \prec \alpha \prec 3\frac{\pi}{2}$ and zero if $\alpha = \frac{\pi}{2}$.

The expression of work depends on the applied external action.

Axial force

A bar of length L is stressed by an axial force *F* (Figure 7.1), which causes a displacement Δ at its point of application.

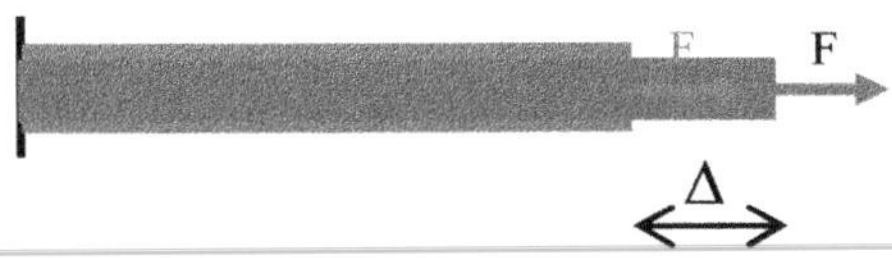

Figure 7.1. *Axial force*[1]

The elementary mechanical work of the axial force is written knowing that the force and displacement are in the same direction.

$$dW = F.d\Delta \quad [7.3]$$

In linear elastic behavior, displacement is proportional to the applied force. It is therefore possible to write

$$F = k_N \Delta \quad [7.4]$$

When we substitute the relationship [7.4] with [7.3], the elementary mechanical work becomes

$$dW = k_N \Delta.d\Delta \quad [7.5]$$

Thus the total mechanical work of force *F* that caused displacement Δ is

$$W = \int_0^{\Delta} k_N \Delta.d\Delta = \frac{1}{2} k_N \Delta^2 = \frac{1}{2} F\Delta \quad [7.6]$$

1 All of the figures in this chapter are available to view in full color at www.iste.co.uk/khalfallah/analysis1.zip.

The total mechanical work of force F is the area between curve F- Δ and the axis of abscissas (Figure 7.2).

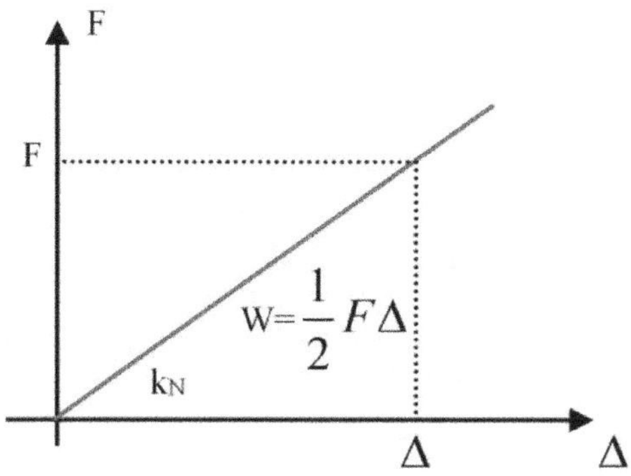

Figure 7.2. *Work of force F*

Bending moment

In the same way, we define the mechanical work of the bending moment by the scalar product of the bending moment M and the elementary slope $d\omega$ (Figure 7.3).

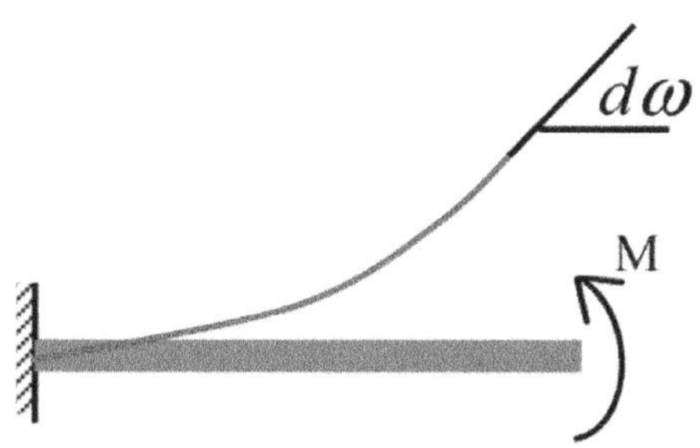

Figure 7.3. *Bending moment*

$$dW = \vec{M}.d\vec{\omega} \quad [7.7]$$

The applied slope and the bending moment rotate in the same direction. The relationship [7.7] is written as

$$dW = M.d\omega \quad [7.8]$$

The bending moment and the slope generated are proportional.

$$M = k_f . \omega \qquad [7.9]$$

Substituting the relationship [7.9] into equation [7.8], we obtain

$$dW = k_f . \omega . d\omega \qquad [7.10]$$

The total work of a bending moment creates a slope ω which becomes

$$W = \int_0^{\omega} k_f \omega . d\omega = \frac{1}{2} k_f \omega^2 = \frac{1}{2} M \omega \qquad [7.11]$$

The total bending moment work of force M is the area between curve M- ω and the axis of abscissas (Figure 7.4).

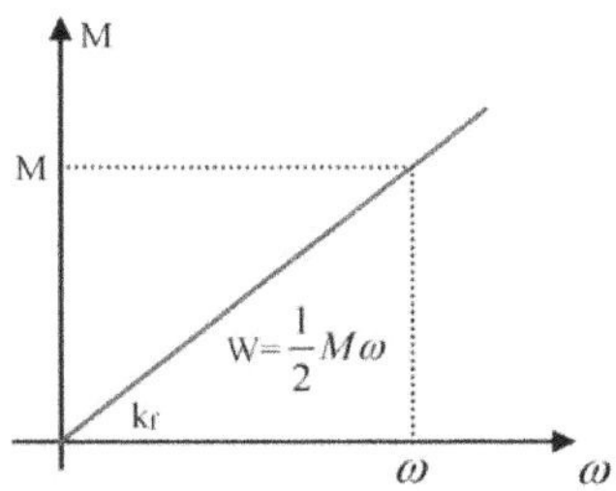

Figure 7.4. *Mechanical work of moment M*

Shear force

The fundamental relationship of the work of external forces [7.1] is always applicable to structures subjected to shear stresses (Figure 7.5).

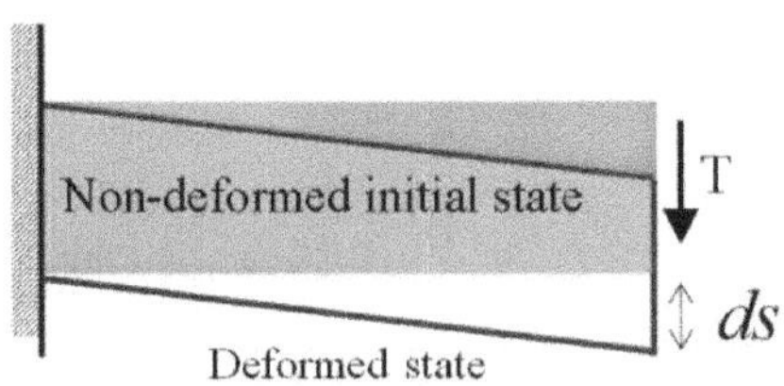

Figure 7.5. *Transversal or shear force*

To generate a transverse slipping, it is necessary to provide an elementary work equal to

$$dW = \vec{T}.d\vec{s} = T.\,ds \qquad [7.12]$$

Knowing that

$$T = k_c.\mathrm{s}, \qquad [7.13]$$

hence

$$dW = k_c\, s.ds \qquad [7.14]$$

The total mechanical work is

$$W = \int_0^s k_c\, s.ds = \frac{1}{2}k_c\, s^2 = \frac{1}{2}Ts \qquad [7.15]$$

The total work of the shear force T is shown schematically in Figure 7.6.

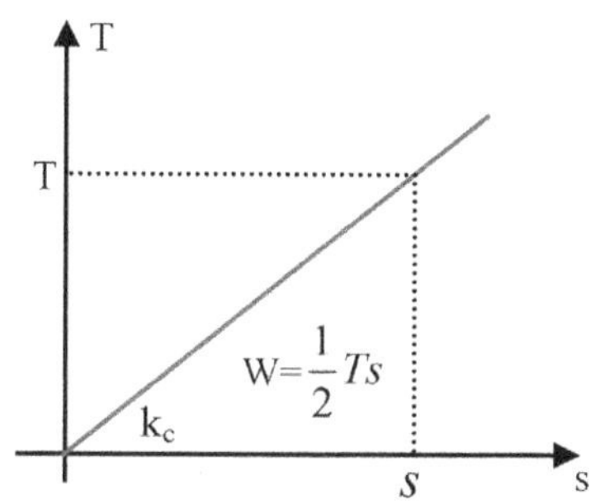

Figure 7.6. *Work of transversal force*

Torque moment

A circular transversal section of a bar is stressed by a torque moment M_t (Figure 7.7).

The torque moment M_t turns on the plane (YZ). The fiber (AB) is horizontal in the initial state. It rotates with the angle $d\varphi$ when the torque moment M_t is applied (Figure 7.7).

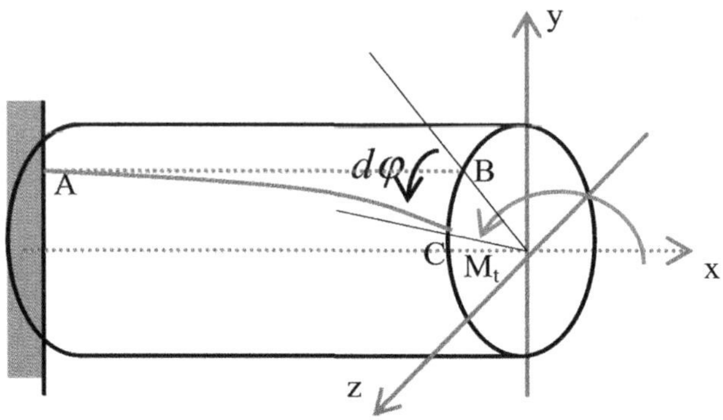

Figure 7.7. *Angular deformation due to torque moment*

The elementary work needed to rotate the section and the plane (YZ) of angle $d\varphi$ is

$$dW = M_T.d\varphi \quad [7.16]$$

Knowing that

$$M_T = k_T.\varphi \quad [7.17]$$

Integrating the relationships [7.16] and [7.17], we can write

$$dW = k_T.\varphi.d\varphi \quad [7.18]$$

The total work is

$$W = \int_0^{\varphi} k_T\varphi.d\varphi = \frac{1}{2}k_T\varphi^2 = \frac{1}{2}M_T\varphi \quad [7.19]$$

In the same way, the total work of torque moment M_t is the area between curve M_T- φ and the axis of abscissas (Figure 7.8).

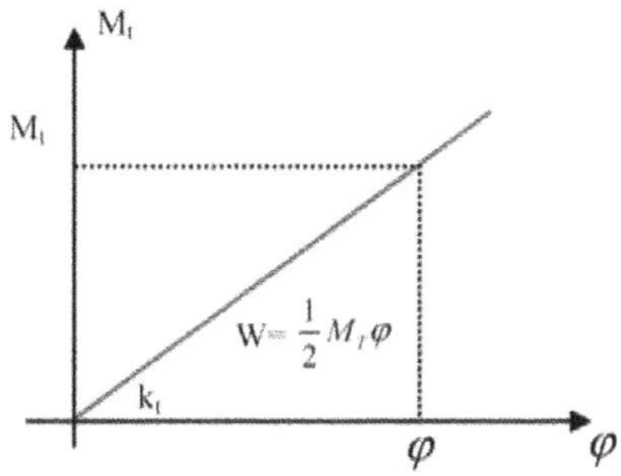

Figure 7.8. *Work of torque moment*

In the case of a structure which is subjected simultaneously to a normal force, a bending moment, a shear stress and a torque moment, the total mechanical work can be deduced using the superposition principle.

$$W=\frac{1}{2}F\Delta+\frac{1}{2}M\omega+\frac{1}{2}Ts+\frac{1}{2}M_T\varphi \quad [7.20]$$

In this area, there are two different cases:

Case 1: When an applied action varies in some way with the corresponding deflection, the mechanical work of this action is written as

$$W=\int_0^\Delta F.d\Delta \quad [7.21a]$$

So the total work of action F is the area between curve F-Δ and the axis of abscissas (Figure 7.9).

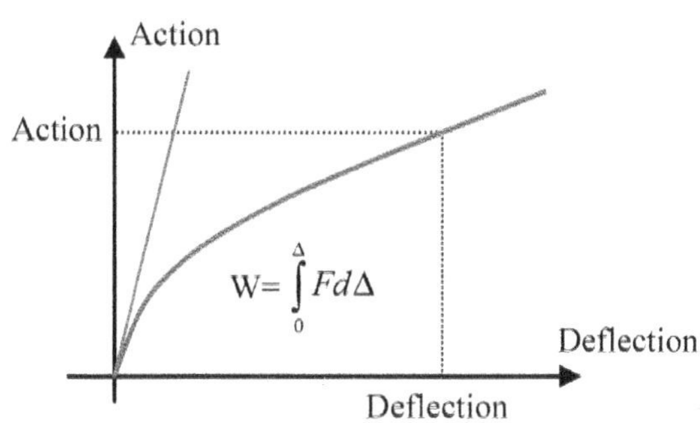

Figure 7.9. *Work of force F*

This action can be a normal force, a bending moment, a shear stress or a torque moment.

Case 2: If applied action F is independent of the deflection which is generated, we deduce the mechanical work by the following expression:

$$W = \int_0^\Delta F.d\Delta = F\int_0^\Delta d\Delta = F.\Delta \quad [7.21b]$$

Figure 7.10 shows the graphical aspect of the mechanical work of the applied action.

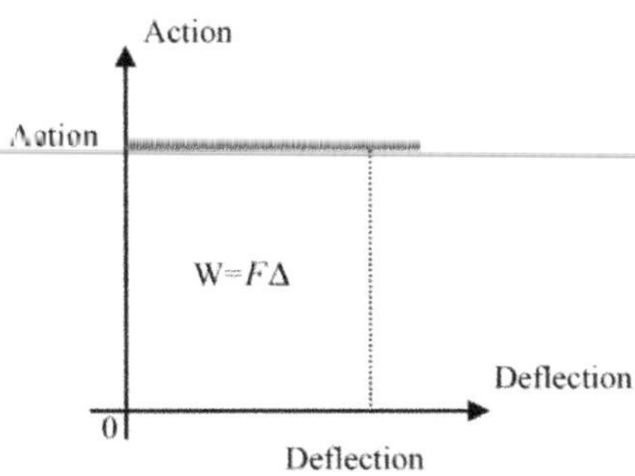

Figure 7.10. *Work of an action independent of deflection*

7.2. Internal or strain energy

The expression of a structure's internal energy depends on the internal action, as well as the geometric and mechanical characteristics of the material constituting the structure. Depending on the action inside the structure, these are as follows.

Normal force

Consider an element of a truss bar of length L that is stressed by an axial internal force (Figure 7.11).

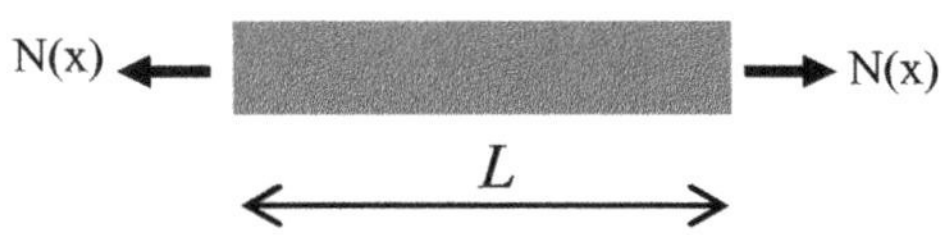

Figure 7.11. *Normal force*

We define the internal elementary work of axial force N(x) by

$$dW_{in} = \vec{N}(x).d(\overrightarrow{\Delta L}) \quad [7.22]$$

The stress–strain relationship is described by Hooke's law.

$$\sigma(x) = E\varepsilon(x) \quad [7.23]$$

Knowing that

$$\sigma(x) = \frac{N(x)}{\Omega} \quad [7.24]$$

$$\varepsilon(x) = \frac{\Delta L}{L} \quad [7.25]$$

With E being the modulus of longitudinal elasticity of the material used and Ω the cross-section area.

Substituting relationships [7.24] and [7.25] in the formula [7.23], we obtain the expression of normal force.

$$N(x) = \frac{E\Omega.\Delta L}{L} \quad [7.26]$$

By substituting the relationship [7.26] into the relationship [7.22], the internal elementary work becomes

$$dW_{in} = \frac{E\Omega}{L}\Delta L.d(\Delta L) \quad [7.27]$$

The internal work required to deflect the bar by a quantity ΔL is written as

$$W_{in} = \frac{E\Omega}{L}\int_{0}^{\Delta L}\Delta L.d(\Delta L) = \frac{E\Omega}{2L}\Delta L^{2} \quad [7.28]$$

Internal work can also be written as

$$W_{in} = \frac{1}{2}N(x)\Delta L \quad [7.29]$$

Using the relationship [7.26], the lengthening of the bar is

$$\Delta L = \frac{N(x)\mathrm{L}}{E\Omega} \qquad [7.30]$$

and the work of internal actions is written as

$$W_{in} = \frac{1}{2}\frac{N^2(x).\mathrm{L}}{E\Omega} \qquad [7.31]$$

In the integral form, the expression [7.31] is written as

$$W_{in} = \frac{1}{2}\int_0^L \frac{N^2(x)}{E\Omega}dx \qquad [7.32]$$

The mechanical work of internal actions is equal to the internal energy of deflections. Hence

$$U_{in} = \frac{1}{2}\int_0^L \frac{N^2(x)}{E\Omega}dx \qquad [7.33]$$

Bending moment

Consider a structural element of length L stressed by a bending moment M(x) (Figure 7.12).

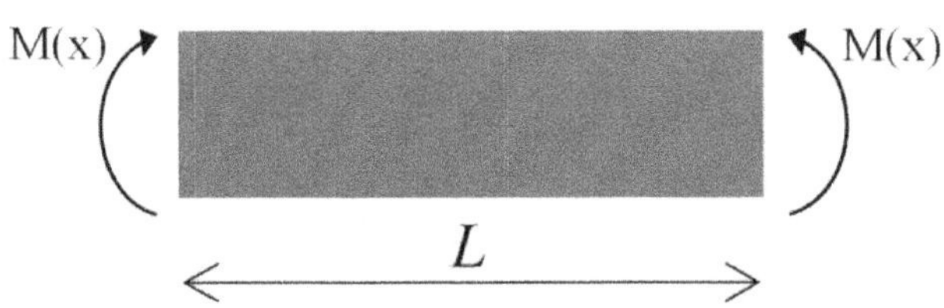

Figure 7.12. *Bending moment*

The elementary work due to the effect of the bending moment M (x) is

$$dW_{in} = M(x).d\omega \qquad [7.34]$$

with

$$M(x) = k.\omega, \qquad [7.35]$$

where k is the rigidity factor of the bar.

Substituting the expression [7.35] into the relationship [7.34], the internal work of the bending moment is

$$W_{in} = \int_0^{\omega} k\omega . d\omega = \frac{1}{2} M(x)\omega \qquad [7.36]$$

Using the relationship [6.14], the differential slope is

$$d\omega = \frac{M(x)}{EI} dx \qquad [7.37]$$

In substituting equation [7.37] with equation [7.34], we obtain

$$dW_{in} = \frac{M^2(x)}{EI} dx \qquad [7.38]$$

The internal work to deflect a bar with angle ω is

$$W_{in} = \frac{1}{2} \int_0^L \frac{M^2(x)}{EI} dx$$

The internal strain energy is

$$U_{in} = \frac{1}{2} \int_0^L \frac{M^2(x)}{EI} dx$$

Shear force

The same procedure is applied to calculate the internal strain energy of a bar subjected to a shear force.

$$U_{in} = \frac{1}{2} \int_0^L \frac{T^2(x)}{G\Omega} dx$$

where G is Young's modulus of transverse elasticity of the material of the bar.

Torque moment

The internal energy of a bar subjected to a torque moment is deduced by

$$U_{in} = \frac{1}{2}\int_0^L \frac{M_T(x)}{GI_0}dx,$$

where I_0 is the moment of polar inertia of the section.

Note that when a section is stressed by a set of actions, the internal energy is equal to the sum of the internal energies of each action, using the principle of superposition.

7.3. Principle of energy conservation

Sections 7.1 and 7.2 above have illustrated the expressions of the work of external forces and the internal energy of a bar stressed by an internal action. In this section, we use the principle of energy conservation to calculate the deflections of elastic structures.

EXAMPLE 7.1.– Bar stressed by a normal force (Figure 7.13)

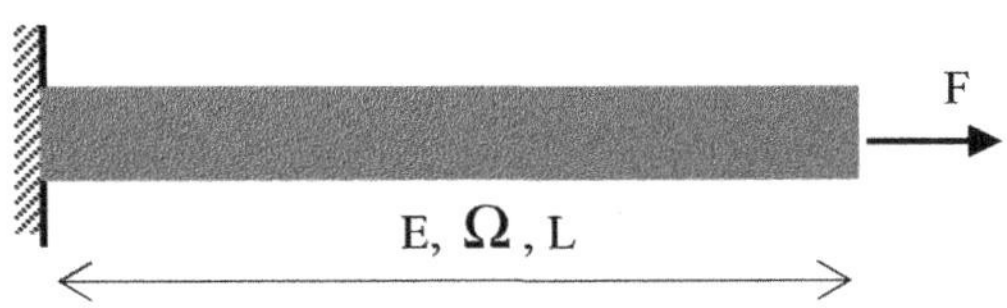

Figure 7.13. *Normal force*

Using the relationship [7.6], the applied force F generates external work equal to

$$W_{ext} = \frac{1}{2}F.\Delta$$

Δ is the displacement that corresponds to the equilibrium limit (Figure 7.14).

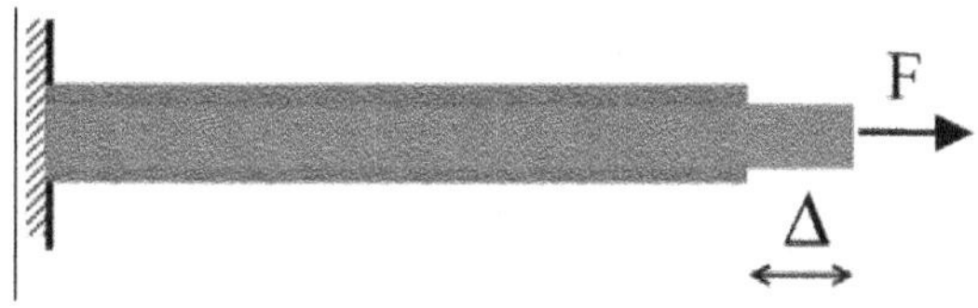

Figure 7.14. *Bar deflection*

The relationship [7.33] allows us to evaluate the energy strain on the bar.

$$U_{in} = \frac{1}{2}\frac{F^2 L}{E\Omega}$$

Applying the principle of energy conservation and taking into account relationships [7.6] and [7.33], we can write

$$W_{ext} = U_{in}$$

This equality makes it possible to calculate the displacement of the application point of the external force.

$$\Delta = \frac{FL}{E\Omega}$$

This result can be deduced using Hooke's fundamental relationship.

EXAMPLE 7.2.– Bending bar (Figure 7.15)

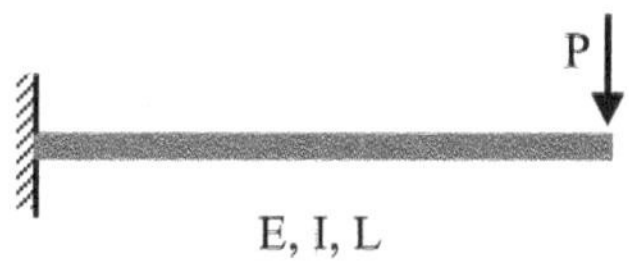

Figure 7.15. *Bending*

The force P generates a mechanical work equal to

$$W_{ext} = \frac{1}{2}F.\Delta$$

Δ is the vertical displacement of the application point of force P (Figure 7.16).

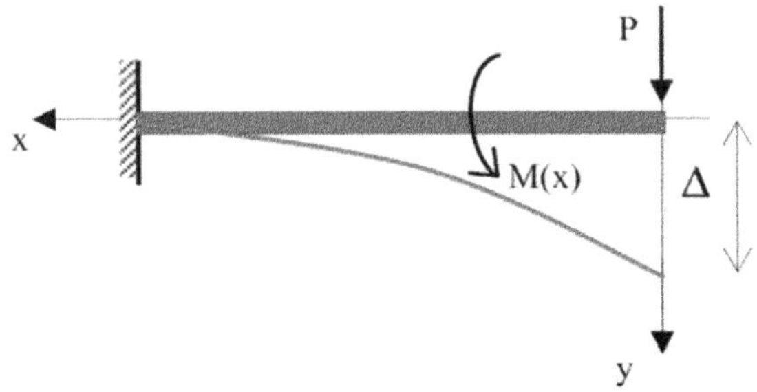

Figure 7.16. *Beam deflection*

The total work of the external force is

$$W_{ext} = \frac{1}{2} P.\Delta$$

The internal strain energy, taking into consideration just the bending effect, is

$$W_{in} = \frac{1}{2}\int_0^L \frac{M^2(x)}{EI} dx$$

Knowing that the bending moment M(x)= Px, the internal energy becomes

$$W_{in} = \frac{1}{2}\int_0^L \frac{P^2x^2}{EI} dx = \frac{P^2L^3}{6EI}$$

Using the principle of energy conservation ($U_{in} = W_{ext}$) allows us to write

$$\frac{1}{2} P.\Delta = \frac{P^2L^3}{6EI}$$

We deduce the vertical displacement of the application point of force P by

$$\Delta = \frac{PL^3}{3EI}$$

EXAMPLE 7.3.– Cantilever beam under pure bending (Figure 7.17)

Figure 7.17. *Pure bending*

The mechanical work of moment M_0 is equal to

$$W_{ext} = \frac{1}{2} M_0 . \omega_0$$

where ω_0 is the slope of the free section of the cantilever beam (Figure 7.18).

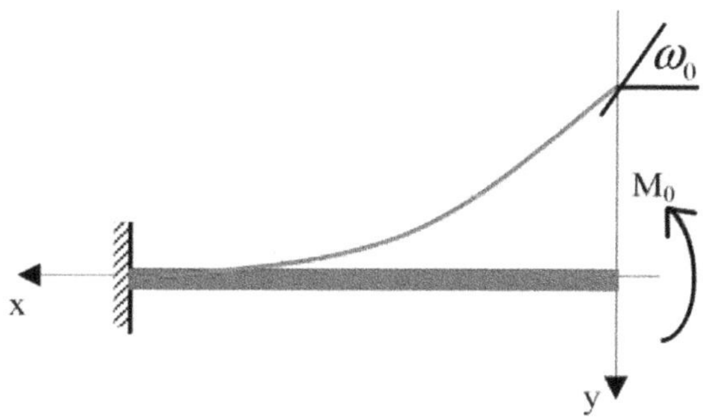

Figure 7.18. *Cantilever deflection*

The internal strain energy of the bar is

$$W_{in} = \frac{1}{2} \int_0^L \frac{M^2(x)}{EI} dx$$

According to the applied moment M_0, internal energy is written as

$$W_{in} = \frac{M_0^2 L}{2EI}$$

Using the principle of energy conservation, the slope of the section at the free end is

$$\omega_0 = \frac{M_0 L}{EI}$$

7.4. Principle of virtual work

John Bernoulli introduced the principle of virtual work in 1717. It is also called the unit action method. This principle is a very powerful tool for calculating deflections at specified points on trusses, beams and plane frames.

In this chapter, we focus on deformable structures. This part presents the development of the principle of virtual forces of deformable structures. The principle of virtual work is one of the general methods of calculating structural deflections.

The principle of virtual forces can be established as follows:

When a set of external actions, F_{ex}, is applied to a deformable structure, internal actions, F_{in}, develop at any point of the structure. Equilibrium equations ensure the relationship between external and internal actions.

External actions cause external displacement Δ at any point on the structure's surface. In the same context, internal actions generate internal displacements δ verifying the compatibility relationships.

In general, the principle of virtual work applies when the work of external actions is equal to that of internal actions.

$$\sum F_{ex}.\Delta = \sum F_{in}.\delta \qquad [7.39]$$

Considering any solid body, we calculate, for example, the displacement at point A on the structure's surface (Figure 7.19).

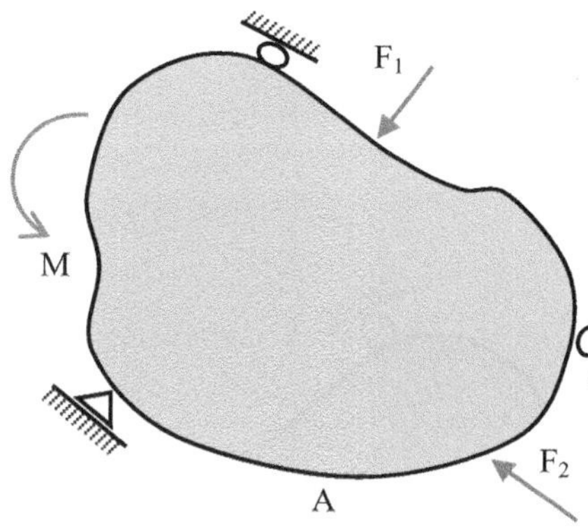

Figure 7.19. *Deformable body*

Under the effect of applied forces, a solid body deforms and obtains a deformed equilibrium configuration, which differs from its initial position (Figure 7.20).

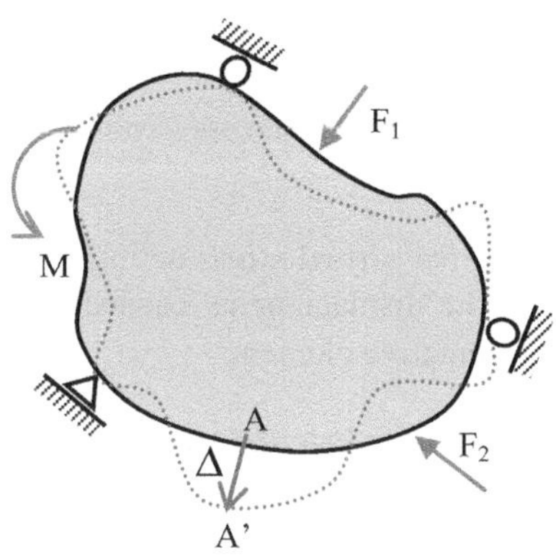

Figure 7.20. *Deformed body*

Now, a virtual unit force is applied at point A in the direction (AA') (Figure 7.21). The unit virtual force generates virtual internal forces N_{in} in any representative element of the structure. The equilibrium relationship is ensured between the unit virtual force and the internal virtual actions.

Once the real actions (M, F_1, F_2) are applied, point A displaces by Δ, generating a displacement dL of the representative point. The reciprocity of the virtual work of internal actions and of external forces through the displacements Δ and dL allows us to write

$$1.\Delta = \sum_{i} N_{in}.dL \qquad [7.40]$$

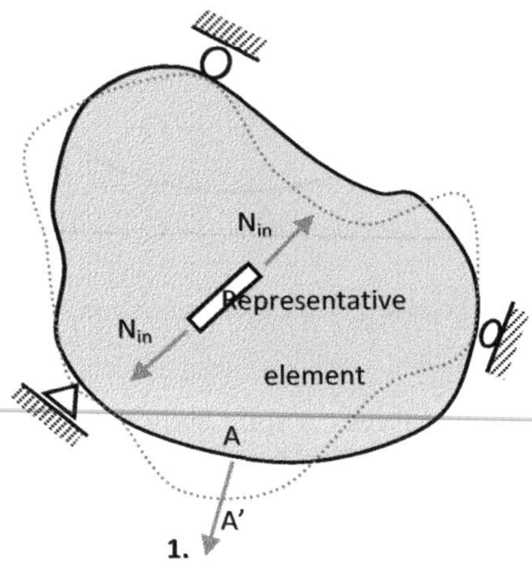

Figure 7.21. *Internal force*

Where $F_{ex}=1$, N_{in} is the internal virtual stress acting on the considered element, Δ and dL are, respectively, the displacements caused by the real forces and the internal action N_{in} of the representative element.

The relationship [7.40] allows us to calculate the displacement at point A.

$$\Delta = \sum N_{in}.\mathrm{dL} \qquad [7.41]$$

The relationship [7.41] can also be used to calculate the slope at point A by applying a unit virtual moment. This moment causes internal virtual forces N_{in} and the slope at point A can be in this case

$$1.\omega = \sum N_{in}.\mathrm{dL} \qquad [7.42]$$

Hence, the slope at point A is calculated as

$$\omega = \sum N_{in}.\mathrm{dL} \qquad [7.43]$$

Where M_{ex}=1, N_{in} is the internal virtual stress acting on the element, ω is the slope caused by real actions and dL is the internal displacement of the representative element.

Relationships [7.41] and [7.43] represent the fundamental equation of virtual work. They can be applied to calculate displacements and slopes.

7.4.1. *Method of virtual work: trusses*

In trusses, the virtual work method is used to calculate the displacement of a joint. It is necessary to calculate the vertical displacement of joint B and the horizontal displacement of joint C of the truss (Figure 7.22). It is assumed that the membrane rigidity $E\Omega$ is constant for all bars.

The forces on the structure's bars can be determined by one of the truss analysis methods (Chapter 4).

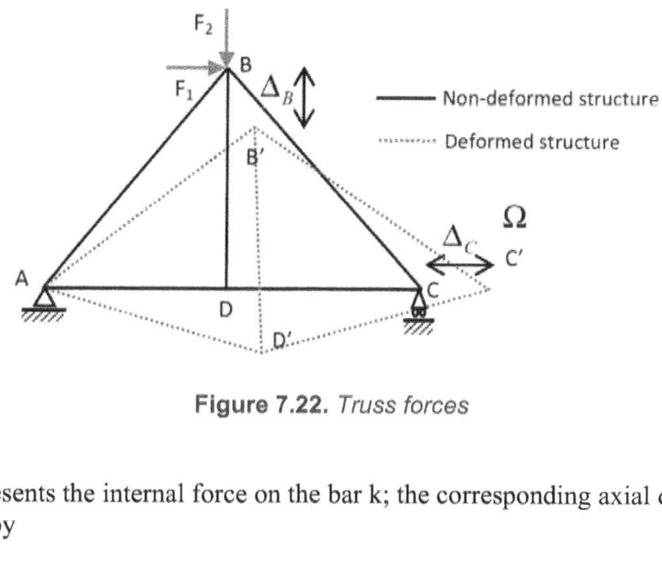

Figure 7.22. *Truss forces*

N_k represents the internal force on the bar k; the corresponding axial deflection is calculated by

$$\delta_k = \frac{N_k L_k}{E\Omega} \quad [7.44]$$

where L_k, $E\Omega$ are, respectively, the length and the axial rigidity of bar k.

To calculate the vertical displacement of joint B, a virtual force is applied to joint B along the positive directions of the axes (Figure 7.23).

The work of the unit virtual force through displacement at the desired point in the real system is

$$W_{ex} = 1.\Delta_B \quad [7.45]$$

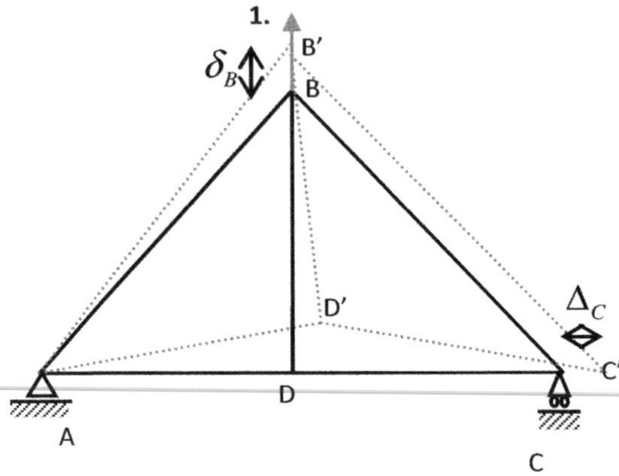

Figure 7.23. *Virtual deflection of the structure*

In the same way, the work of the internal virtual forces through the displacements of the bars of the real system is written as

$$W_{in} = \sum_k n_k.\delta_k \quad [7.46]$$

where n_k is the virtual force in bar k due to the unit force.

By applying the fundamental principle of virtual work, we can write

$$1.\Delta = \sum_k n_k.\delta_k \quad [7.47]$$

Substituting the relationship [7.44] into relationship [7.47], the displacement at point B is given as

$$\Delta_B = \sum_k n_k.\frac{N_k.L_k}{E\Omega} \quad [7.48]$$

7.4.1.1. *Analysis procedure*

The steps for calculating the displacement of statically determinate trusses using the virtual work method are as follows:

– the internal forces N_k in the bars of the statically determinate truss are calculated by one of the truss analysis methods;

– in the virtual system, where we want to calculate the displacement in a positive direction, we apply a unit virtual force to the joint. Calculating internal forces due to the unit virtual action is feasible by using another method of truss analysis;

– the displacement can be calculated by applying the relationship [7.48]. The expression of the virtual work due to the real and virtual forces evaluated in 1 and 2 allows us to calculate the displacement. If the displacement obtained is positive, it means that the displacement and the unit force follow the same direction.

EXAMPLE 7.4.–

Calculate the vertical displacement in joint B of the truss (Figure 7.24). $E\Omega$ is assumed to be constant.

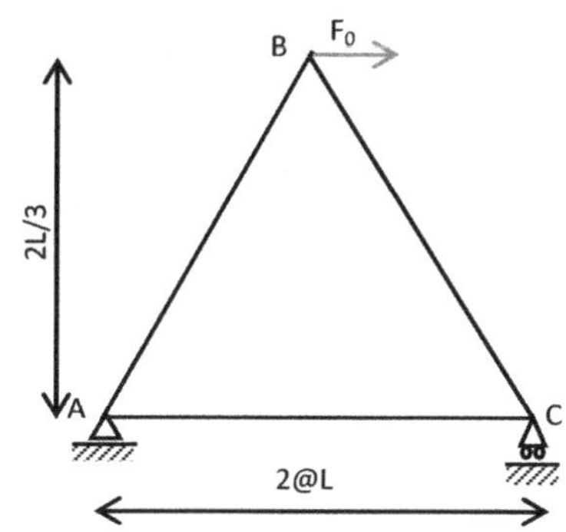

Figure 7.24. *Given structure*

Study of the real structure

Calculating internal forces (Figure 7.24) is carried out using the method of joint equilibrium. The results obtained are grouped in Table 7.1.

Bar	AB	BC	AC
N	$\frac{\sqrt{13}}{6}F_0$	$-\frac{\sqrt{13}}{6}F_0$	$\frac{1}{2}F_0$
Nature of N	Traction	Compression	Traction

Table 7.1. *Forces on the bars*

Study of the virtual structure

A unit virtual force is applied to joint B (Figure 7.25) and then the internal forces in the bars are calculated by the joint equilibrium method.

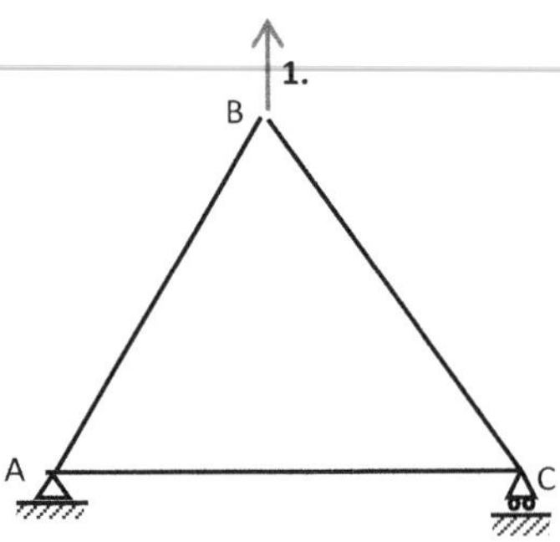

Figure 7.25. *Given structure*

The forces on the bars are grouped in Table 7.2.

Bar	AB	BC	AC
n	$\frac{\sqrt{13}}{4}$	$\frac{\sqrt{13}}{4}$	$-\frac{3}{4}$
Nature of n	Traction	Traction	Compression

Table 7.2. *Unit forces*

Table 7.3 groups the forces on the bars of the virtual and real structure (Figure 4.13). We can calculate the displacement using the relationship [7.48].

Bar	N	n	L	$\frac{nNL}{E\Omega}$
AB	$\frac{\sqrt{13}}{4}F_0$	$\frac{\sqrt{13}}{6}$	$\frac{\sqrt{13}}{2}L$	$\frac{13\sqrt{13}}{48}\frac{F_0 L}{E\Omega}$
BC	$-\frac{\sqrt{13}}{4}F_0$	$\frac{\sqrt{13}}{6}$	$\frac{\sqrt{13}}{2}L$	$-\frac{13\sqrt{13}}{48}\frac{F_0 L}{E\Omega}$
AC	$\frac{1}{2}F_0$	$-\frac{1}{3}$	$2L$	$-\frac{1}{3}\frac{F_0 L}{E\Omega}$
$\sum\frac{nNL}{E\Omega}$				$-\frac{1}{3}\frac{F_0 L}{E\Omega}$

Table 7.3. *Displacements on the bars*

From Table 7.3, vertical displacement of point B is given as

$$\Delta_B^V = -\frac{1}{3}\frac{F_0 L}{E\Omega}$$

The negative sign shows that joint B is displaced in the opposite direction to that of the unit force.

EXAMPLE 7.5.–

Repeat Example 7.4 to calculate the horizontal displacement in support C.

A unit virtual force is applied to joint C in a positive direction (Figure 7.26). Calculations of internal forces in the bars of the virtual structure are grouped in Table 7.4.

Bar	AB	BC	AC
n	0	0	1
Nature of n	Neutral	Neutral	Traction

Table 7.4. *Forces due to unit action*

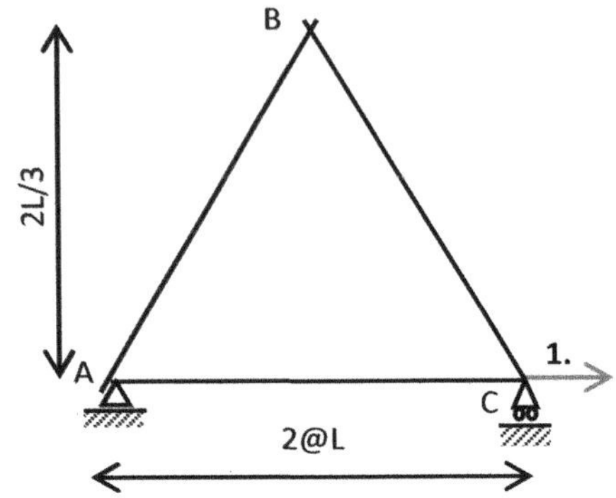

Figure 7.26. *Given structure*

The internal forces on the bars are grouped in Table 7.5.

Bar	N	n	L	$\frac{nNL}{E\Omega}$
AB	$\frac{\sqrt{13}}{4}F_0$	0	$\frac{\sqrt{13}}{2}L$	0
BC	$-\frac{\sqrt{13}}{4}F_0$	0	$\frac{\sqrt{13}}{2}L$	0
AC	$\frac{1}{2}F_0$	1	$2L$	$\frac{F_0 L}{E\Omega}$
$\sum\frac{nNL}{E\Omega}$				$\frac{F_0 L}{E\Omega}$

Table 7.5. *Forces on the bars*

Hence, the horizontal displacement at point C is

$$\Delta_C^H = \frac{F_0 L}{E\Omega}$$

The deformed structure takes the following configuration (Figure 7.27).

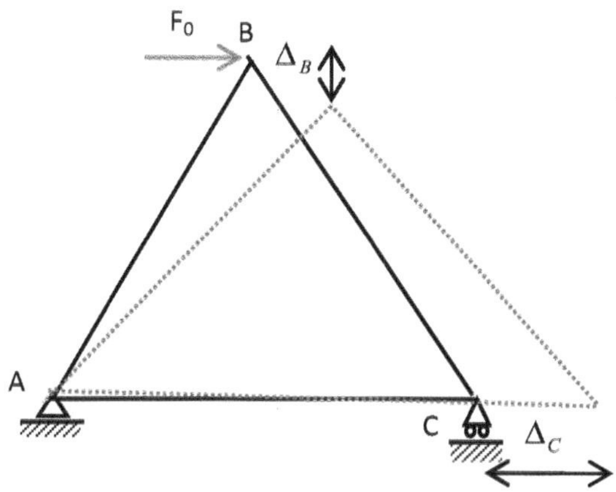

Figure 7.27. *Deformation of the structure*

7.4.2. *Method of virtual work: beams*

The expression of the virtual work method is used to calculate the deflections of the determine beams. In this context, we consider a beam subjected to any load q(x) (Figure 7.28) where we want to calculate the vertical displacement at point B of the beam.

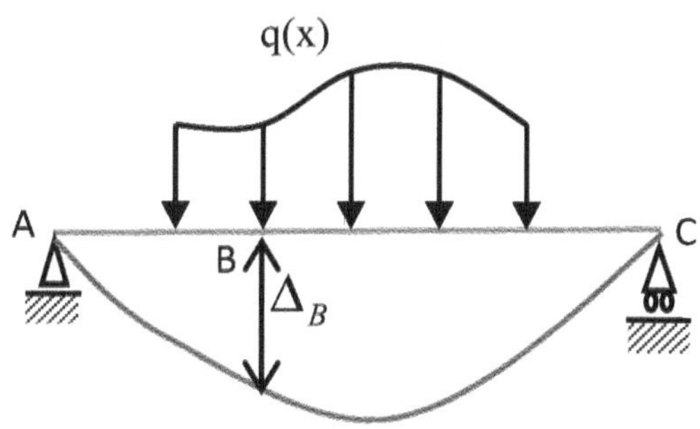

Figure 7.28. *Given beam*

To develop the expression giving the vertical displacement Δ_B using the virtual work method, we apply a unit virtual force in a positive direction (Figure 7.29).

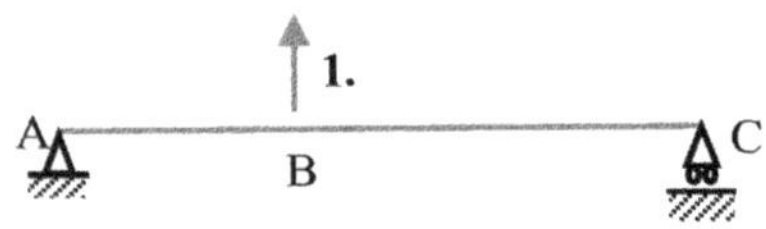

Figure 7.29. *Unit action*

The work of the unit virtual force through the deflection generated in point B is

$$W_{ex} = 1.\Delta_B \qquad [7.49]$$

Similarly, the elementary work of virtual internal actions is written as

$$dW_{in} = m(x)\,\mathrm{d}\omega(x), \qquad [7.50]$$

where m(x) is the virtual moment along the beam.

The variation of slope $d\omega(x)$ along element dx [7.16] is written as

$$d\omega(x) = \frac{\mathrm{M}(x)}{EI}dx \qquad [7.51]$$

M(x) is the bending moment due to real loads.

Substituting the relationship [7.51] with [7.50], we obtain

$$dW_{in} = m(x)\frac{M(x)}{EI}dx \qquad [7.52]$$

The total virtual work of the whole beam can be deduced by integrating relationship [7.52].

$$W_{in} = \int_0^L m(x)\frac{M(x)}{EI}dx \qquad [7.53]$$

Applying the principle of virtual work leads to formulating the expression of vertical displacement in point B.

$$1.\Delta_B = \int_0^L m(x)\frac{M(x)}{EI}dx \qquad [7.54]$$

or

$$\Delta_B = \int_0^L \frac{m(x)M(x)}{EI} dx \qquad [7.55]$$

The relationship [7.55] is used to calculate the slope of the section passing through point B.

$$\omega_B = \int_0^L \frac{m(x)M(x)}{EI} dx \qquad [7.56]$$

where $m(x)$ is the bending moment due to applying a unit virtual moment at the desired point.

7.4.2.1. *Analysis procedure*

To calculate the deflection at a point on a statically determinate beam using the virtual work method, it is necessary to follow the following steps:

– formulate the expression of the bending moment along the beam subjected to the force of real loads;

– formulate the expression of the unit virtual moment. If we want to find a displacement, a unit force must be applied. If we want to find a slope, a unit virtual moment must be applied to the point chosen in a positive direction;

– the deflection at a point is calculated using one of equations [7.55] or [7.56].

EXAMPLE 7.6.– Beam

Calculate the deflection and slope at point A of the cantilevered beam (Figure 7.30). We assume that thc flexural rigidity EI is constant.

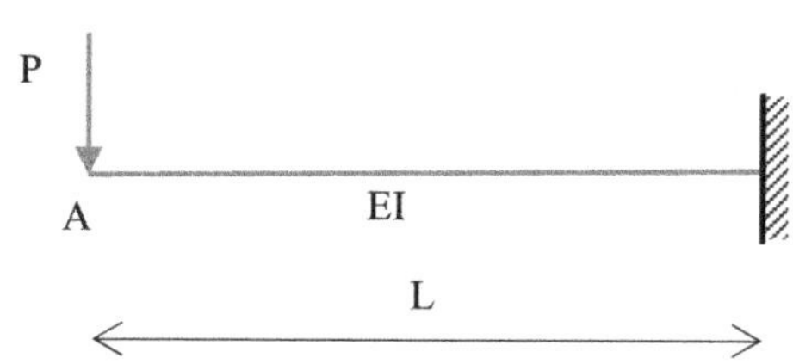

Figure 7.30. *Given beam*

The variation of the bending moment on reference (Axy) is

$$M(x) = -Px$$

A unit virtual force is applied at point A (Figure 7.31).

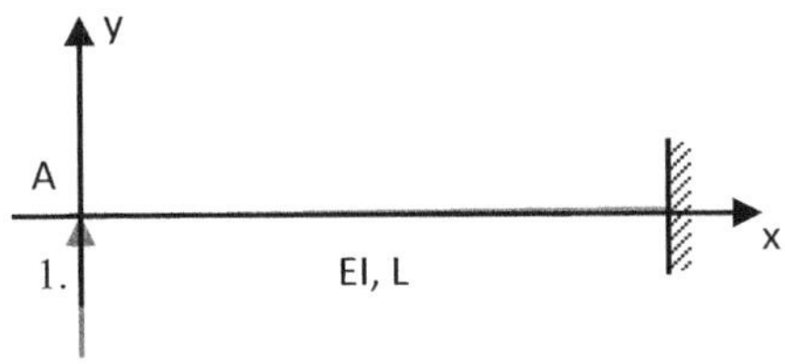

Figure 7.31. *Virtual force*

The variation of the unit virtual moment is

$$m(x) = x$$

The vertical displacement of point A can be deduced by

$$\Delta_A = \int_0^L x\frac{-Px}{EI}dx = \frac{-PL^3}{3EI}$$

The slope of the section passing through point A can be obtained by applying a unit virtual moment at the same point (Figure 7.32).

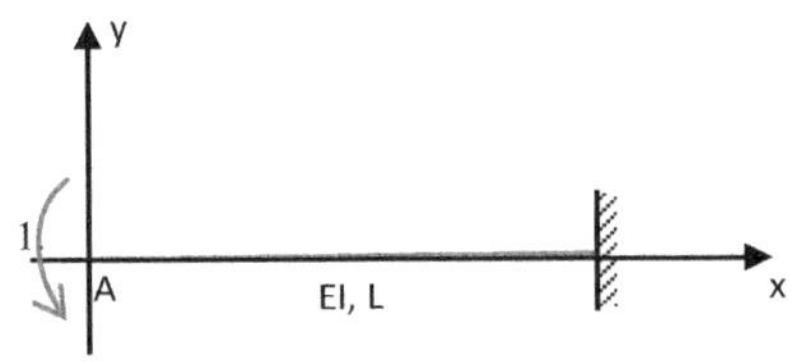

Figure 7.32. *Unit structure*

The variation of the virtual moment along the cantilever is

$$m(x) = -1$$

In the same way, the slope of the section passing through point A is

$$\omega_A = \int_0^L (-1)\frac{(-Px)}{EI}dx = \frac{PL^2}{2EI}$$

The deformed cantilever is shown in Figure 7.33.

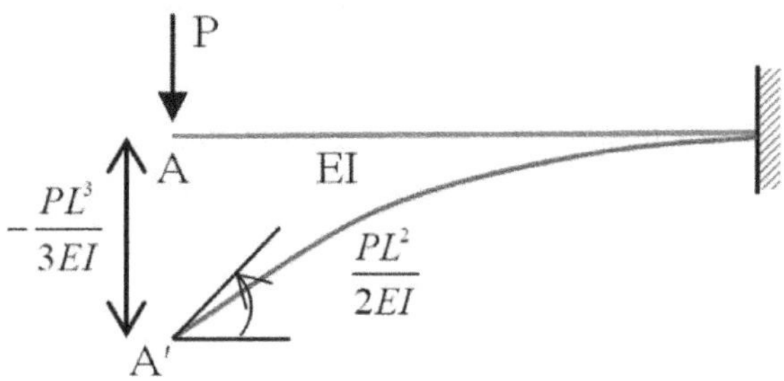

Figure 7.33. *Deformed cantilever*

EXAMPLE 7.7.– A beam with many loads

Calculate the deflection and slope at point A of the cantilever (Figure 7.34). We assume that the flexural rigidity EI is constant.

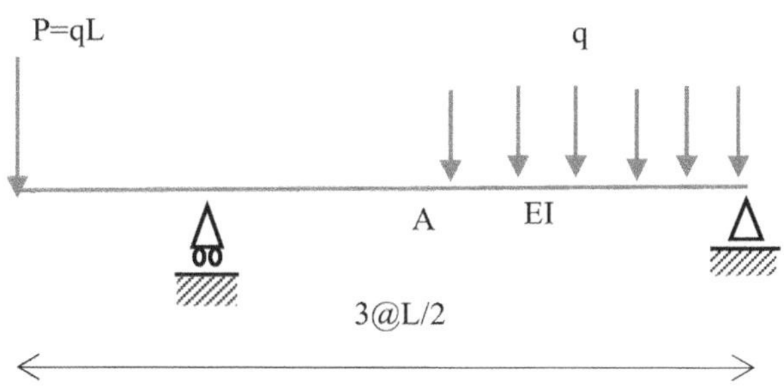

Figure 7.34. *Given beam*

We associate reference (Bxy) with the beam (Figure 7.35).

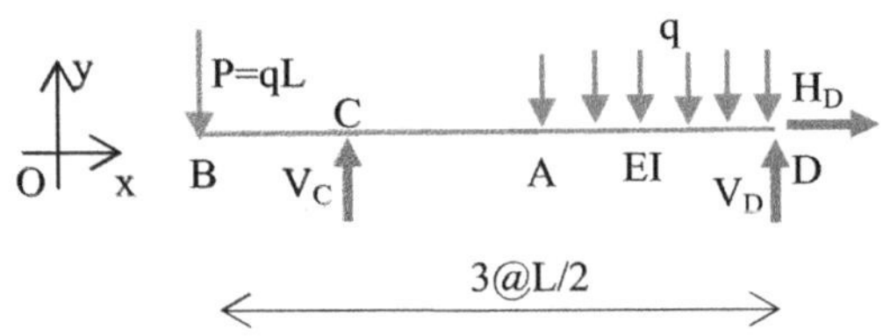

Figure 7.35. *Free-body beam*

The variation of the bending moment of external actions is

$0 \le x \le \frac{L}{2}$:

$$M(x) = -Px = -qLx$$

$\frac{L}{2} \le x \le L$:

$$V_C = \frac{13}{8}qL$$

$$M(x) = \frac{13}{8}qL(x - \frac{L}{2}) - qLx = \frac{5}{8}qLx - \frac{13}{16}qL^2$$

$L \le x \le 3\frac{L}{2}$:

$$M(x) = \frac{13}{8}qL(x - \frac{L}{2}) - qLx - \frac{1}{2}q(x-L)^2 = \frac{13}{8}qLx - \frac{1}{2}qx^2 - \frac{21}{16}qL^2$$

A unit virtual force is exerted at point A (Figure 7.36).

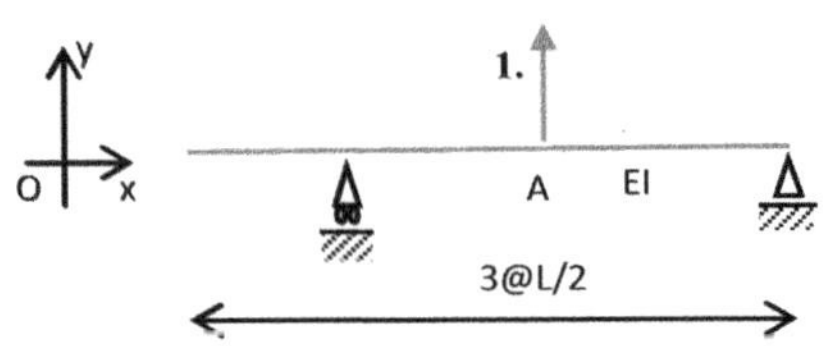

Figure 7.36. *Unit force*

The expression of the unit virtual moment is

$0 \leq x \leq \frac{L}{2}$:

$$\mathrm{m}(x) = 0$$

$\frac{L}{2} \leq x \leq L$:

$$V_C = -\frac{1}{2}$$

$$\mathrm{m}(x) = -\frac{1}{2}x$$

$L \leq x \leq 3\frac{L}{2}$:

$$\mathrm{m}(x) = \frac{1}{2}x - \frac{L}{2}$$

The vertical displacement of point A can be deduced by

$$\Delta_A = \int_0^{L/2} 0.\frac{-qLx}{EI}dx + \int_{L/2}^{L} (-\frac{1}{2}x)\frac{\frac{5}{8}qLx - \frac{13}{16}qL^2}{EI}dx + \int_L^{3L/2} (\frac{x}{2} - \frac{L}{2})\frac{\frac{13}{8}qLx - \frac{1}{2}qx^2 - \frac{21}{16}qL^2}{EI}dx = \frac{361}{768}\frac{qL^4}{EI}$$

To calculate the slope at point A, we apply the unit virtual moment at the same point (Figure 7.37).

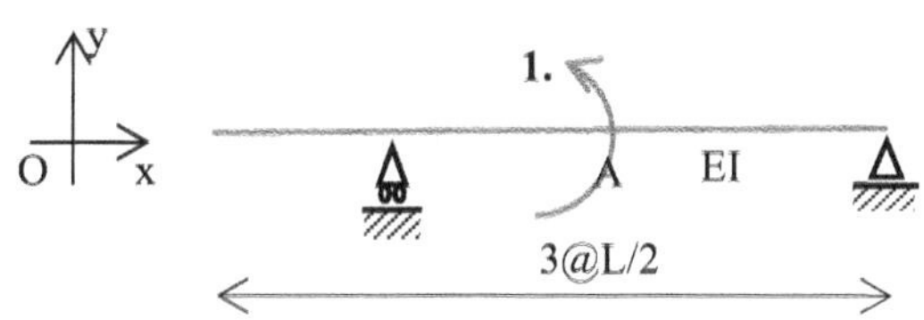

Figure 7.37. *Unit moment*

The variation of the virtual moment along the cantilever is calculated as

$$0 \leq x \leq \frac{L}{2}:$$

$$\mathrm{m}(x) = 0$$

$$\frac{L}{2} \leq x \leq L:$$

$$V_C = \frac{1}{L}$$

$$\mathrm{m}(x) = \frac{1}{L}x$$

$$L \leq x \leq 3\frac{L}{2}:$$

$$\mathrm{m}(x) = \frac{1}{L}x - 1$$

The slope of the section passing through point C is obtained as

$$\omega_A = \int_0^{L/2} 0.\frac{-qLx}{EI}dx + \int_{L/2}^{L} (\frac{1}{L}x)\frac{\frac{5}{8}qLx - \frac{13}{16}qL^2}{EI}dx + \int_{L}^{3L/2} (\frac{x}{L} - 1)\frac{\frac{13}{8}qLx - \frac{1}{2}qx^2 - \frac{21}{16}qL^2}{EI}dx = \frac{qL^3}{3EI} \qquad [7.57]$$

The deformed cantilever is shown in Figure 7.38.

$$\omega_A = \frac{qL^3}{3EI}$$

$$\Delta_A = \frac{361}{768}\frac{qL^4}{EI}$$

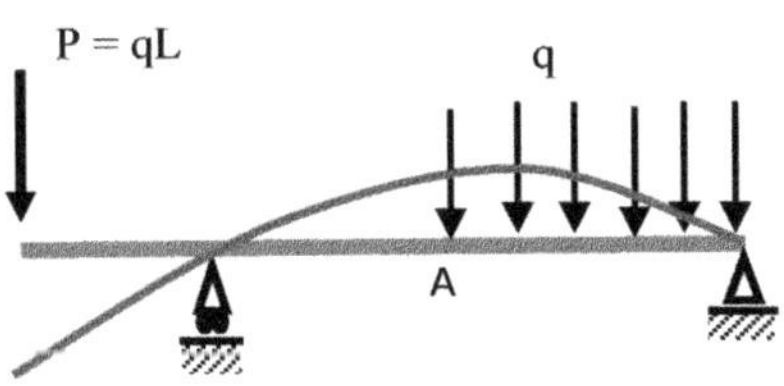

Figure 7.38. *Deformed beam*

7.4.3. *Method of virtual work: frames*

The virtual work method applies to frames in the same way to that of beams. Therefore, to calculate the displacement Δ or the slope ω at a point of a statically determinate frame, it is necessary to apply a unit virtual force or a unit virtual moment respectively. The corresponding external virtual work is

Force applied:

$$W_{ex} = 1.\Delta \quad [7.58]$$

Moment applied:

$$W_{ex} = 1.\omega \quad [7.59]$$

Frames are usually subjected to a normal force, a tangential force and a bending moment. In this case, the virtual work of internal actions is equal to the sum of the internal virtual work.

Internal force:

$$W_{in} = \sum_{i} \int_{L_i} n_i(x) \frac{N_i(x)}{E\Omega} dx \quad [7.60]$$

Internal bending moment:

$$W_{in} = \sum_{i} m_i(x) \frac{M_i(x)}{EI} dx \quad [7.61]$$

By neglecting the effect of the shear force, the expression of the total internal virtual work can be obtained by integrating relationships [7.60] and [7.61].

$$W_{in} = \sum_{i} \int_{L_i} n_i(x) \frac{N_i(x)}{E\Omega} dx + \sum_{i} m_i(x) \frac{M_i(x)}{EI} dx \quad [7.62]$$

Using the principle of virtual work allows us to write:

Displacement:

$$\Delta = \sum_i \int_{L_i} n_i(x) \frac{N_i(x)}{E\Omega} dx + \sum_i m_i(x) \frac{M_i(x)}{EI} dx \quad [7.63]$$

Slope:

$$\omega = \sum_i \int_{L_i} n_i(x) \frac{N_i(x)}{E\Omega} dx + \sum_i m_i(x) \frac{M_i(x)}{EI} dx \quad [7.64]$$

Axial deflections in the bars of a frame are generally very small compared to bending deflections. The relationships [7.63] are [7.64] are written as:

Displacement:

$$\Delta = \sum_i m_i(x) \frac{M_i(x)}{EI} dx \quad [7.65]$$

Slope:

$$\omega = \sum_i m_i(x) \frac{M_i(x)}{EI} dx, \quad [7.66]$$

where $m_i(x)$ is the expression of the bending moment due to a unit virtual force [7.65] and a unit virtual moment [7.66].

7.4.3.1. *Analysis procedure*

To calculate the deflection and/or the slope at a point on a statically determinate frame using the virtual work method, it is necessary to follow the following steps:

– formulate the expression of the bending moment on each bar of a frame subjected to the force of real loads;

– formulate the unit virtual moment. If we are considering a displacement, a unit force must be applied at the desired point. In the case of a slope, a unit virtual moment is applied to the selected point in a positive direction;

– the displacement or slope at the requested point is calculated using relationships [7.65] or [7.66] respectively.

EXAMPLE 7.8.–

Calculate the horizontal displacement and the slope at point C of the frame (Figure 7.39). It gives P = 2.25 qL.

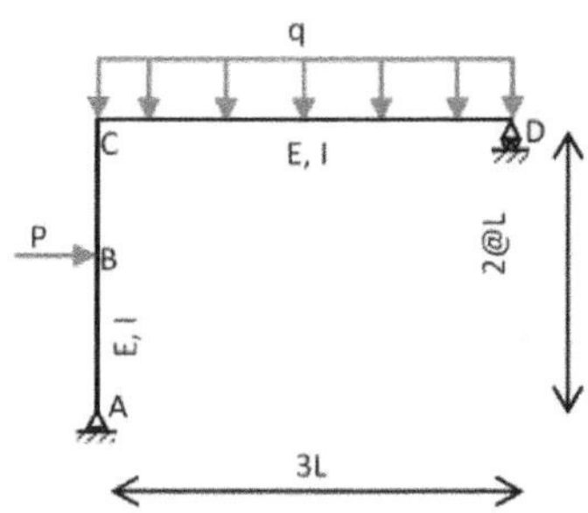

Figure 7.39. *Given frame*

Calculation of the slope at point C

Static analysis of the frames allows us to calculate the support reactions.

$$V_A = 0.75qL$$
$$H_A = -2.25qL$$
$$V_D = 2.25qL$$

Bar (AC)

$0 \leq x \leq L$:

$$\mathrm{M}(x) = 2.25qLx$$

$L \leq x \leq 2L$:

$$\mathrm{M}(x) = 2.25qL^2$$

Bar (CD)

$0 \leq x \leq 3L$:

$$M(x) = 2.25qL^2 + 0.75qLx - \frac{1}{2}qx^2$$

To calculate the slope at point C, we apply the unit virtual moment at point C (Figure 7.40).

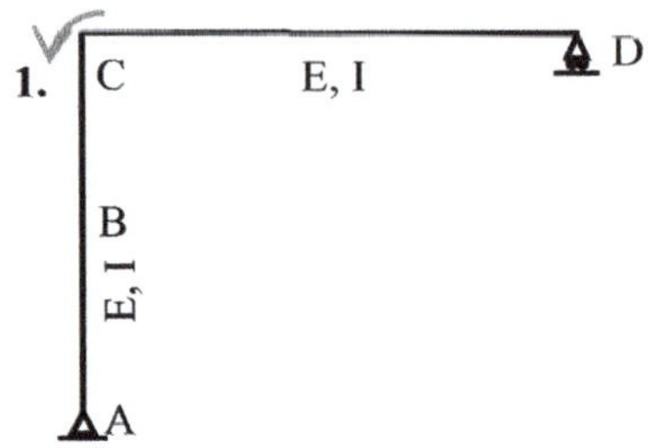

Figure 7.40. *Unit moment*

Under the effect of the unit moment, static analysis of the frame makes it possible to deduce the variation of the bending moment on the bars of the frame.

$$V_A = \frac{1}{3L}$$

$$H_A = 0$$

$$V_D = -\frac{1}{3L}$$

Bar (AC)

$0 \le x \le 2L$:

$$\mathrm{m}(x) = 0$$

Bar (CD)

$0 \le x \le 3L$:

$$m(x) = \frac{1}{3L}x - 1$$

The external work of the unit moment is

$$W_{ex} = 1.\omega_C$$

In another part, the internal virtual work, taking into account the bending effect, is given as

$$W_{in} = \sum \int_{L_i} m_i(x)\frac{M_i(x)}{EI}dx = \int_0^L m_{AB}(x)\frac{M_{AB}(x)}{EI}dx + \int_L^{2L} m_{BC}(x)\frac{M_{BC}(x)}{EI}dx + \int_0^{3L} m_{CD}(x)\frac{M_{CD}(x)}{EI}dx$$

We substitute the expressions of the bending moments of each bar. We obtain

$$W_{in} = -3.375\frac{qL^3}{EI}$$

Applying relationship [7.67] allows us to deduce

$$\omega_C = -3.375\frac{qL^3}{EI}$$

Horizontal displacement calculation at point C

A unit virtual force is applied at point C (Figure 7.41).

Static analysis of the frame is given as

$$V_A = -\frac{2}{3}$$

$$H_A = -1$$

$$V_D = \frac{2}{3}$$

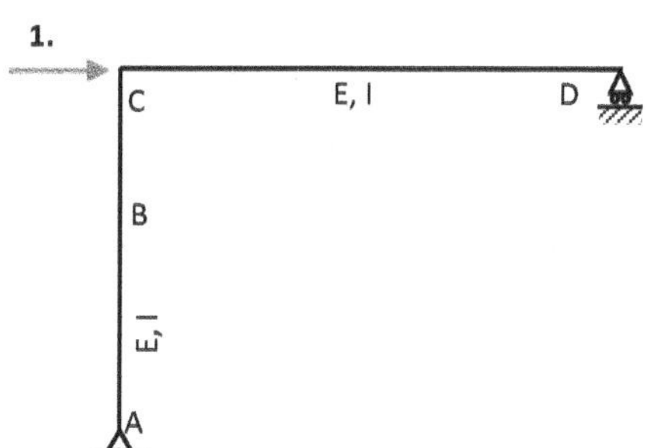

Figure 7.41. *Effect of the virtual force*

Bar (AC)

$$0 \le x \le 2L:$$

$$\mathrm{m}(x) = x$$

Bar (CD)

$$0 \le x \le 3L:$$

$$\mathrm{m}(x) = 2L - \frac{2}{3}x$$

The external work of the unitary moment is given as

$$W_{ex} = 1.\Delta_C$$

The internal virtual work, considering the bending stress, is given as

$$W_{in} = \sum \int_{L_i} m_i(x)\frac{M_i(x)}{EI}dx = \int_0^L m_{AB}(x)\frac{M_{AB}(x)}{EI}dx + \int_L^{2L} m_{BC}(x)\frac{M_{BC}(x)}{EI}dx + \int_0^{3L} m_{CD}(x)\frac{M_{CD}(x)}{EI}dx$$

Substituting the expressions of the bending moments of each bar, we obtain

$$W_{in} = 10.875\frac{qL^4}{EI}$$

The horizontal displacement at point C can be deduced from relationship [7.65].

$$\Delta_C = 10.875\frac{qL^4}{EI}$$

The deformed frame under the applied load system is shown in Figure 7.42.

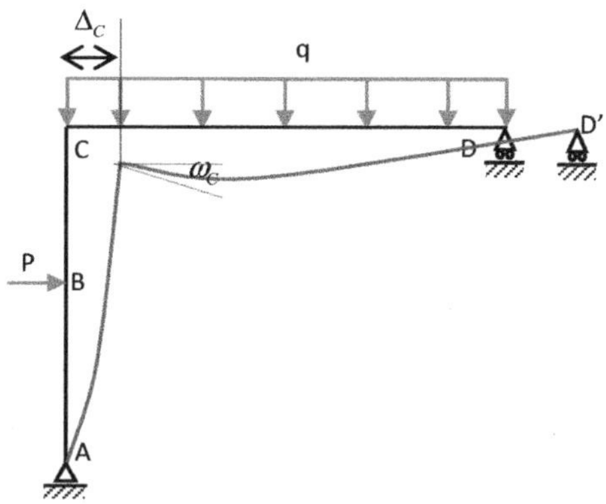

Figure 7.42. *Deformed frame*

7.5. Conservation of energy and strain energy

Internal energy is the ability of a structure to provide internal work when it is subject to external actions. Strain energy is the energy stored in a structure due to its deflection. The statement of the principle of energy conservation is "the external work of an elastic structure is equal to the work of internal actions or the strain energy stored in the structure".

The principle of energy conservation allows us to write

$$W_{ex} = W_{in} \qquad [7.67]$$

or

$$W_{ex} = U \qquad [7.68]$$

where U is the internal or strain energy.

In another way, we can define the internal or strain energy according to the work of the external actions [7.1], which is equal to that of the internal actions.

7.6. Castigliano's theorem

Another method of calculating structural deflections based on the concept of energy is described in this section. This method only applies to elastic structures bearing the name of its author (the Castigliano method).

7.6.1. *Displacement theorem statement*

The partial derivative of internal energy in relation to an applied force is equal to the displacement of the applied force point in its direction.

$$\frac{\partial U}{\partial P_i} = \Delta_i \qquad [7.69]$$

7.6.2. *Slope theorem statement*

The partial derivative of internal energy in relation to an applied moment is equal to the slope of the section passing through this point in the direction of the slope of the moment.

$$\frac{\partial U}{\partial M_i} = \omega_i \qquad [7.70]$$

To derive Castigliano's theorem, we consider a structure subject to n forces; $F_1, F_2, F_3, \ldots, F_i, \ldots, F_{n-1}, F_n$ (Figure 7.43).

The virtual work of external forces is

$$W_{ex} = \frac{1}{2}\sum_i F_i\,\Delta_i \qquad [7.71]$$

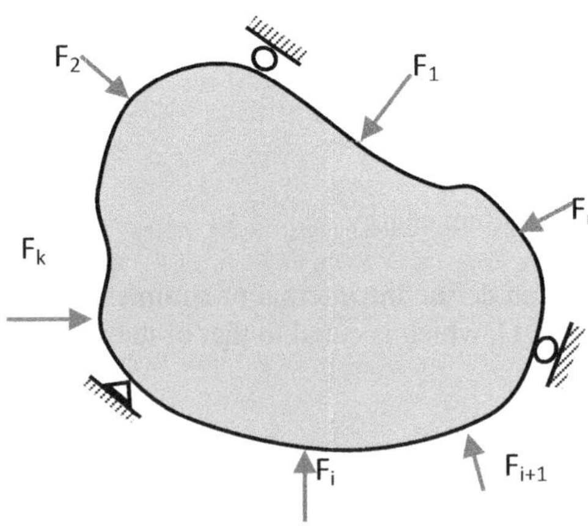

Figure 7.43. *Solid body*

By applying the principle of energy conservation, internal energy is equal to the work of external forces.

$$U = W_{ex} \qquad [7.72]$$

The internal energy depends on the applied forces. If the external force F_i increases by one increment dF_i, the internal energy increases by quantity dU. The expression of internal energy is written in this case as.

$$U + dU = U + \frac{\partial U}{\partial F_i} dF_i \qquad [7.73]$$

The differential quantity of the force dF_i generates a deflection that corresponds to an increase in displacement $d\Delta_i$. The increment of external work is

$$dW_{ex} = \frac{1}{2} dF_i \, .d\Delta_i \qquad [7.74]$$

This corresponds to an increase in internal energy such that

$$dU = \frac{1}{2} dF_i \, .d\Delta_i \qquad [7.75]$$

The second order differential of dU can be neglected.

The set of applied forces F_1, F_2, F_3, …, F_i, …, F_{n-1}, F_n, respectively generates displacements $\Delta_1, \Delta_2, \Delta_3, \ldots, \Delta_i, \ldots, \Delta_{n-1}, \Delta_n$, which leads to the provision of internal energy equal to:

$$U + dU = U + dF_i \, \Delta_i \qquad [7.76]$$

The quantity $dU = dF_i \, \Delta_i$ is the additional internal energy generated by the force increment dF_i.

The strain energy is independent of the application order of external forces. Equation [7.76] allows us to deduce that

$$\Delta_i = \frac{\partial U}{\partial F_i} \qquad [7.77]$$

The analysis method can be applied to set out Castigliano's second theorem relating to slopes. It is

$$\omega_i = \frac{\partial U}{\partial M_i} \quad [7.78]$$

7.6.3. *Application of Castigliano's theorem to truss analysis*

The internal energy expression of a truss is

$$U = \frac{1}{2}\sum_i \frac{N_i^2 L}{E\Omega} \quad [7.79]$$

The displacement of a joint in the direction of an applied force is equal to

$$\Delta_i = \sum_i N_i \left(\frac{\partial N_i}{\partial F_i}\right)\frac{L_i}{E\Omega} \quad [7.80]$$

EXAMPLE 7.9.–

Repeat Example 7.4 to determine vertical displacement in joint C.

We exert force F in the direction of the requested displacement (Figure 7.44).

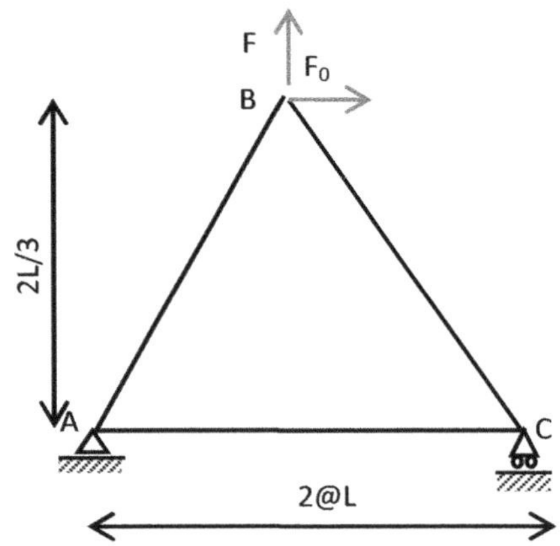

Figure 7.44. *Given structure*

The stresses on the bars of a truss under the effect of force F_0 have already been established in Example 7.4.

Under the effect of force F, the stresses on the bars of the structure are grouped in Table 7.6 using the superposition principle.

Bar	$N^{(F_0)}$	$N^{(F)}$	L	$N^{(F_0)}+N^{(F)}$
AB	$\frac{\sqrt{13}}{4}F_0$	$\frac{\sqrt{13}}{6}F$	$\frac{\sqrt{13}}{2}L$	$\frac{\sqrt{13}}{4}F_0+\frac{\sqrt{13}}{6}F$
BC	$-\frac{\sqrt{13}}{4}F_0$	$\frac{\sqrt{13}}{6}F$	$\frac{\sqrt{13}}{2}L$	$-\frac{\sqrt{13}}{4}F_0+\frac{\sqrt{13}}{6}F$
AC	$\frac{1}{2}F_0$	$-\frac{1}{3}F$	$2L$	$\frac{1}{2}F_0-\frac{1}{3}F$

Table 7.6. *Forces on the bars*

The internal strain energy of the truss is

$$U=\frac{1}{2}\sum_i \frac{N_i^2 L_i}{E\Omega}=\frac{1}{2}\frac{1}{E\Omega}(N_{AB}^2.L_{AB}+N_{BC}^2.L_{BC}+N_{AC}^2.L_{AC})$$

Substituting the expressions of the stresses in the internal energy formula and taking the partial derivative with respect to force F, we obtain

$$\frac{\partial U}{\partial F}=-\frac{2}{3}(\frac{F_0}{2}-\frac{F}{3})\frac{L}{E\Omega}$$

Given that applied force F is not a real force, we can write

$$\frac{\partial U}{\partial F}=-\frac{1}{3}\frac{F_0 L}{E\Omega}$$

Vertical displacement of point B is given as

$$\Delta_B=\frac{\partial U}{\partial F}=-\frac{1}{3}\frac{F_0 L}{E\Omega}$$

The result obtained is the same as in Example 7.4.

7.6.4. *Using Castigliano's theorem to analyze beams and frames*

The internal energy of a frame composed of several bars can be deduced from the internal energy formulated from a structural bar.

$$U = \frac{1}{2}\sum_i \int_0^{L_i} \frac{M_i^2(x)}{EI}dx \qquad [7.81]$$

Applying Castigliano's theorem allows us to calculate the displacement Δ_i at a point on the bar.

$$\Delta_i = \frac{\partial U}{\partial F_i} = \frac{\partial}{\partial F_i}\left[\int_0^{L} \frac{M^2(x)}{2EI}dx\right] = \int_0^{L} \frac{M(x)}{EI}\left(\frac{\partial M(x)}{\partial F_i}\right)dx \qquad [7.82]$$

At a point on a frame, the displacement is

$$\Delta_i = \frac{\partial U}{\partial F_i} = \sum \frac{\partial}{\partial F_i}\left[\int_0^{L_i} \frac{M_i^2(x)}{2EI}dx\right] = \sum \int_0^{L_i} \frac{M_i(x)}{EI}\left(\frac{\partial M(x)}{\partial F_i}\right)dx \qquad [7.83]$$

In the same way, the slope at a point on a bar can be deduced by

$$\omega_i = \frac{\partial U}{\partial M_i} = \frac{\partial}{M_i}\left[\int_0^{L} \frac{M^2(x)}{2EI}dx\right] = \int_0^{L} \frac{M(x)}{EI}\left(\frac{\partial M(x)}{\partial M_i}\right)dx \qquad [7.84]$$

If the desired point is on a frame, the corresponding slope is given as

$$\omega_i = \frac{\partial U}{\partial M_i} = \sum \frac{\partial}{\partial M_i}\left[\int_0^{L_i} \frac{M_i^2(x)}{2EI}dx\right] = \sum \int_0^{L_i} \frac{M_i(x)}{EI}\left(\frac{\partial M(x)}{\partial M_i}\right)dx \qquad [7.85]$$

EXAMPLE 7.10.– Cantilever

Repeat Example 7.6 to calculate the displacement and slope at point A of the cantilever (Figure 7.30). At the same point, a virtual force is exerted (Figure 7.45).

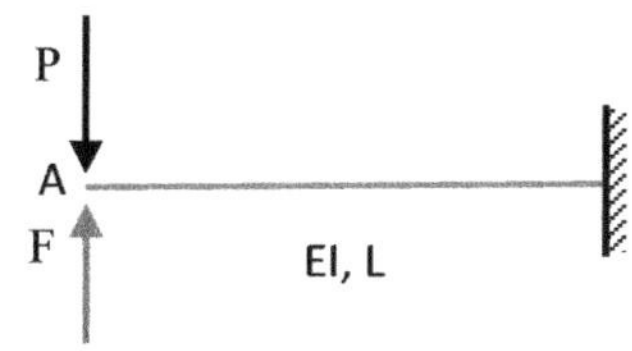

Figure 7.45. *Given cantilever*

The expression of the bending moment due to applied forces is given as

$$M(x) = (F - P)x$$

The strain energy of the bar is written as

$$U = \frac{1}{2}\int_0^L \frac{M^2(x)}{EI} dx$$

Applying Castigliano's theorem, the displacement at point A is written as

$$\Delta_A = \frac{\partial U}{\partial F} = \frac{\partial U}{\partial M(x)} \frac{\partial M(x)}{\partial F}$$

It can also be written as

$$\Delta_A = \int_0^L \frac{M(x)}{EI}.(x).dx = \frac{(F-P)L^3}{3EI}$$

Force F is zero, so the previous relationship is written as

$$\Delta_A = -\frac{PL^3}{3EI}$$

Similarly, to calculate the slope at point A, we apply a virtual moment M_A (Figure 7.46).

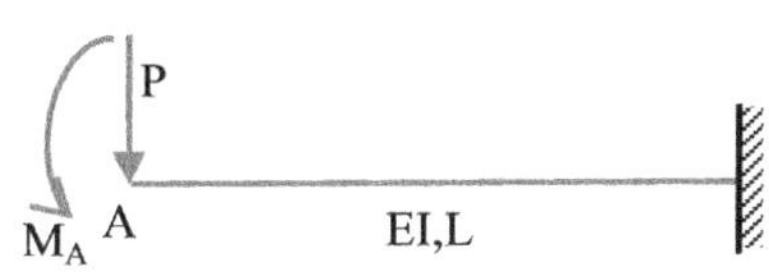

Figure 7.46. *Given cantilever beam*

The expression of the bending moment is

$$M(x) = -Px - M_A$$

The slope at point A using Castigliano's theorem is written as

$$\omega_A = \frac{\partial U}{\partial M_A} = \frac{\partial U}{\partial M(x)} \frac{\partial M(x)}{\partial M_A}$$

or

$$\omega_A = \int_0^L \frac{M(x)}{EI}.(-1).dx = \frac{1}{EI}\left[\frac{PL^2}{2} + M_A L\right]$$

Knowing that $M_A = 0$, the previous relationship is written as

$$\omega_A = \frac{PL^2}{2EI}$$

The deformed cantilever beam is shown in Figure 7.33.

EXAMPLE 7.11.– Frame

Repeat Example 7.8 to calculate the horizontal displacement and slope at point C of the frame (Figure 7.39).

Calculation of the slope at point C

We apply a virtual moment M_C at point C (Figure 7.47).

The equilibrium equations allow us to calculate the support reactions.

$$V_A = 0.75qL + \frac{M_C}{3L}$$

$$H_A = -2.25qL$$

$$V_D = 2.25qL - \frac{M_C}{3L}$$

The variation of the bending moment in each bar is

Bar (AC)

$0 \le x \le L:$

$$M(x) = 2.25qLx$$

$L \le x \le 2L:$

$$M(x) = 2.25qL^2$$

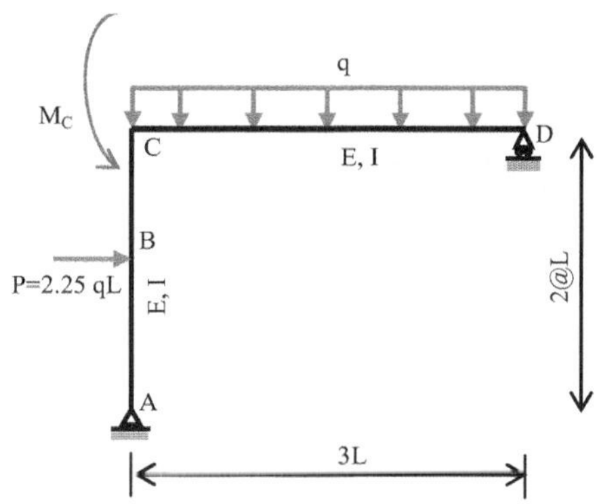

Figure 7.47. *Application of MC*

Bar (CD)

$0 \le x \le 3L$

$$M(x) = 2.25qL^2 + 0.75qLx - \frac{1}{2}qx^2 + (\frac{1}{3L}x - 1)M_C$$

The internal energy of the frame is written in the following form:

$$U = \frac{1}{2}\sum_i \int_0^{L_i} \frac{M_i^2(x)}{EI}dx = \frac{1}{2}\int_0^{L} \frac{M_{AB}^2(x)}{EI}dx + \frac{1}{2}\int_L^{2L} \frac{M_{BC}^2(x)}{EI}dx + \frac{1}{2}\int_0^{3L} \frac{M_{CD}^2(x)}{EI}dx$$

The slope of point C can be deduced using the presented method.

$$\omega_C = \frac{\partial U}{\partial M_C} = \frac{\partial U}{\partial M(x)}\frac{\partial M(x)}{\partial M_C} = \int_0^{3L} \frac{M_{CD}(x)}{EI}\frac{\partial M(x)}{\partial M_C}dx$$

$$\omega_C = -3.375\frac{qL^3}{EI} - \frac{3}{2}\frac{M_C L}{EI}$$

Given that $M_C = 0$, the slope of joint C is

$$\omega_C = -3.375\frac{qL^3}{EI}$$

Calculation of horizontal displacement at point C

This time, the force F is applied to joint C (Figure 7.48).

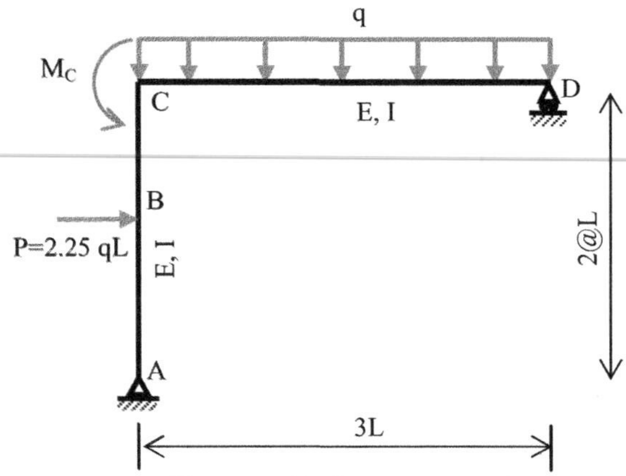

Figure 7.48. *Force at joint C*

Static analysis of the frame makes it possible to determine the support reactions and formulate the bending moment expressions.

$$V_A = 0.75qL - \frac{2}{3}F$$

$$H_A = -(2.25qL + F)$$

$$V_D = 2.25qL + \frac{2}{3}F$$

Bar (AB)

$0 \le x \le L$:

$$M(x) = (2.25qL + F)x$$

Bar (BC)

$$L \leq x \leq 2L:$$

$$M(x) = (2.25qL + F)x - 2.25qL(x - L) = Fx + 2.25qL^2$$

Bar (CD)

$$0 \leq x \leq 3L:$$

$$M(x) = 0.75qLx + 2.25qL^2 - \frac{1}{2}qx^2 + F(2L - \frac{2}{3}x)$$

The internal energy of the frame is written as

$$U = \frac{1}{2}\sum_i \int_0^{L_i} \frac{M_i^2(x)}{EI} dx = \frac{1}{2}\int_0^{L} \frac{M_{AB}^2(x)}{EI} dx + \frac{1}{2}\int_L^{2L} \frac{M_{BC}^2(x)}{EI} dx + \frac{1}{2}\int_0^{3L} \frac{M_{CD}^2(x)}{EI} dx$$

The horizontal displacement at point C can be deduced from Castigliano's method.

$$\Delta_C = \frac{\partial U}{\partial F} = \frac{\partial U}{\partial M(x)}\frac{\partial M(x)}{\partial F} =$$
$$\frac{1}{EI}\left[\int_0^{L} (2.25qL + F)x^2 dx + \int_L^{2L} (Fx + 2.25qL^2).xdx\right] +$$
$$\frac{1}{EI}\left[\int_0^{3L} (0.75qL - \frac{2}{3}F)x + 2.25qL^2 - 2FL - \frac{1}{2}qx^2)(-\frac{2}{3}x - 2L)dx\right]$$

Since F = 0, we can deduce the horizontal displacement at point C as

$$\Delta_C = 10.875\frac{qL^4}{EI}$$

7.7. Maxwell–Betti law

The statement of the Maxwell–Betti law is: "for any elastic structure, the virtual work of force F_1 (moment M_1) through the displacement (the slope) caused by force

F_2 (moment M_2) is equal to the virtual work of force F_2 (moment M_2) through the displacement (the slope) generated by force F_1 (moment M_1)".

To show the validity of the Maxwell–Betti law, we consider a deformable body stressed by two forces F_1 and F_2, respectively, generating displacements Δ_1 and Δ_2 (Figure 7.49).

The application of the force is carried out in the following process: force F_1 is applied, the body takes any deformed configuration. We then apply force F_2 and the body is deformed once more and takes another deformed position. The total work due to applying forces F_1 and F_2 is

$$W(F_1, F_2) = \frac{1}{2} F_1 \Delta_1 + \frac{1}{2} F_2 \Delta_2 + F_1 \Delta_{12} \quad [7.86]$$

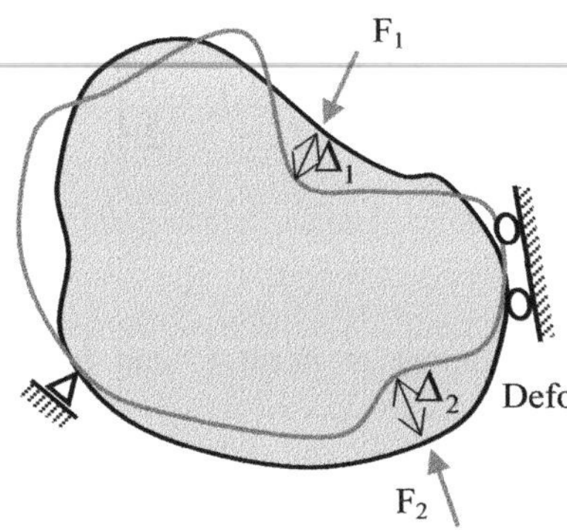

Figure 7.49. *Deformable body*

Force F_1 causes additional work through the displacement Δ_{12} caused by force F_2, independent of the effect of force F_1.

We reverse the order of applying forces; we first apply force F_2 then force F_1. The total external work due to applying forces F_2 and F_1 is

$$W(F_2, F_1) = \frac{1}{2} F_2 \Delta_2 + \frac{1}{2} F_1 \Delta_1 + F_2 \Delta_{21} \quad [7.87]$$

In the same way, force F_2 causes additional work through the displacement Δ_{21} caused by applying force F_1, independent of the effect of force F_2.

The total work is independent of the order of applying forces F_1 and F_2. This allows us to write

$$F_1 \Delta_{12} = F_2 \Delta_{21} \tag{7.88}$$

Equation [7.88] shows that the work of force F_1 through displacement Δ_{12} caused by force F_2 is equal to the external work of F_2 through displacement Δ_{21} caused by force F_1 (reciprocity of external work).

Relationship [7.88] also applies to equal forces ($F_1 = F_2$). The reciprocity of deflections becomes

$$\Delta_{12} = \Delta_{21} \tag{7.89}$$

EXAMPLE 7.12.–

Verify the Maxwell–Betti law applied to the beam in Figure 7.50.

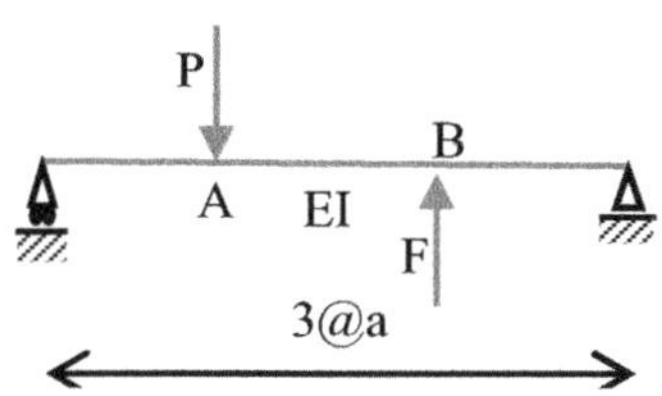

Figure 7.50. *Given beam*

The deflection expressions of each section using the differential equation of the elastic line method are

$0 \leq x \leq a$:

$$y(x) = \frac{-P(a-L)}{6EIL}(x^3 + a(a-2L)x)$$

$a \leq x \leq 3a$:

$$y(x) = \frac{-Pa}{6EIL}(x^3 - 3Lx^2 + (a^2 + 2L^2)x - a^2 L)$$

With L = 3a.

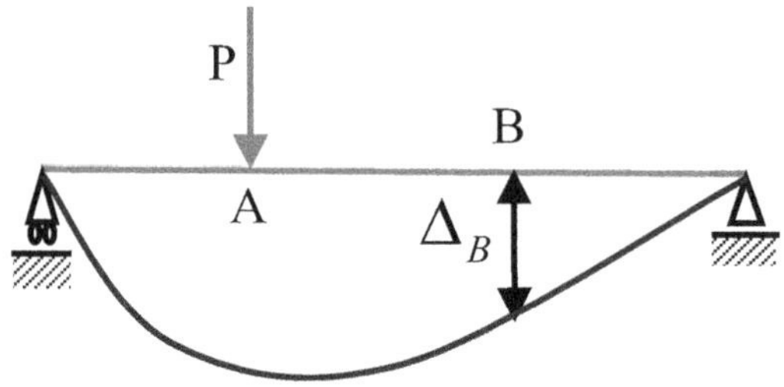

Figure 7.51. *Application of force in A*

We first apply force P to point A (Figure 7.51). The displacement of point B is given by

$$\Delta_B = y(2a) = \frac{-Pa}{6EI(3a)}(8a^3 - 3(3a)4a^2 + (a^2 + 2.9a^2)(2a) - a^2(3a))$$

$$\Delta_B = \frac{-7Pa^3}{18EI}$$

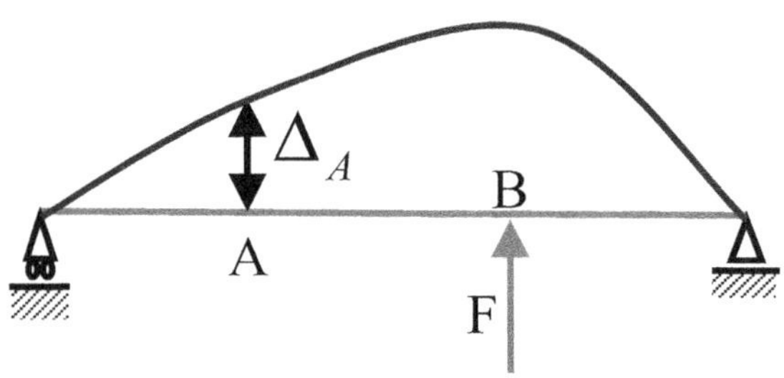

Figure 7.52. *Application of force in B*

We then apply force F (Figure 7.52). Point A moves by a value of Δ_A .

$$\Delta_A = y(a) = \frac{+F(2a-3a)}{6EI(3a)}(a^3 + 2a(2a-6a)a)$$

$$\Delta_A = \frac{+7Fa^3}{18EI}$$

The external work of force F through displacement Δ_B is equal to

$$W(\text{F}) = F.\Delta_B = \frac{7F.Pa^3}{18EI}$$

Similarly, the external work of force P through displacement Δ_A is

$$W(\text{P}) = P.\Delta_A = \frac{7P.Fa^3}{18EI}$$

Hence, $W(\text{P}) = W(F)$

7.8. Conclusion

First, we presented the formulation of the work of the external forces and the formulation of the strain energy of a structure subjected to an axial force, a bending moment, a shear stress and a torque moment. These expressions lead to the calculation of displacements and/or slopes by applying the principle of energy conservation.

Second, we examined the principle of virtual work. This principle is a powerful tool for calculating displacements and/or slopes at distinctive points. It was introduced to analyze trusses, beams and frames.

In the same context, the calculation of the displacement (slope) using Castigliano's theorem can be carried out by deriving the internal energy expression in relationship to the applied force (at the applied moment) at the desired point.

Finally, the Maxwell–Betti law allows us to quantify the displacement and/or the slope: "the displacement (slope) at point i due to a force (moment) applied at point j is equal to the displacement (at the slope) at point j when a force (moment) is applied at point i".

7.9. Problems

Exercise 1

– Express the work of force P and thc internal energy of the following structures. Consider that $E\Omega$ and EI are constant.

– Calculate the vertical displacement at point A in each case.

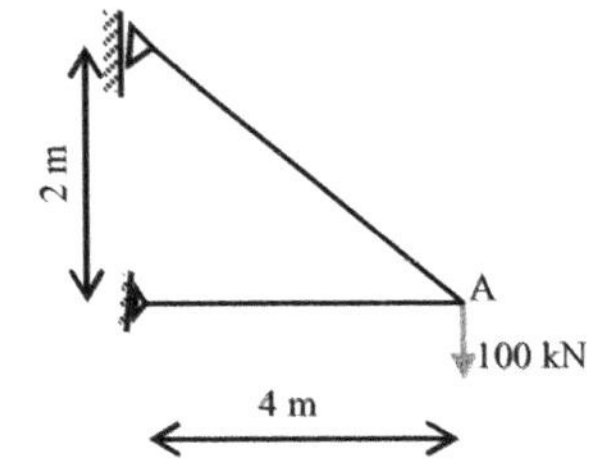

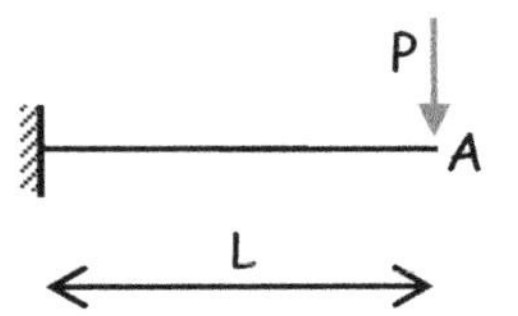

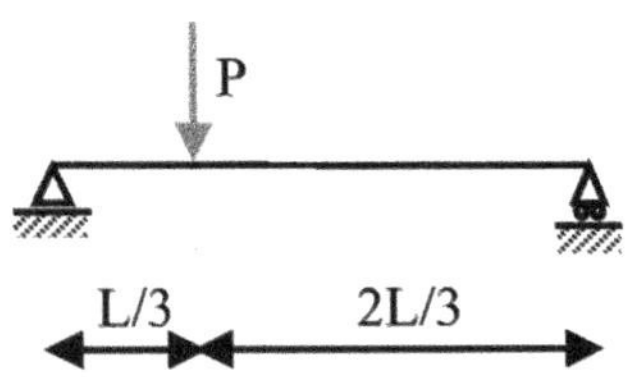

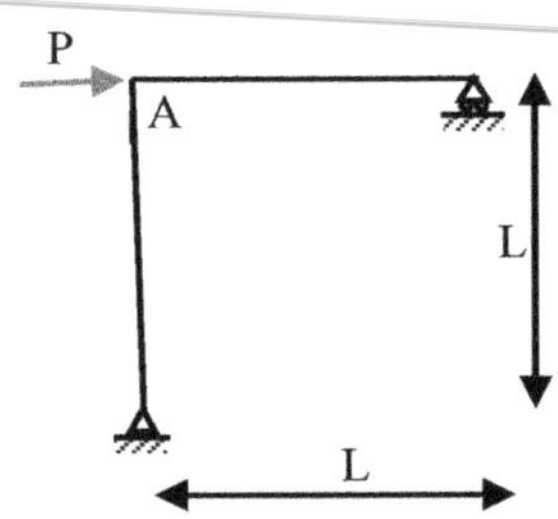

Exercise 2

Calculate the slope in the direction of the slope of moment M_0. EI is assumed to be constant.

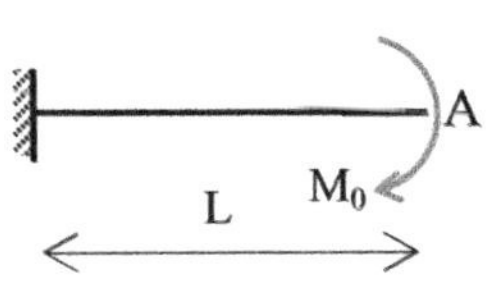

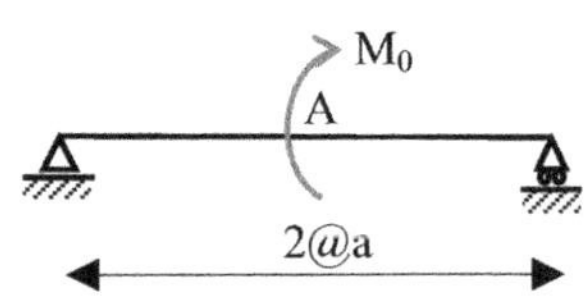

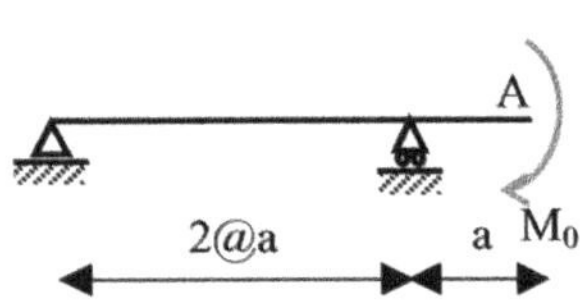

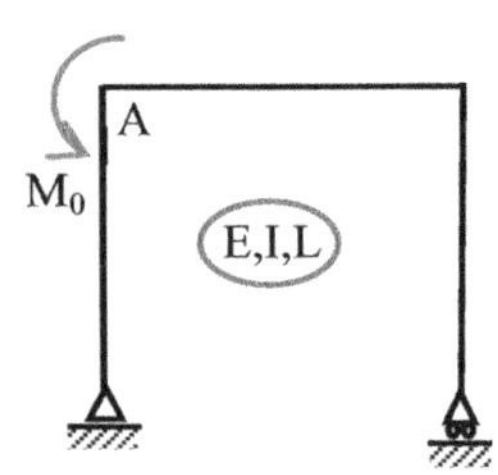

Exercise 3

Using the principle of virtual work, calculate the vertical and horizontal displacements at point A of the following structures: EI and EΩ are constant.

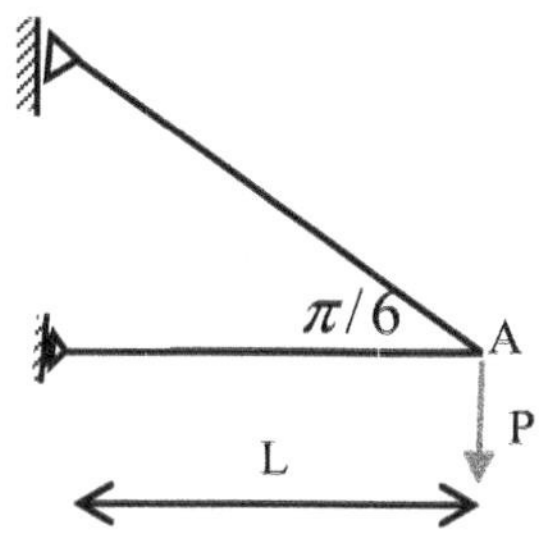

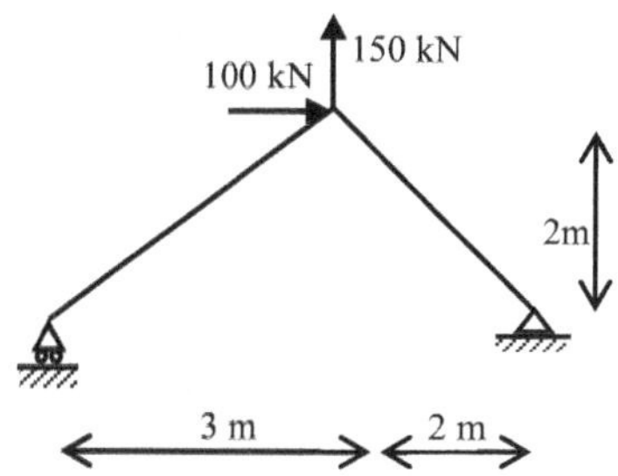

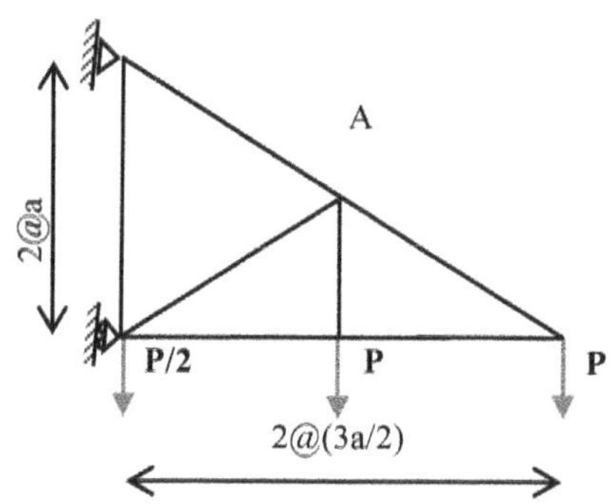

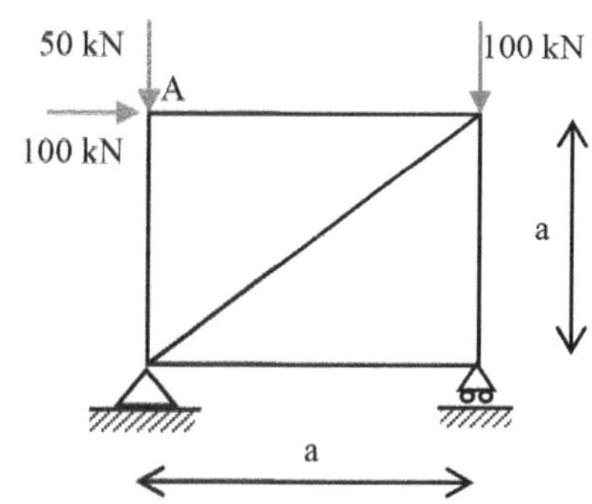

Exercise 4

Using the principle of virtual work, calculate the vertical and horizontal displacement and the slope at point A of the following structures: EI is assumed to be constant.

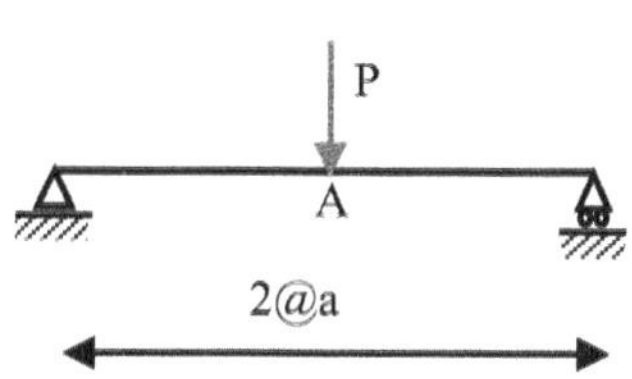

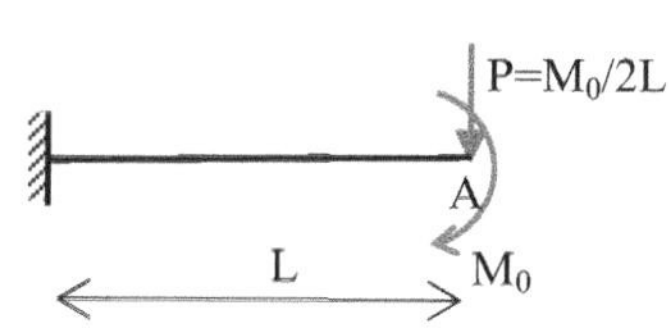

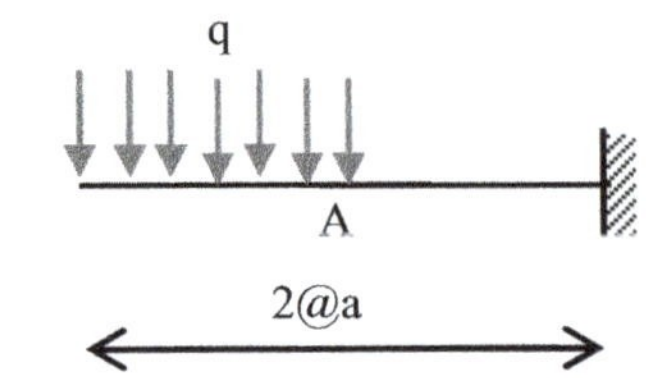

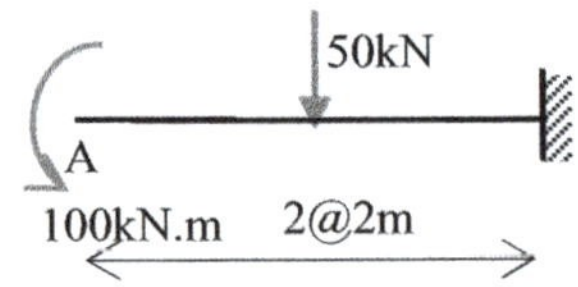

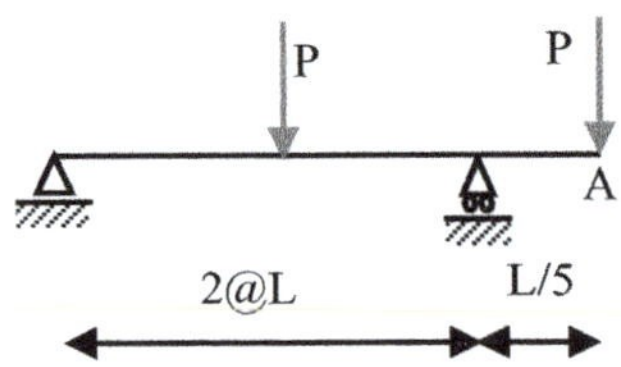

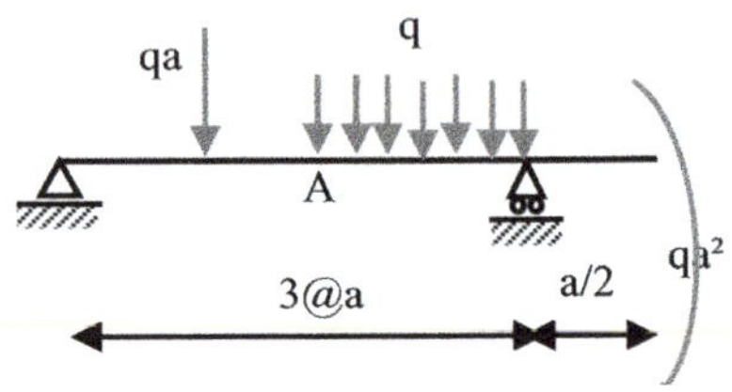

Exercise 5

Using the principle of virtual work, calculate the vertical and horizontal displacement and the slope at point A of the following structures: EI is assumed to be constant.

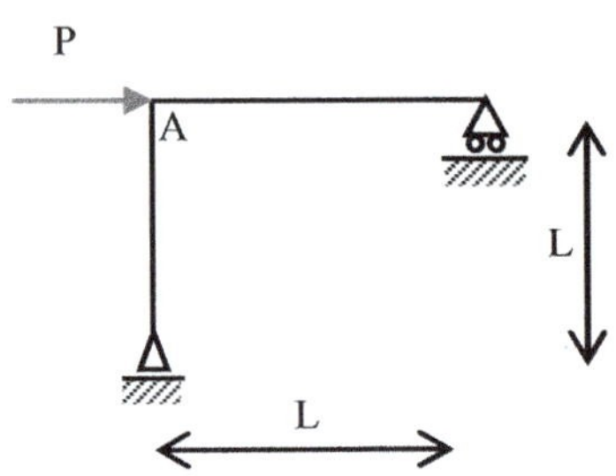

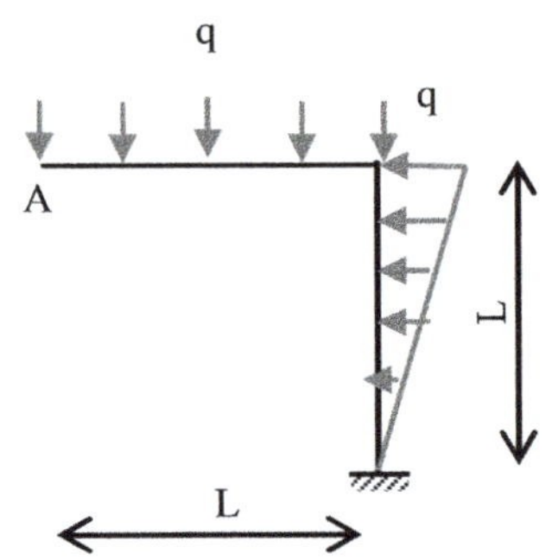

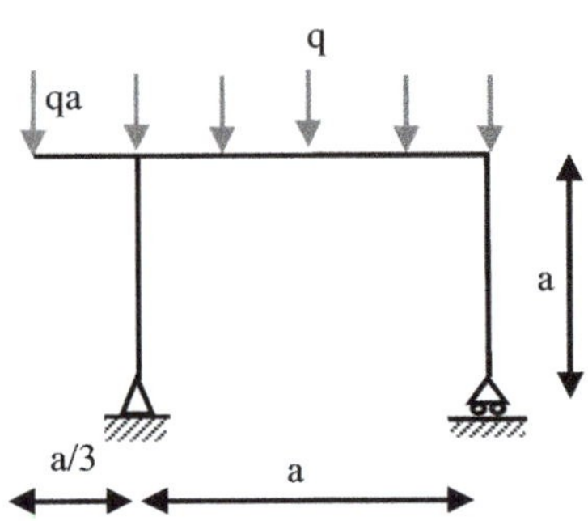

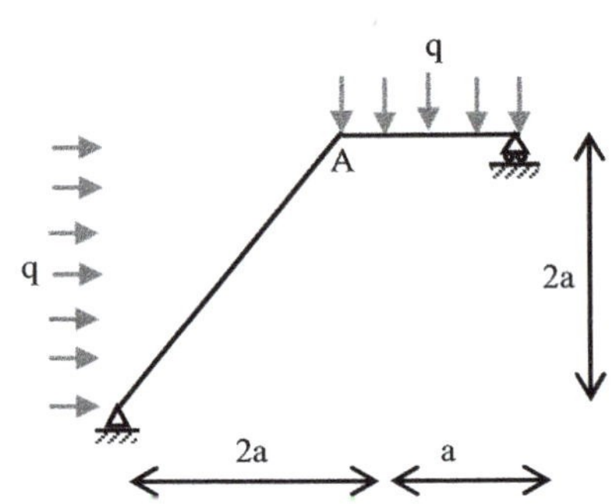

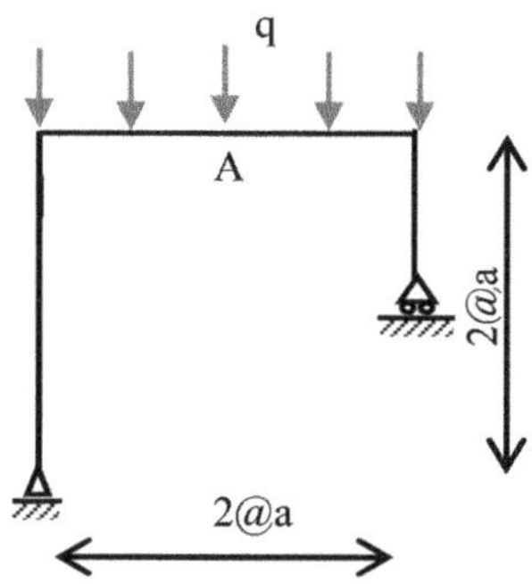

Exercise 6

Determine the vertical and horizontal displacement and the slope at point A of the following structures.

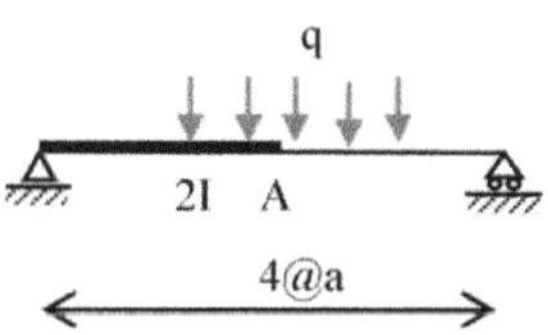

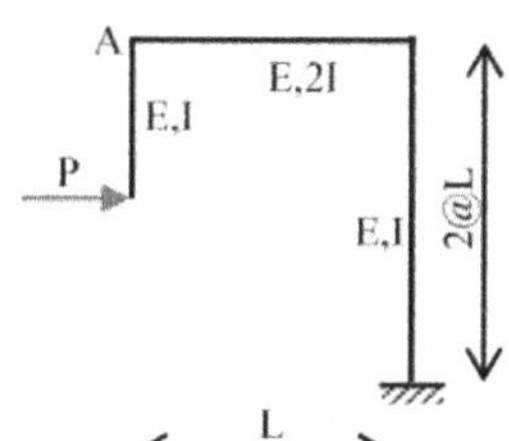

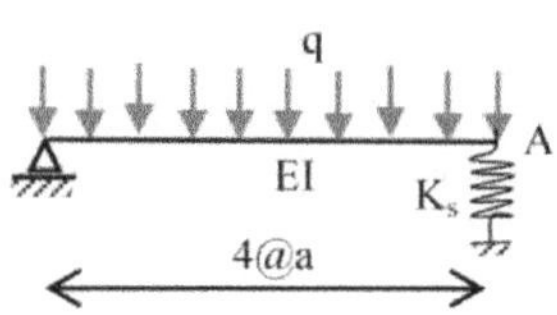

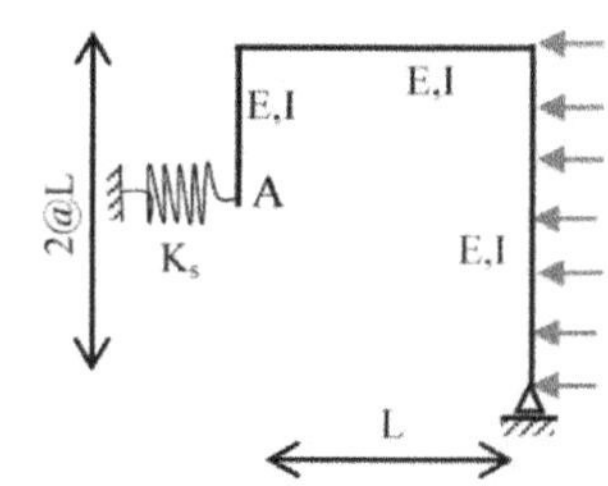

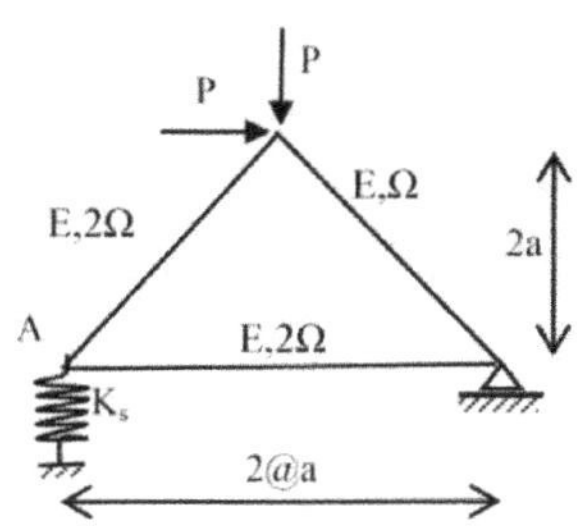

Exercise 7

Using Castigliano's method, calculate the vertical displacement in A and the slope at point B in the following structures: EI is assumed to be constant.

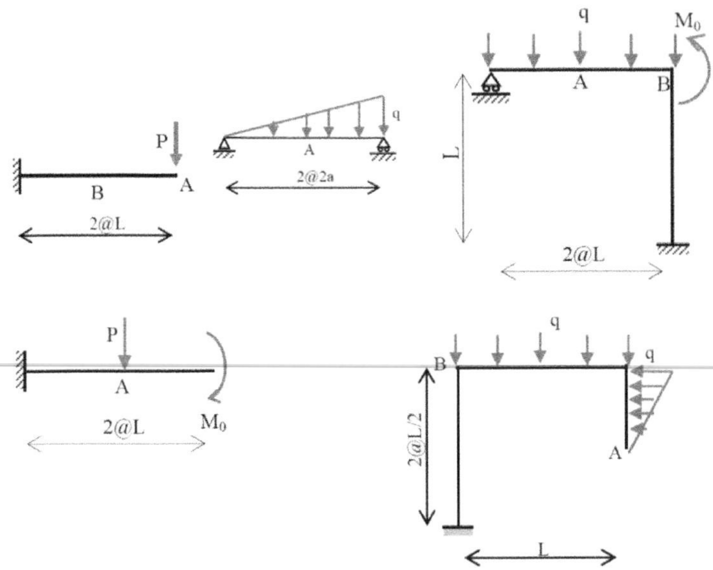

Exercise 8

Using Castigliano's method, calculate the unknown, represented by for each structure.

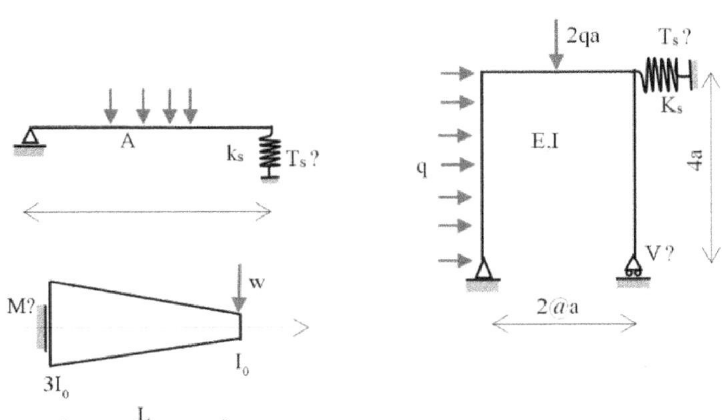

Exercise 9

Using Castigliano's method, remove the static indeterminacy of the following structures and deduce the corresponding unknown.

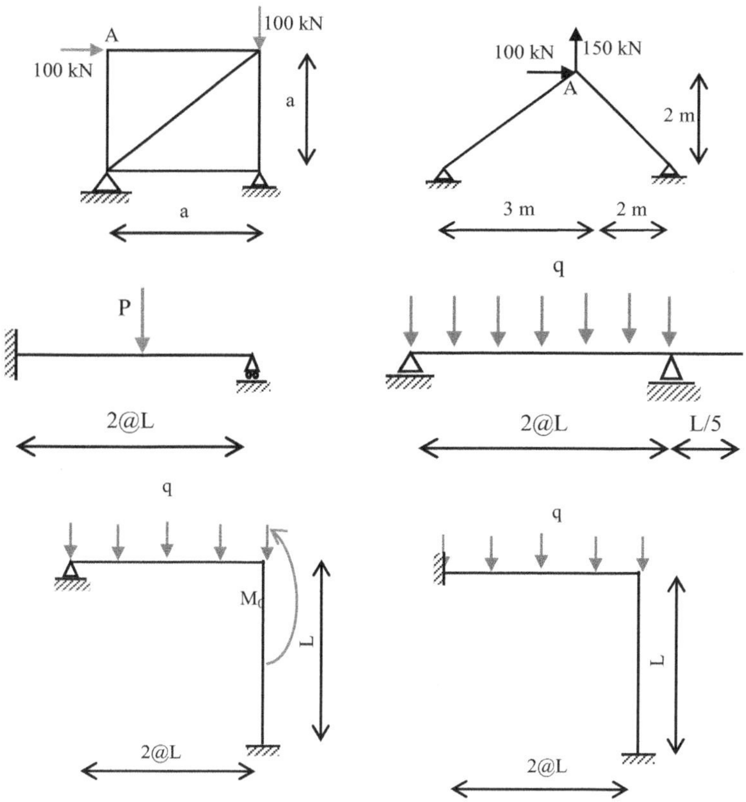

Exercise 10

Verify Maxwell's law for the following structures.

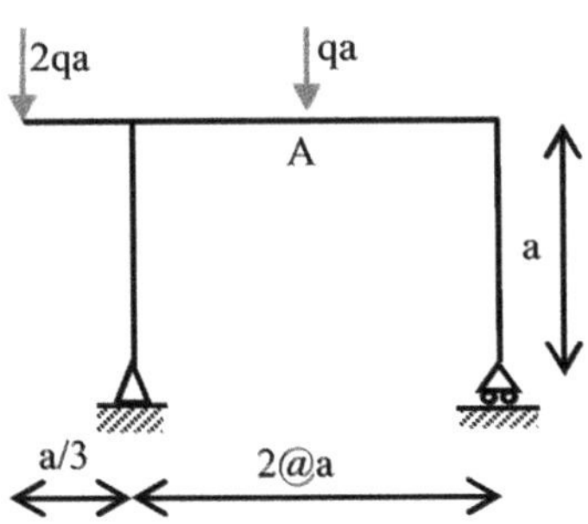
2qa
qa
A
a
a/3
2@a

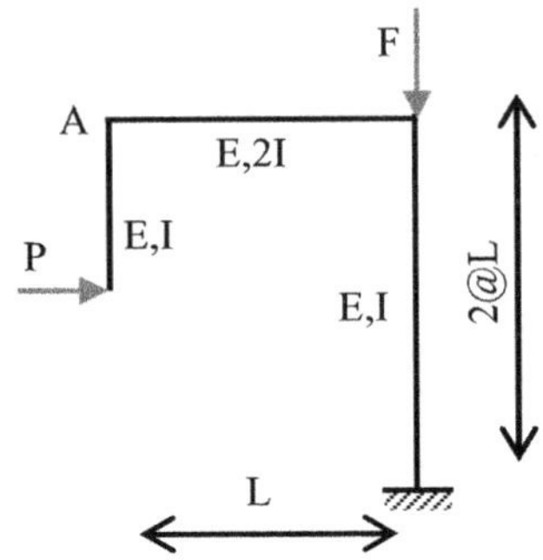
F
A
E,2I
E,I
P
E,I
2@L
L

8

Cable Analysis

After studying this chapter, the student will be able to:

– distinguish between rigid and deformable cable structures;

– be familiar with the funicular shape of a cable structure;

– analyze cable structures subjected to concentrated and distributed forces;

– calculate the thrust of the support system and the maximum tension.

In the first section, we analyze cable structures subjected to different loads, which includes concentrated forces, uniform forces and other forces. For each load, we evaluate the thrust, tension and deflection of the cables. In the second section, we describe how to analyze a cable when the inflection point is located outside the cable. Finally, we present an overview of the analysis of suspension bridges.

8.1. Introduction

Cables have been used for centuries and are chosen because of their structural performance, light weight and aesthetic appearance. Cable structures can be divided into two categories: (1) rigid cables and (2) deformable or flexible cables. This classification is based on the cable's deflection and geometry when loaded. Rigid cable structures can bear external loads without any change to their geometry. Conversely, deformable cable structures have an observable deflection when loaded. However, it is necessary to assume that the deflections are small.

The role of cable structures is to bear the loads from their own weight (dead weight) and the live loads of the arches, suspended roofs and bridge structures as they are transmitting loads from one element to another. The dead weight of relatively short cables is neglected, but when cables are used in electric or transmission lines the weight must be considered.

Cables are structural elements used in various constructions including suspension bridges, support wires, transmission lines, electrical networks, stadiums and covered sports halls (Figure 8.1). The cables are intended to withstand traction forces but they have no resistance to bending, shear stress, compression or torsion. Their mode of resistance is traction, generally directed along the tangent of the cable from a given point.

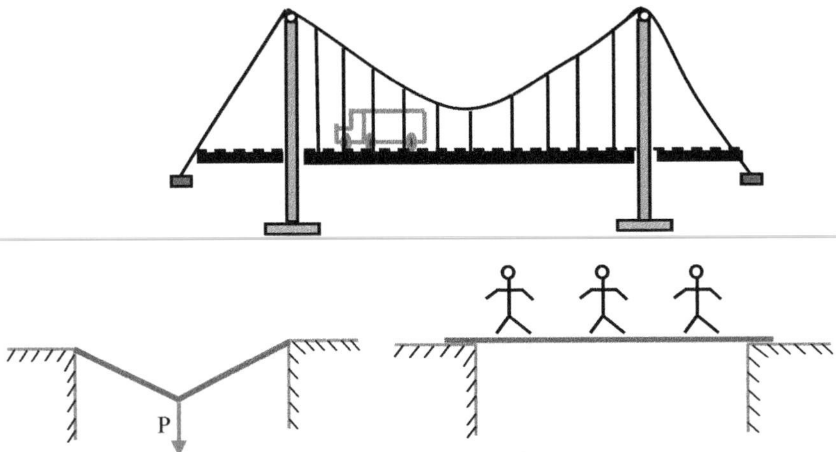

Figure 8.1. *Cables*[1]

In addition, cables are key elements of structural behavior; they are currently used in modern structures. They can be temporarily used as ropes to stabilize ships and boats, for example.

8.2. Mechanical characteristics of cables

The cables used in the field of construction are characterized by high mechanical strength and high fire resistance. They are flexible and coiled to increase their mechanical efficiency. Cables can cover large-span structures. Their efficiency increases with increasing length and they can reach hundreds of meters.

The cables are characterized by their shape (curvature), which depends on several parameters, such as the nature of the force, the point of the force's application, the length of the cable and the placement of the supports.

1 All of the figures in this chapter are available to view in full color at www.iste.co.uk/khalfallah/analysis1.zip.

Cable structures are classified according to their treatment of the dead weight:

– cables with the dead weight neglected;

– cables with the dead weight considered.

Depending on the loads applied, the cables take on different geometric shapes.

8.3. Hypothesis of cable analysis

To analyze cable structures, it is necessary to consider the following simplifying assumptions:

– the cables are inextensible and perfectly flexible;

– the cables have no resistance to bending or shearing;

– the internal force in the cables is traction, directed along the tangent to the curve at a given point;

– the horizontal component of the tension force at the support is called the system thrust.

8.4. Cable analysis

The role of cable structures is similar to that of other types of structures. The primary goal is to transfer actions from one element to another (on the ground). In a guyed bridge, for example, the internal forces are transferred from the slab to the transverse cables and then to the longitudinal cables.

Depending on the loads applied, we can distinguish two categories: (1) cables stressed by concentrated forces and (2) cables subject to distributed loads. In cable structure analysis, the flexibility of the cables makes it possible to neglect the shear and flexural strength while the inextensibility makes the cable length constant before and after the loading system is applied. This allows us to state that the geometry of cable structures retains its initial configuration after external loads are applied.

8.4.1. *Cables subject to concentrated force*

The cable (Figure 8.2) is stressed by a concentrated force P and each portion is a straight segment.

1) *Direct method*

Determine the shape of the cable if the system thrust is H.

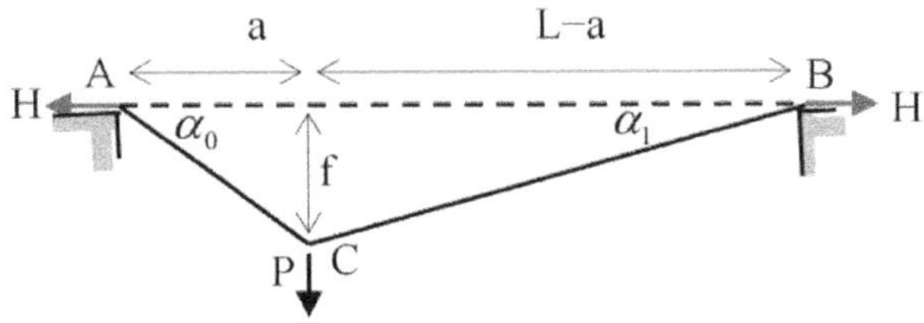

Figure 8.2. *Given cable*

The support reactions are calculated using equilibrium equations (Figure 8.3).

$$V_A = P(1-\frac{a}{L}) \quad [8.1]$$

$$V_B = P\frac{a}{L} \quad [8.2]$$

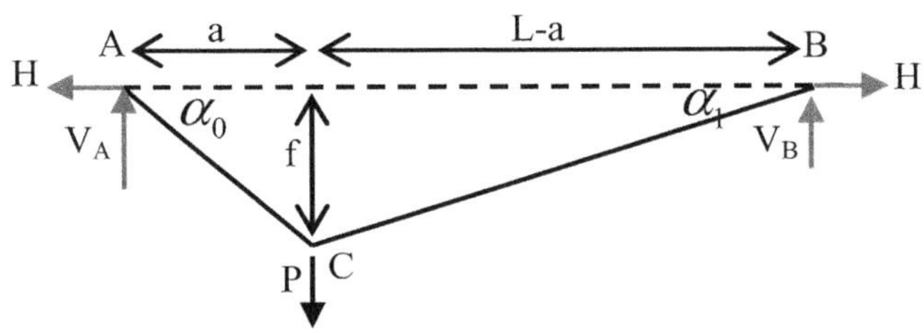

Figure 8.3. *Free-body diagram*

Given that $tg\alpha_0 = -\frac{V_A}{H}$, the inclinations of the cable are given as

$$\cos\alpha_0 = \frac{H}{\sqrt{V_A^2 + H^2}} \quad [8.3]$$

$$\cos\alpha_1 = \frac{H}{\sqrt{V_B^2 + H^2}} \quad [8.4]$$

The tension forces in the portions of the cable can be deduced by

$$T_{AC} = \frac{H}{\cos\alpha_0} = \sqrt{V_A^2 + H^2} \qquad [8.5]$$

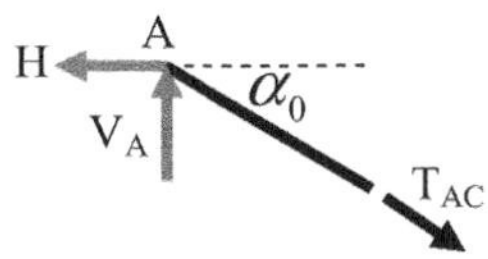

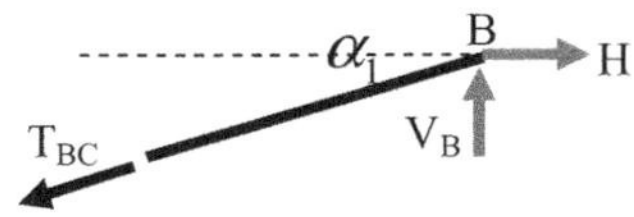

$$T_{BC} = \frac{H}{\cos\alpha_1} = \sqrt{V_B^2 + H^2} \qquad [8.6]$$

EXAMPLE 8.1.–

For H = 24 kN, P = 20 kN, a = 10 m and L = 25 m, the internal forces and maximum deflection (Figure 8.2) are given as

$$V_A = P(1 - \frac{a}{L}) = 20(1 - \frac{10}{25}) = 12\,\text{kN}$$

$$V_B = P\frac{a}{L} = 20\frac{10}{25} = 8\,\text{kN}$$

The forces in the portions of the cable are

$$T_{AC} = \sqrt{V_A^2 + H^2} = \sqrt{12^2 + 24^2} = 26.83\,\text{kN}$$

$$T_{BC} = \sqrt{V_B^2 + H^2} = \sqrt{8^2 + 24^2} = 25.30\,\text{kN}$$

The inclinations of the segments are

$$\cos\alpha_0 = \frac{H}{\sqrt{V_A^2 + H^2}} = \frac{24}{26.83} = 0.894\,,\ \alpha_0 = 26.62°$$

$$\cos\alpha_1 = \frac{H}{\sqrt{V_B^2 + H^2}} = \frac{24}{25.30} = 0.948\,,\ \alpha_1 = 18.56°$$

The deflection of the point of application of force P is

$$f = a.tg\alpha_0 = a.(-\frac{V_A}{H}) = 10.(-\frac{12}{24}) = -5\,\text{m}$$

2) *Inverse method*

The inverse method evaluates the system thrust when the total length of the cable is known.

The total length can be deduced by

$$L_t = \frac{a}{\cos\alpha_0} + \frac{L-a}{\cos\alpha_1} \qquad [8.7]$$

Substituting equations [8.3] and [8.4] in the relationship [8.7], we obtain

$$L_t = a\sqrt{1+\frac{P^2}{H^2}(1-\frac{a}{L})^2} + (L-a)\sqrt{1+\frac{P^2}{H^2}(\frac{a}{L})^2} \qquad [8.8]$$

After setting $\delta = \frac{a}{L}$ and $\lambda = \frac{L_t}{L}$, equation [8.8] is written as

$$\lambda = \delta\sqrt{1+\frac{P^2}{H^2}(1-\delta)^2} + (1-\delta)\sqrt{1+\frac{P^2}{H^2}\delta^2} \qquad [8.9]$$

Resolving equation [8.9] means we can evaluate the system thrust.

$$H = \frac{2.\lambda.\delta.(1-\delta)}{\sqrt{\lambda^4 - 2\lambda^2(2\delta^2 - 2\delta + 1) + 4\delta^2 - 4\delta + 1}}P \qquad [8.10]$$

EXAMPLE 8.2.–

Example 8.1 is repeated again.

The total length of the cable is given by the relationship [8.7].

Where $L_t = 30\,\text{m}$

Therefore, $\delta = 0.40$ and $\lambda = 1.20$.

The system thrust is given as

$$H = 0.734P = 14.68\,\text{kN}$$

We find a gain on the system thrust of 26.6%.

EXAMPLE 8.3.– Cables subject to multiple concentrated forces

The cable structure (Figure 8.4) is solicited by concentrated forces. The deflections of points B, C and D are, respectively, y_B, h and y_D. It is necessary to determine the support reactions, the tensions in the cable sections and the height differences y_B and y_D.

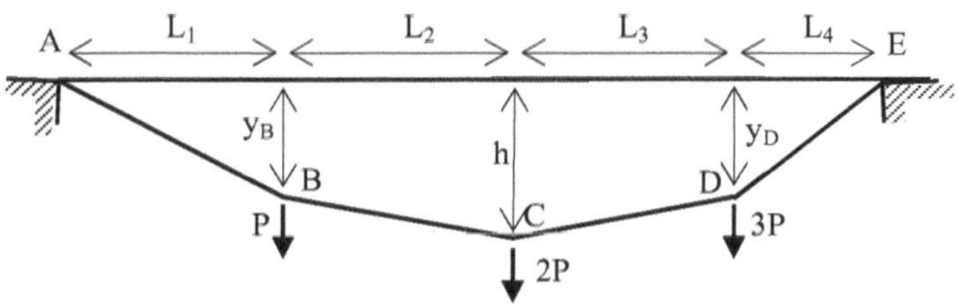

Figure 8.4. *Cable structure under concentrated forces*

Evaluating the support reactions is carried out using the static equations (Figure 8.5).

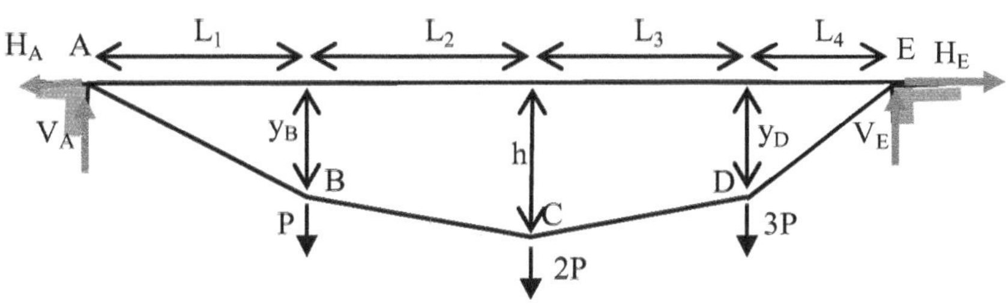

Figure 8.5. *Support reactions*

$$\sum \overrightarrow{M_E} = \vec{0}$$

$$V_A.(L_1 + L_2 + L_3 + L_4) - P.(L_2 + L_3 + L_4) - 2P.(L_3 + L_4) - 3P.(L_4) = 0$$

Hence

$$V_A = P.\frac{(L_2 + L_3 + L_4)}{(L_1 + L_2 + L_3 + L_4)} + 2P.\frac{(L_3 + L_4)}{(L_1 + L_2 + L_3 + L_4)} + 3P.\frac{L_4}{(L_1 + L_2 + L_3 + L_4)}$$

It can also be written as

$$V_A = P.\frac{(L_2 + 3L_3 + 6L_4)}{(L_1 + L_2 + L_3 + L_4)}$$

The equation of vertical equilibrium leads to

$$V_E = P.\frac{(6L_1 + 5L_2 + 4L_3)}{(L_1 + L_2 + L_3 + L_4)}$$

In the same way, the sum of the moments relating to point C is

$$\sum \overrightarrow{M_C} = \vec{0}$$

$$V_A.(L_1 + L_2) - H_A.\text{h} - P.L_2 = 0$$

$$H_A = V_A \frac{L_1 + L_2}{h} - P\frac{L_2}{h}$$

The thrust to support E is

$$H_E = H_A = V_A \frac{L_1 + L_2}{h} - P\frac{L_2}{h}$$

The deflections of joints B and D are

$$\sum \overrightarrow{M}_B = \vec{0} : V_A.L_1 - H_A.y_B = 0$$

$$y_B = \frac{V_A}{H_A} L_1$$

$$\sum \overrightarrow{M}_D = \vec{0}: V_E.\mathrm{L}_4 - H_E.y_c = 0$$

$$y_D = \frac{V_E}{H_E} L_4$$

To determine the tensions in cables (AB), (BC), (CD) and (DE), we use the equilibrium equations for each cable.

Segment (AB)

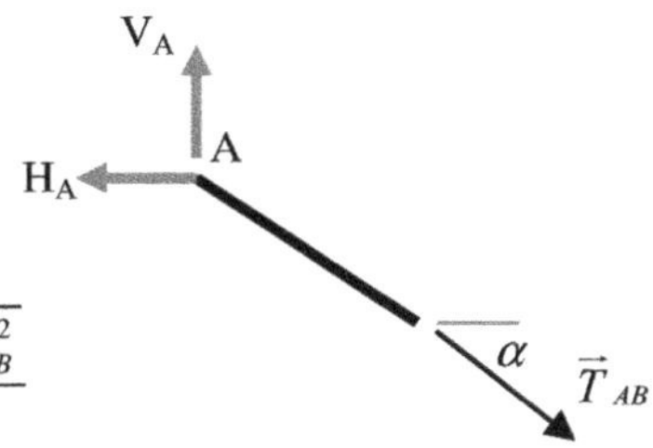

$$T_{AB}.\cos\alpha - H_A = 0$$

$$T_{AB} = \frac{H_A}{\cos\alpha} = (V_A \frac{L_1 + L_2}{h} - P\frac{L_2}{h})\frac{\sqrt{L_1^2 + y_B^2}}{L_1}$$

Segment (BC)

$$T_{BC} = \frac{H_A}{\cos\beta} = (V_A \frac{L_1 + L_2}{h} - P\frac{L_2}{h}).\frac{\sqrt{L_2^2 + (y_C - y_B)^2}}{L_2}$$

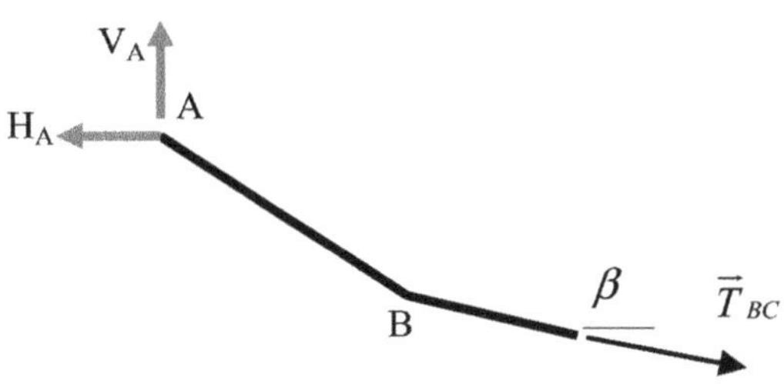

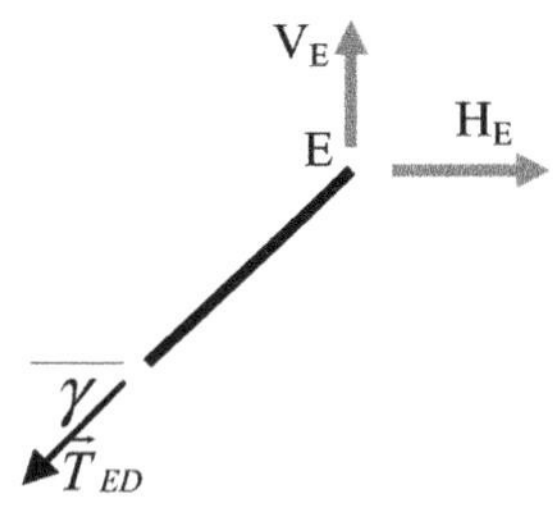

The same method applies to bars ED and DC.

Segment (ED)

$$T_{ED}.\cos\gamma - H_D = 0$$

$$T_{ED} = \frac{H_E}{\cos\gamma} = (V_A \frac{L_1 + L_2}{h} - P\frac{L_2}{h})\frac{\sqrt{L_4^2 + y_D^2}}{L_4}$$

Segment (DC)

$$T_{DC}=\frac{H_E}{\cos\psi}=(V_A\frac{L_1+L_2}{h}-P\frac{L_2}{h}).\frac{\sqrt{L_3^2+(y_C-y_D)^2}}{L_3}$$

In this particular case: $L_1=L_4=4\,\text{m}$, $L_2=L_3=3\,\text{m}$ and $h=2\,\text{m}$.

The formulas above lead to the determination of the reactions

$$V_A=\frac{18}{7}P$$

$$V_E=\frac{24}{7}P$$

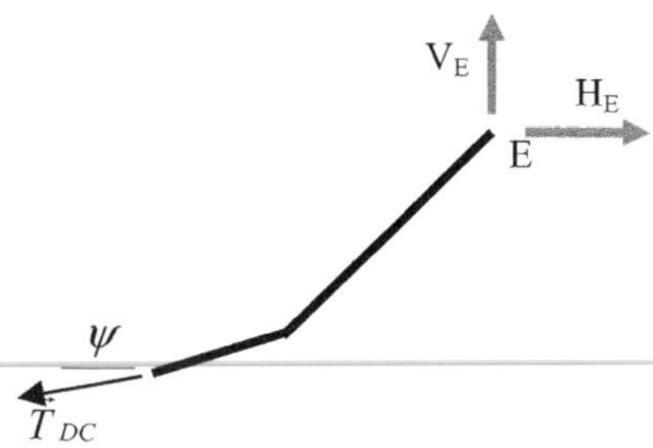

and $H_E=-H_A=\frac{15}{2}P$

The deflections of points B and D can be deduced by

$$y_B=\frac{V_A}{H_A}L_1=1.37\text{ m}$$

$$y_D=\frac{V_E}{H_E}L_4=1.82\,\text{m}$$

The tensions in the cable sections are

$$T_{AB}=\frac{H_A}{\cos\alpha}=(V_A\frac{L_1+L_2}{h}-P\frac{L_2}{h})\frac{\sqrt{L_1^2+y_B^2}}{L_1}=7.92.P$$

$$T_{BC}=\frac{H_A}{\cos\beta}=(V_A\frac{L_1+L_2}{h}-P\frac{L_2}{h})\frac{\sqrt{L_2^2+(y_C-y_B)^2}}{L_2}=7.66\,\text{P}$$

$$T_{ED}=\frac{H_E}{\cos\gamma}=(V_A\frac{L_1+L_2}{h}-P\frac{L_2}{h})\frac{\sqrt{L_4^2+y_D^2}}{L_4}=8.23P$$

$$T_{DC}=\frac{H_E}{\cos\psi}=(V_A\frac{L_1+L_2}{h}-P\frac{L_2}{h})\frac{\sqrt{L_3^2+(y_C-y_D)^2}}{L_3}=7.51P$$

The diagram of the tensions in the cable segments is shown in Figure 8.6.

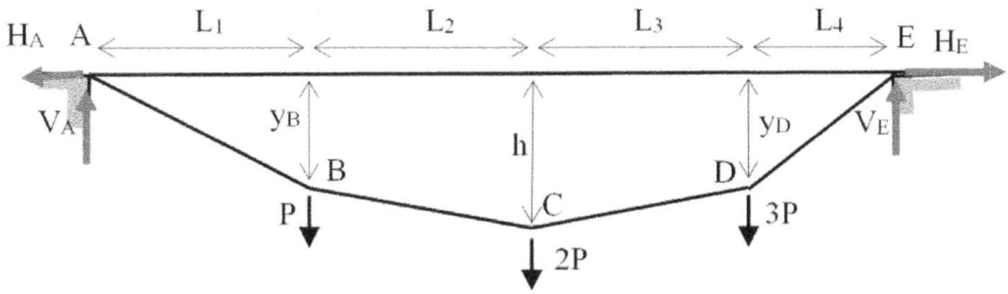

Figure 8.6. *Normal force in cable sections*

8.4.2. *Cables subject to distributed forces*

The cables are used to bear permanent loads and live loads. In the case of a guyed bridge, for example, the longitudinal cables are held by a series of transverse cables, which are generally regularly spaced (Figure 8.7).

In this case, it is necessary to determine the curve of the passage of the longitudinal cables, which depends on the number of transverse cables (that is, the number of concentrated forces). For ease of analysis, the concentrated forces applied to the cables are replaced by a uniformly distributed load (Figure 8.8).

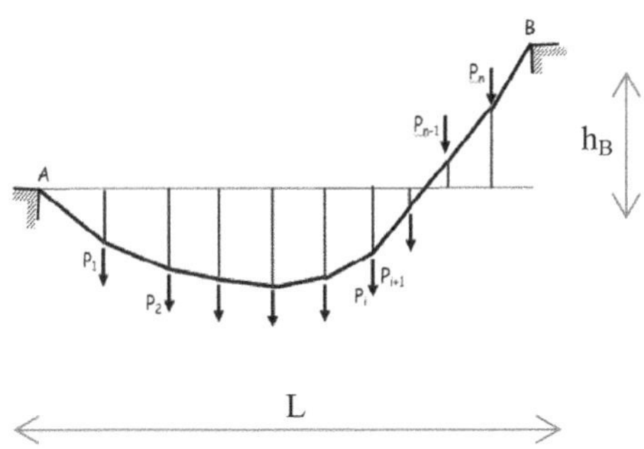

Figure 8.7. *Cable under concentrated force*

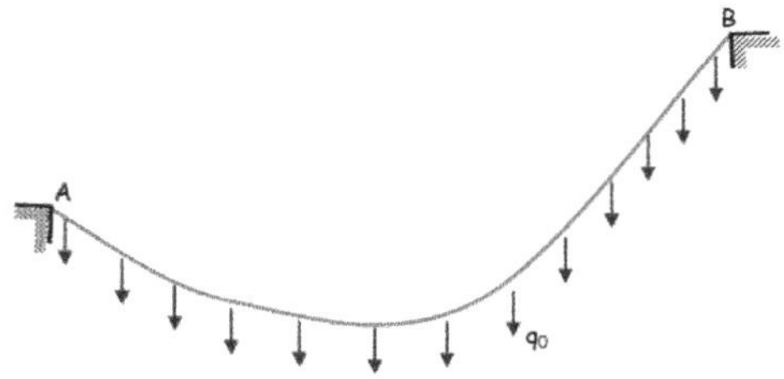

Figure 8.8. *Cable under distributed force*

Generally, a portion of cable of length Δx is acted upon by a variable load q(x) whose ends are subjected to tensions $T(x)$ and $T(x)+\Delta T(x)$ (Figure 8.9).

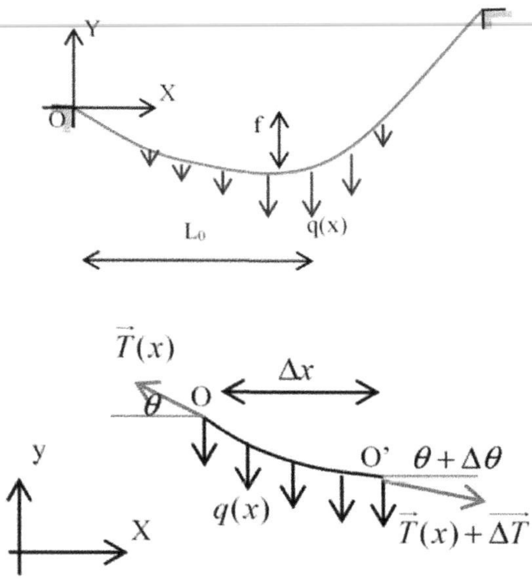

Figure 8.9. *Cable portion*

The static equations allow us to construct the following equation system:

$$\sum \vec{F}_{xx} = \vec{0} \; : \; -T(x)\cos\theta + (T(x)+\Delta T)\cos(\theta+\Delta\theta) = 0 \qquad [8.11]$$

$$\sum \vec{F}_{yy} = \vec{0} \;: T(x)\sin\theta - (T(x) + \Delta T)\sin(\theta + \Delta\theta) - q(x)\Delta x = 0 \qquad [8.12]$$

$$\sum \overrightarrow{M}_{0'} = \vec{0} \;: T(x)\sin\theta.\Delta x - T(x)\cos\theta.\Delta y - q(x)\Delta x \frac{\Delta x}{2} = 0 \qquad [8.13]$$

Equations [8.11], [8.12] and [8.13] can be simplified when $\Delta x \to 0$, $\Delta y \to 0$, $\Delta\theta \to 0$ and $\Delta T \to 0$.

$$\frac{d(T(x)\cos\theta)}{dx} = 0 \qquad [8.14]$$

$$\frac{d(T(x)\sin\theta)}{dx} = q(x) \qquad [8.15]$$

$$\frac{dy}{dx} = tg\theta \qquad [8.16]$$

The successive integration of equations [8.14] to [8.16] leads to

$$T(x)\cos\theta = C_1 \qquad [8.17]$$

$$T(x)\sin\theta = q(x)x + C_2, \qquad [8.18]$$

where q(x) is assumed to be constant along Δx. The constants C_1 and C_2 can be deduced by applying the limit conditions.

$$T(x = 0)\cos(\theta = 0) = H_A \;:\; C_1 = H_A$$

$$T(x)\cos\theta = H_A \qquad [8.19]$$

$$T(x = 0)\sin(\theta = 0) = 0 :\; C_2 = 0$$

$$T(x)\sin\theta = q(x)x \qquad [8.20]$$

Using equation [8.16], and given that $\cos\theta = \dfrac{H_A}{T(x)}$ and $\sin\theta = \dfrac{q(x)x}{T(x)}$, we can write

$$\frac{dy}{dx} = \frac{q(x)x}{H_A} \qquad [8.21]$$

If we suppose that $q(x) = q_0$, the curve equations of the cable become

$$y(x) = \frac{1}{2}\frac{q_0 x^2}{H_A} + C_3$$

Knowing that $y(L_0) = f$: $C_3 = f - \frac{q_0 L_0^2}{2H_A}$.

Finally, the parabola equation is

$$y(x) = \frac{q_0}{2H_A}(x^2 - L_0^2) + f \qquad [8.22]$$

Knowing that $y(0) = 0$, the system thrust is

$$H_A = \frac{1}{2}\frac{q_0 L_0^2}{f}$$

Using equations [8.19] and [9.20] allows us to determine the tension in the cable.

$$T(x) = \sqrt{q_0^2 x^2 + H_A^2} = \sqrt{q_0^2 x^2 + \frac{q_0^2 L_0^4}{4f^2}} \qquad [8.23]$$

The maximum tension in the cable can be determined using the relationship [8.23].

$$T_{max} = T(L_0) = q_0 L_0 \sqrt{1 + \frac{L_0^2}{4f^2}} \qquad [8.24]$$

1) *Direct method*

The cable is acted by a uniformly distributed load with an intensity of q_0 (Figure 8.8). It is customary to write

$$q(x)\Delta x = q_0 \Delta s, \qquad [8.25]$$

where Δs is the curvilinear length of the cable.

$$\Delta s = \sqrt{\Delta x^2 + \Delta y^2} \qquad [8.26]$$

Substituting relationship [8.26] into [8.25], we obtain

$$q(x) = q_0 \sqrt{1 + (\frac{\Delta y}{\Delta x})^2} \quad [8.27]$$

The length of the studied element is very small, thus equation [8.27] is written in the following form

$$q(x) = q_0 \sqrt{1 + (\frac{dy}{dx})^2} \quad [8.28]$$

Substitute the relationship [8.28] with the relationship [8.21], we obtain

$$\frac{dy}{dx} = \frac{q_0 x \sqrt{1 + (\frac{dy}{dx})^2}}{H_A} \quad [8.29]$$

By the resolution of equation [8.21], the cable slope expression is given as

$$\frac{dy}{dx} = \frac{q(x).x}{H_A} = \frac{T(x)}{H_A} \quad [8.30]$$

We derive equation [8.30] and use equation [8.15] to obtain

$$\frac{dT(x)}{dx} = H_A.\frac{d^2 y}{dx^2} = q(x) \quad [8.31]$$

Knowing that $\rho = \frac{dy}{dx}$, the relationship [8.31] is written as

$$\frac{d\rho}{dx} = \frac{q_0 \sqrt{1 + \rho^2}}{H_A} \quad [8.32]$$

Equation [8.32] is written as

$$\frac{d\rho}{\sqrt{1 + \rho^2}} = \frac{q_0 dx}{H_A} \quad [8.33]$$

Resolution of the differential equation with separate variables [8.33] is written as

$$\rho = \sinh(\frac{q_0}{H_A} x + C_1) \quad [8.34]$$

Substituting the expression ρ with [8.34], the ordinate of the cable is given as

$$y(x) = \frac{H_A}{q_0} \cosh(\frac{q_0}{H_A} x + C_1) + C_2 \quad [8.35]$$

The limit conditions allow us to determine the integration constants C_1 and C_2.

$$y(x=0) = 0 \; : \; \frac{H_A}{q_0} \cosh(C_1) + C_2 = 0 \quad [8.36]$$

$$y(x=L) = 0 \; : \; \frac{H_A}{q_0} \cosh(\frac{q_0}{H_A} L + C_1) + C_2 = 0 \quad [8.37]$$

Subtraction of relationships [8.36] and [8.37] leads to

$$\frac{H_A}{q_0} \cosh(\frac{q_0}{H_A} L + C_1) - \frac{H_A}{q_0} \cosh(C_1) = 0 \quad [8.38]$$

Equation [8.38] is valid if and only if

$$C_1 = -\frac{q_0 L}{2H_A}$$

$$C_2 = -\frac{H_A}{q_0} \cosh(\frac{q_0 L}{2H_A})$$

The parabola equation of the curve of the cable is

$$y(x) = \frac{H_A}{q_0} \cosh(\frac{q_0}{H_A} x - \frac{q_0 L}{2H_A}) - \frac{H_A}{q_0} \cosh(\frac{q_0 L}{2H_A}) \quad [8.39]$$

EXAMPLE 8.4.–

Determine the tension in the cable at points A, B and C solicited by a uniform load $q_0 = 15$ kN/m.L (Figure 8.10).

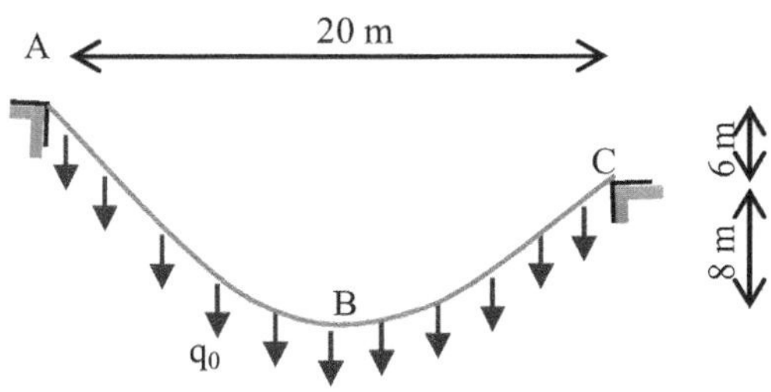

Figure 8.10. *Given cable*

A calculation reference is associated with point B (Figure 8.11).

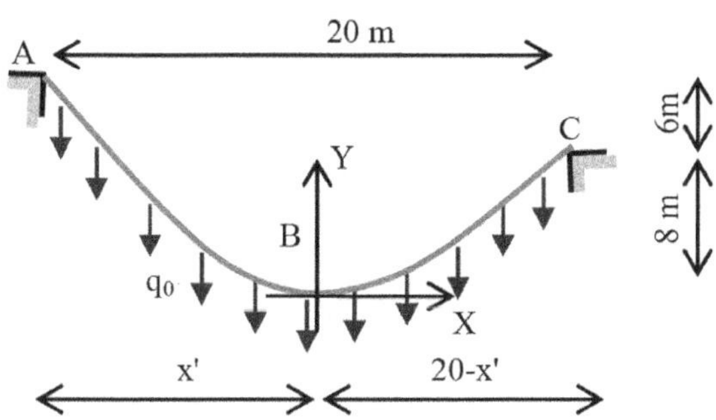

Figure 8.11. *Spotting the structure*

The parabola equation is given by

$$y(x) = \frac{q_0}{2H_A} x^2$$

Substituting the coordinates of points A and C in the preceding relationship makes it possible to formulate

$$6x'^2 - 14x' + 280 = 0$$

The accepted solution is x'=11.39 m.

The system thrust at support A is given as

$$H_A = \frac{q_0}{28} x'^2 = \frac{15}{28}(11.39)^2 = 69.50\,\text{kN}$$

The tension at point B is $(\theta = 0)$.

$$T_B = T(x') = H_A = 69.50kN$$

We have $tg\theta(x) = \dfrac{dy}{dx} = \dfrac{q_0}{H_A}x$

$$tg\theta_A = \frac{dy}{dx} = \frac{q_0}{H_A}(-x') = \frac{15}{69.50}(-11.39) = -2.458$$

$$\theta_A = -67.864\,°$$

Hence, the tension in the cable at point A is

$$T_A = \frac{H_A}{\cos\theta_A} = \frac{69.50}{\cos(67.864)} = 184.444\,\text{kN}$$

At point C, we have

$$tg\theta_C = \frac{dy}{dx} = \frac{q_0}{H_A}(20 - x') = \frac{15}{69.50}(8.61) = 1.858$$

Hence $\theta_C = 61.71°$

$$T_C = \frac{H_A}{\cos\theta_C} = \frac{69.50}{\cos(61.71)} = 146.644\,\text{kN}$$

The curve equation of the cable can be established using the relationship [8.39] and substituting the expression of thrust H_A.

2) *Inverse method*

The expression, given the total length of the cable, is given as

$$L_t = \int_0^L \sqrt{1 + (\frac{dy}{dx})^2}\, dx \qquad [8.40]$$

We substitute expression [8.15], and by introducing the deflection in relationship [8.40], we obtain

$$L_t = \int_0^L \sqrt{1+(\frac{fL}{L_0^2})^2(\frac{2x}{L}-1)^2}\ dx \tag{8.41}$$

The relationship [8.41] can be broken down into a Taylor series.

$$L_t = \int_0^L (1+\frac{1}{2}(\frac{fL}{L_0^2})^2(\frac{2x}{L}-1)^2 - \frac{1}{8}(\frac{fL}{L_0^2})^4(\frac{2x}{L}-1)^4 + ...)dx \tag{8.42}$$

The total length can also be written as

$$L_t = L + \frac{8L}{3}(\frac{f}{L_0})^2 - \frac{32}{5}(\frac{f}{L_0})^4 + ... \tag{8.43}$$

If $L_0=\frac{L}{2}$, we obtain

$$L_t = L(1+\frac{32}{3}(\frac{f}{L})^2 - \frac{512}{5}(\frac{f}{L})^4 + ...) \tag{8.44}$$

The relationship [8.44] allows us to calculate the thrust H using the deflection f, the cable length L and the applied force q_0.

EXAMPLE 8.5.–

A flexible cable is attached to points A and B at different levels (Figure 8.12). It is stressed by a uniformly distributed load q_0, calculating the total length of the cable and the maximum tension.

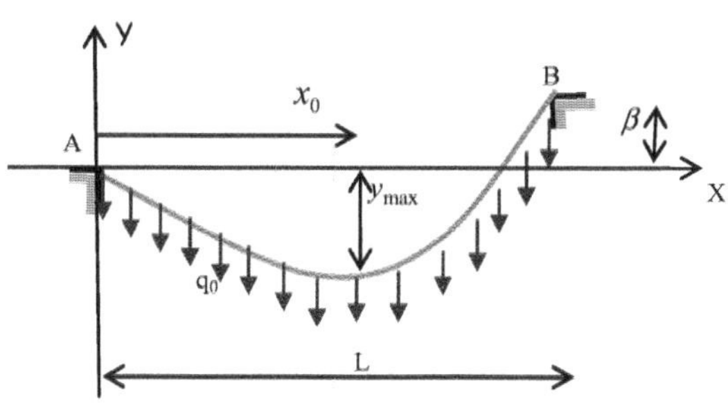

Figure 8.12. *Given cable*

The successive integration of the deflection of relationship [8.15] makes it possible to write

$$\frac{dy}{dx} = \frac{q_0}{H_A} x + C_1$$

$$y(x) = \frac{q_0}{2H_A} x^2 + C_1 x + C_2$$

Integration constants can be deduced by

$$C_1 = \frac{\beta}{L} - \frac{q_0 L}{2H_A} \quad \text{and} \quad C_2 = 0$$

The curve of the cable is

$$y(x) = \frac{q_0}{2H_A} x^2 + \left(\frac{\beta}{L} \frac{q_0 L}{2H_A}\right) x$$

The maximum curve is at the position $x_0 = \frac{L}{2} - \beta \frac{H_A}{q_0 L}$ and the deflection is

$$y_{\max} = \left(\frac{\beta}{L} - \frac{q_0 L}{2H_A}\right)\left(\frac{L}{2} - \frac{\beta H_A}{q_0 L}\right)$$

In the relationship H_A is assumed to be known, and the maximum deflection y_{max} can be determined using q_0 and L.

8.4.3. *Cables subject to any force*

In the previous sections, we have presented the analysis of cable structures subjected to concentrated forces and uniformly distributed forces. This analysis can also be conducted for other loads, such as an operating load that can accompany a permanent load. This phenomenon causes a change in the cable's geometry and possibly a change in the system thrust.

Consider an inextensible cable supported at points A and B and acted upon by any load q(x) (Figure 8.13).

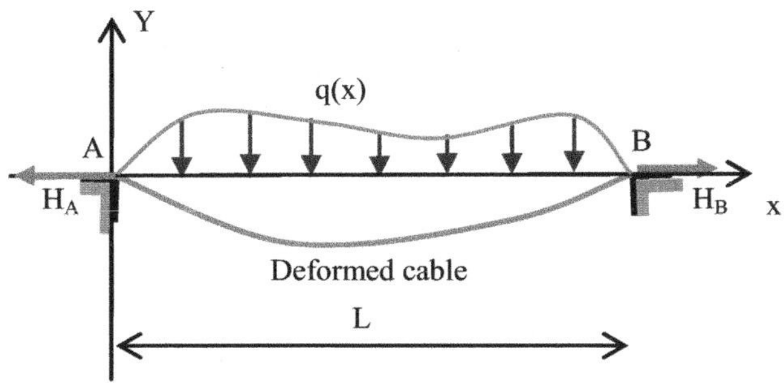

Figure 8.13. *Given structure*

Applying the relationship [8.40] makes it possible to calculate the total length of the cable in a deformed state.

$$L_t = \int_0^L \sqrt{1+(\frac{dy}{dx})^2}\, dx \cong \int_0^L (1+\frac{1}{2}(\frac{dy}{dx})^2)dx \tag{8.45}$$

The slope expression in all sections is

$$\frac{dy(x)}{dx} = \frac{T(x).\sin\theta}{H_A} \tag{8.46}$$

Substituting the relationship [8.46] into the relationship [8.45], the total length is

$$L_t = L + \frac{\sin^2\theta}{2H_A^2}\int_0^L T^2(x)dx \tag{8.47}$$

We deduce the system thrust at support A by

$$H_A = \sin\theta\sqrt{\frac{\int_0^L T^2(x)dx}{2L(\lambda-1)}}, \tag{8.48}$$

with $\lambda = \dfrac{L_t}{L}$

The system thrust [8.48] is deduced using the applied loading. Then, we distinguish the following cases.

Case 1: Concentrated load

The cable (Figure 8.2) takes on a concentrated load P applied at a distance a from support A.

$$x \leq a: \quad T(x) = \frac{P}{\sin\theta}(1 - \frac{a}{L}) \quad [8.49a]$$

$$a \leq x \leq L: \quad T(x) = -\frac{P}{\sin\theta}\frac{a}{L} \quad [8.49b]$$

The system thrust can be obtained by substituting the relationship [8.49] into [8.48].

$$H_A = \sqrt{\frac{\int_0^a P^2(1-\frac{a}{L})^2 dx + \int_a^L P^2(\frac{a}{L})^2 dx}{2L(\lambda-1)}} = P\sqrt{\frac{\frac{a}{L}(1-\frac{a}{L})}{2(\lambda-1)}} \quad [8.50]$$

Particularly if $\frac{a}{L} = 0.4$ and $\lambda = 1.2$, we find $H_A = 0.774P$.

With the inverse method, the thrust is H_A = 0.733P. An error of 5.44% is observed between the two calculation methods.

Case 2: Uniformly distributed load

The cable is stressed by a uniformly distributed load q_0. In this case, the variation of the tension in the cable is

$$x \leq L: T(x) = \frac{1}{\sin\theta}(\frac{q_0 L}{2}x - \frac{q_0 x^2}{2}) \quad [8.51]$$

The system thrust at support A is obtained by substituting relationship [8.51] into [8.48].

$$H_A = \sqrt{\frac{\int_0^L (\frac{q_0 L}{2} - q_0 x)^2 dx}{2L(\lambda-1)}} = q_0 L\sqrt{\frac{1}{24(\lambda-1)}} \quad [8.52]$$

Knowing that $\lambda = \frac{L_t}{L} \simeq 1 + \frac{8}{3}\frac{f^2}{L^2}$, the system thrust becomes

$$H_A = \frac{q_0 L^2}{8f} \qquad [8.53]$$

The deflection of the cable is obtained in Example 8.5. It is given by

$$y(x) = \frac{q_0}{2H_A}x^2 + (\frac{\beta}{L}\frac{q_0 L}{2H_A})x \qquad [8.54]$$

If the given structure is symmetrical $y(\frac{L}{2}) = f$ and $\beta = 0$, the deflection is

$$f = \frac{q_0 L^2}{8H_A} \qquad [8.55]$$

The system thrust at support A is

$$H_A = \frac{q_0 L^2}{8f} \qquad [8.56]$$

8.5. Cables with an inflection point outside the cable

Consider cable (AB) with a length of L_0 and attached to points A and B. The cable is funicular, $tg\theta = 0$, located outside the cable at a distance a from the axis (xx). The geometric dimensions are shown in Figure 8.14.

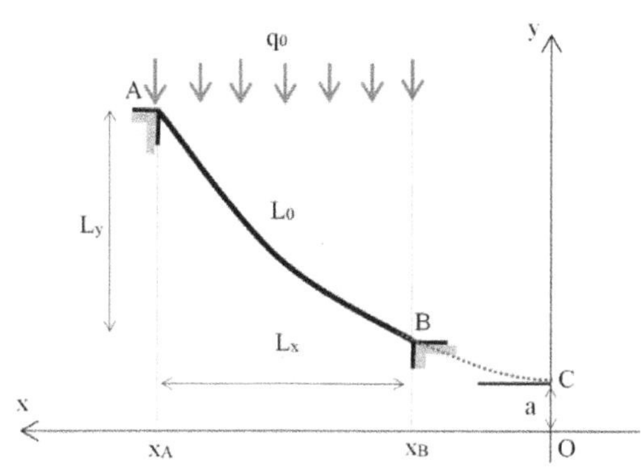

Figure 8.14. *Given cable*

The distance of the x-axis section from point C following the curve of the cable is

$$L_{cx} = a\sinh(\frac{x}{a}) \quad [8.57]$$

At points A and B, we obtain

$$L_{cA} = a\sinh(\frac{x_A}{a}) \quad [8.58a]$$

$$L_{cB} = a\sinh(\frac{x_B}{a}) \quad [8.58b]$$

The curved length L_0 is deduced by

$$L_0 = a\left(\sinh(\frac{x_A}{a}) - \sinh(\frac{x_B}{a})\right) \quad [8.59]$$

In the same way, the height at the point of the y-axis is

$$y(x) = a\cosh(\frac{x}{a}) \quad [8.60]$$

The heights at points A and B are deduced by

$$y_A = a\cosh(\frac{x_A}{a}) \quad [8.61a]$$

$$y_B = a\cosh(\frac{x_B}{a}) \quad [8.61b]$$

The vertical difference between points A and B is

$$L_y = y_A - y_B = a\left(\cosh(\frac{x_A}{a}) - \cosh(\frac{x_B}{a})\right) \quad [8.62]$$

Similarly, the horizontal difference between points A and B may be written as

$$L_x = x_A - x_B \quad [8.63]$$

The relationships [8.59], [8.62] and [8.63] describe the relationship between L_x, L_y and L_0. Therefore, for cable geometry L_x, L_y and L_0, resolving the equations leads to

$$x_B = arctg(\frac{L_y}{L_0}) \qquad [8.64]$$

$$x_A = arctg(\frac{L_y}{L_0}) + L_x \qquad [8.65]$$

$$\cosh(\frac{L_x}{a}) = 1 + (\frac{L_0^2 - L_y^2}{2a^2}) \qquad [8.66]$$

The expression of the deflection of the relationship [8.60] is

$$y'(x) = tg\theta(x) = \sinh(\frac{x}{a}) \qquad [8.67]$$

At the support level, we can write

$$tg\theta_A = tg\theta(x = x_A) = \sinh(\frac{x_A}{a}) \qquad [8.68a]$$

$$tg\theta_B = tg\theta(x = x_B) = \sinh(\frac{x_B}{a}) \qquad [8.68b]$$

The forces at points A and B are shown in Figure 8.15.

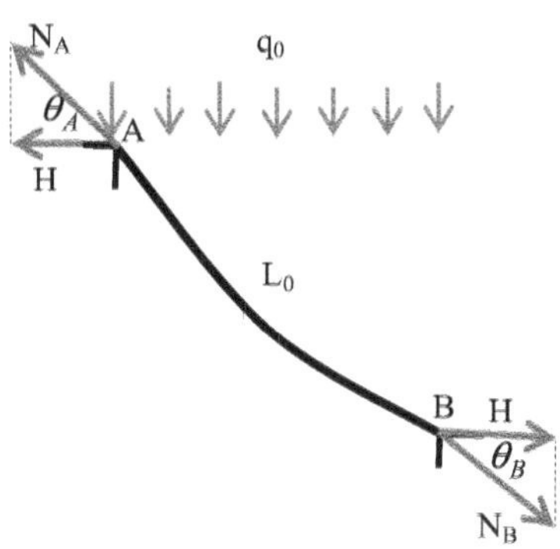

Figure 8.15. *Support forces*

$$tg\theta_A = \frac{\sqrt{N_A^2 - H^2}}{H} \qquad [8.69]$$

Combining the relationship [8.68] with [8.69], we obtain the tensions at the supports

$$N_A = H\sqrt{(1+\sinh^2(\frac{x_A}{a})} \qquad [8.70a]$$

$$N_B = H\sqrt{(1+\sinh^2(\frac{x_B}{a})} \qquad [8.70b]$$

EXAMPLE 8.6.–

A cable is stressed by a uniformly distributed load of intensity q_0 = 15 N/mL. The geometry is given by Figure 8.16.

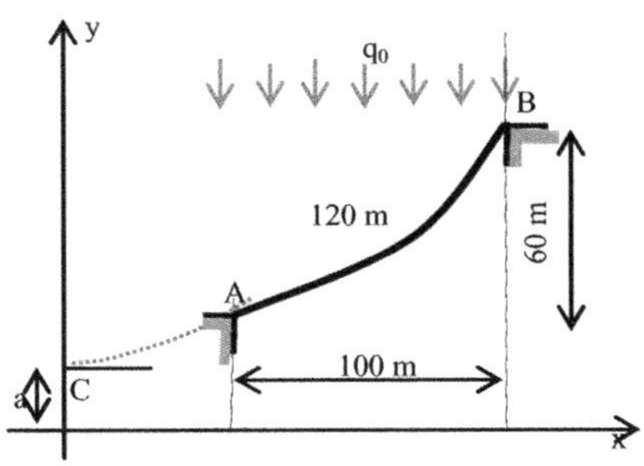

Figure 8.16. *Given cable*

Calculate the system thrust, the support tensions and deduce the curve of the cable.

The relationship [8.66] allows us to calculate the distance *a*:

$$\cosh(\frac{L_x}{a}) = 1 + (\frac{L_0^2 - L_y^2}{2a^2}) = 1 + \frac{10,800}{2a^2}$$

$$\cosh(\frac{100}{a}) = 1 + \frac{10,800}{2a^2}$$

The solution to this equation is $a = 130$ m.

The x-axis of points A and B are

$$x_A = a\, tgh^{-1}(\frac{L_y}{L_0}) - \frac{L_x}{2} = 21.41\,\text{m}$$

$$x_B = x_A + L_x = 121.41\,\text{m}$$

The slope at support A is

$$tg\theta_A = \sinh(\frac{x_A}{a}) = \sinh(\frac{21.41}{130}) = 0.165\,,\ \theta_A = 9.37°$$

The thrust of the cable is

$$H = q_0.a = 1.95\,\text{kN}$$

The expression of support tensions are

$$N_A = H\sqrt{(1+\sinh^2(\frac{x_A}{a})} = 1.95\sqrt{(1+\sinh^2(\frac{21.41}{130})} = 1.976\,\text{kN}$$

$$N_B = H\sqrt{(1+\sinh^2(\frac{x_B}{a})} = 1.95\sqrt{(1+\sinh^2(\frac{121.41}{130})} = 2.864\,\text{kN}$$

The ordinates of the supports can be deduced by

$$y_A = a\cosh(\frac{x_A}{a}) = 130\cosh(\frac{21.41}{130}) = 131.76\,\text{m}$$

$$y_B = a\cosh(\frac{x_B}{a}) = 130\cosh(\frac{121.41}{130}) = 190.936\,\text{m}$$

8.6. Suspension bridges

A suspension bridge deck is generally suspended by hangers held by longitudinal cables passing over pulleys at the ends of each pylon and extending to the blocks anchored in the ground (the anchor) (Figure 8.17).

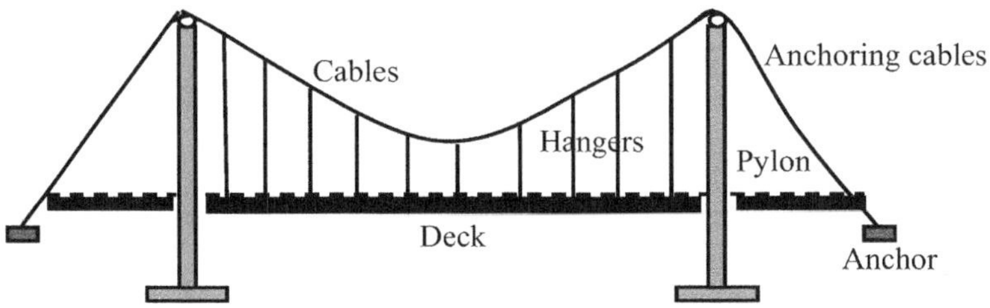

Figure 8.17. *Suspension bridge*

Suspension bridges are used in small span bridges (river bridge, etc.). The load borne by the cables is made up of the dead weight of the deck and vehicles crossing the suspension bridge. In general, suspension bridge cables bear a load, which is uniformly distributed along the bridge. The shape of the cables in most cases is parabolic.

The cables pass over frictionless pulleys, meaning that the tension in the anchoring cables T_A is equal to the tension in the main cables T_C at the end of the pylon (Figure 8.18).

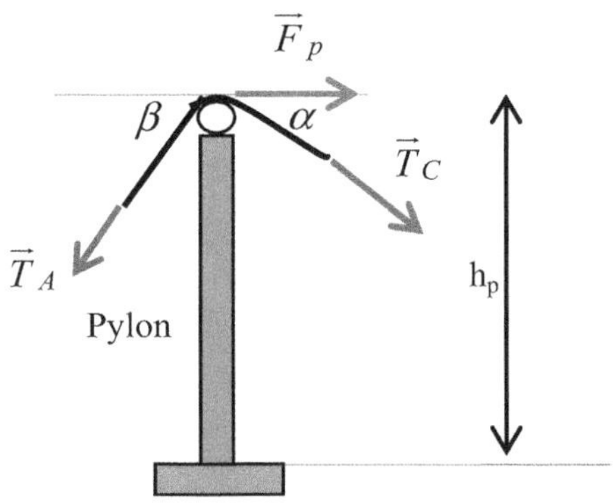

Figure 8.18. *Forces on the pulley*

In general, the inclination of the anchoring cables β is constant and cannot be equal to that of the cables α . There is a horizontal force at the pulley F_p, which is equivalent to

$$F_p = T_c \cos\alpha - T_A \cos\beta \qquad [8.71]$$

The mechanical continuity of the cable makes it possible to consider $T_C = T_A$; the tension force becomes

$$F_p = T_c (\cos\alpha - \cos\beta) \qquad [8.72]$$

This force generates a bending moment along the pylon whose maximum value at the base is

$$M_p^{\max} = F_p .h_p = T_c (\cos\alpha - \cos\beta) h_p \qquad [8.73]$$

where h_p is the height of the pylon.

In addition, the concavity of the cable generates a compression force V_P in the pylon equal to

$$N_p = T_c (\sin\alpha + \sin\beta) \qquad [8.74]$$

To eliminate the force on the pulley, a roller or cylinder is placed at the end of the pylon (Figure 8.19).

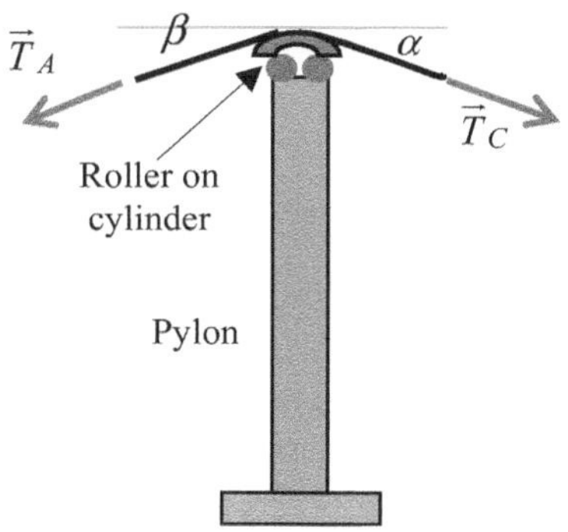

Figure 8.19. *Arrangement at the end of the pylon*

Knowing that the resulting force due to the effect of the cables is eliminated, the stresses in the cable and the pylon are

$$T_c \cos\alpha = T_A \cos\beta \qquad [8.75]$$

$$M_p = 0. \qquad [8.76]$$

$$N_p = T_c \sin\alpha + T_A \sin\beta \qquad [8.77]$$

EXAMPLE 8.7.–

A cable on a suspension bridge passes over two frictionless pulleys placed at the end of each pylon (Figure 8.20). We assume that the load generated by the deck is 110 kN/mL. We must calculate:

– the required diameter of the cable;

– the bending moment in the pylon;

– the normal force at the base of the pylon;

– the weight of the anchor block.

The stress limit of the cables is 540 MPa.

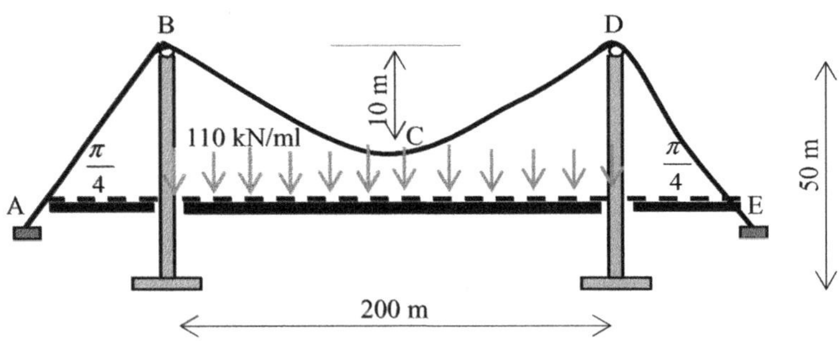

Figure 8.20. *Given suspension bridge*

The ends of the pylons are at the same level. The tensions in the cable at points B and D are equal and maximum. The maximum tensions in the cable can be deduced by

$$T_{\max} = \frac{wL}{2}\sqrt{1+\left(\frac{L}{4f}\right)^2} = 56{,}089.21\text{kN}$$

The resistance condition is

$$\sigma_{max} = \frac{T_{max}}{\frac{\pi D^2}{4}} \leq \overline{\sigma}$$

$$D \geq 363.66 \text{ mm} \;\text{ or }\; D = 363.70 \text{ mm}$$

The inclination of the suspension cables from horizontal is

$$tg\alpha = \frac{wL}{2H}$$

Knowing that $H = \frac{wL^2}{8f} = 55{,}000 \text{ kN}$, $tg\alpha = \frac{4f}{L} = \frac{4.10}{200} = 0.20$

$$\alpha = 11.31°.$$

The maximum bending moment is

$$M_p^{max} = T_c\,(\cos\alpha - \cos\beta)h_p = 56{,}089.2\,(\cos 11.31 - \cos\frac{\pi}{5})50 = 767{,}245.41 \text{ kN.m}$$

$$N_p = T_{max}(\sin\alpha + \sin\beta) = 50{,}661.11 \text{ kN}$$

The necessary counterweight is

$$W_A = T_{max}\cos\alpha = 39{,}655.06 \text{ kN}$$

8.7. Conclusion

In this chapter, we have presented the analysis of cables, which are structures working with forces of traction. In general, the funicular shape of the cables depends on the applied load. In this chapter, we described the analysis of cable structures subjected to concentrated and uniformly distributed loads.

The second part of this chapter focused on the analysis of cables with an inflection point located outside the cable, as well as the analysis of suspension bridges. Each chapter is accompanied by examples to clarify the steps taken.

8.8. Problems

Exercise 1

– Determine the tension in each portion of the following cables.

– Calculate the thrust of each system.

– Calculate the inclinations of the cable to the supports.

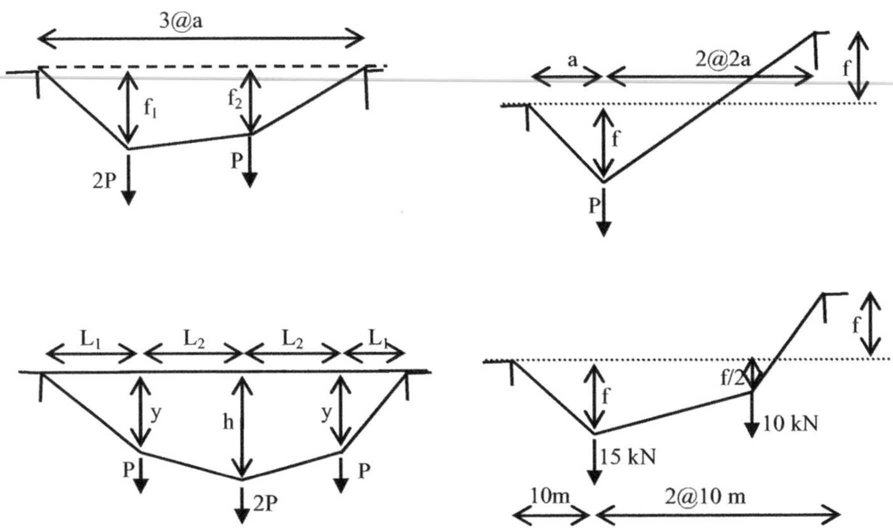

Exercise 2

Using the indirect method, calculate the thrust of each system of cable structures given in Exercise 1 for $\lambda = 1.25$.

Exercise 3

For each cable structure,

– determine the deformation of the cable and deduce the slopes at the supports;

– draw the diagram of tensions in the cable and deduce its maximum value and specify the corresponding position;

– calculate the thrust of each system.

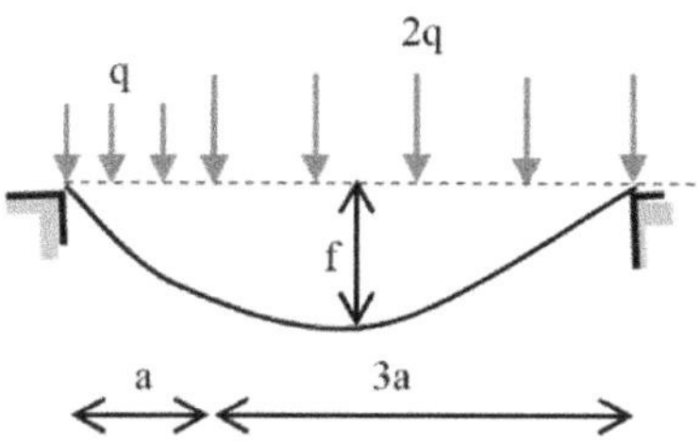

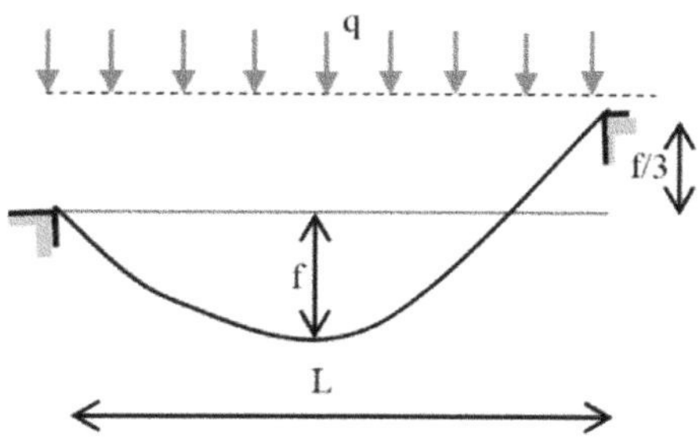

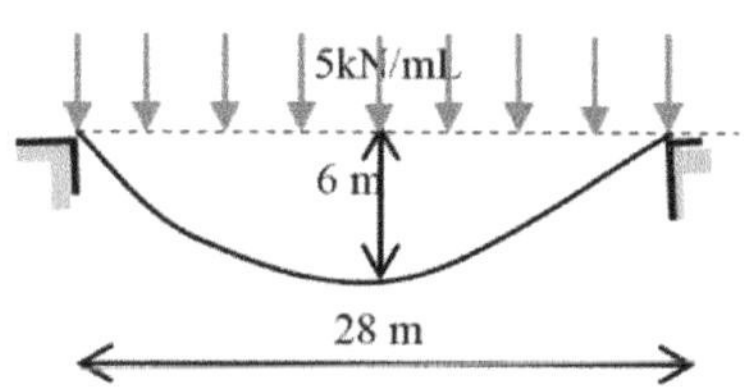

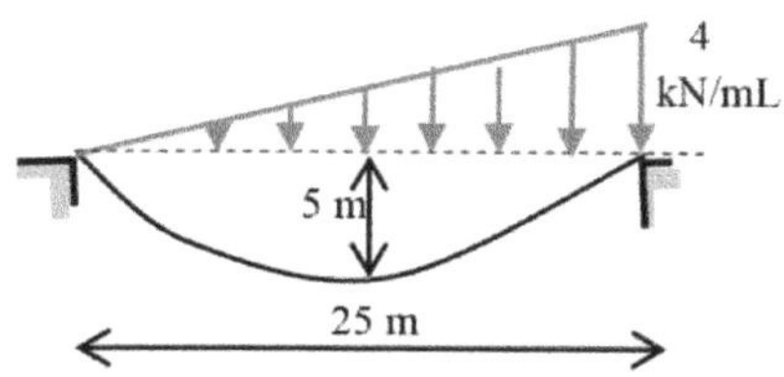

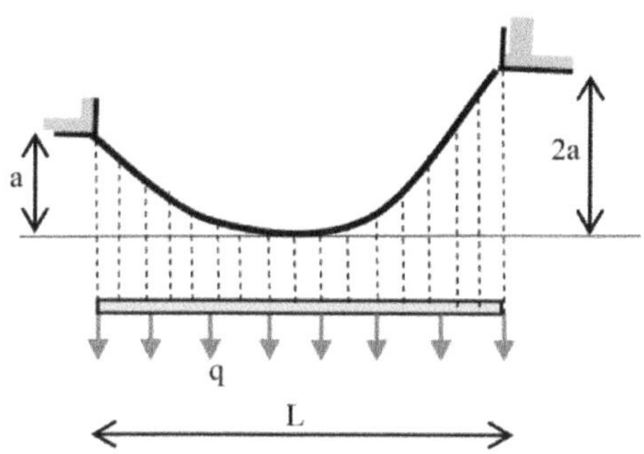

Exercise 4

Using the indirect method, calculate the thrust of each system of cable structures given in Exercise 3 for $\lambda = 1.25$.

Exercise 5

Calculate the thrust of each of the following systems:

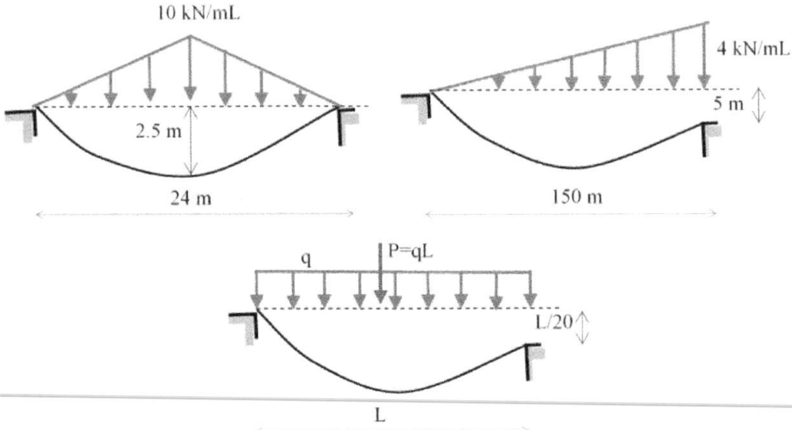

Exercise 6

Calculate the thrust and the expression of tension and deduce the tensions in the supports of the structures in the following cables:

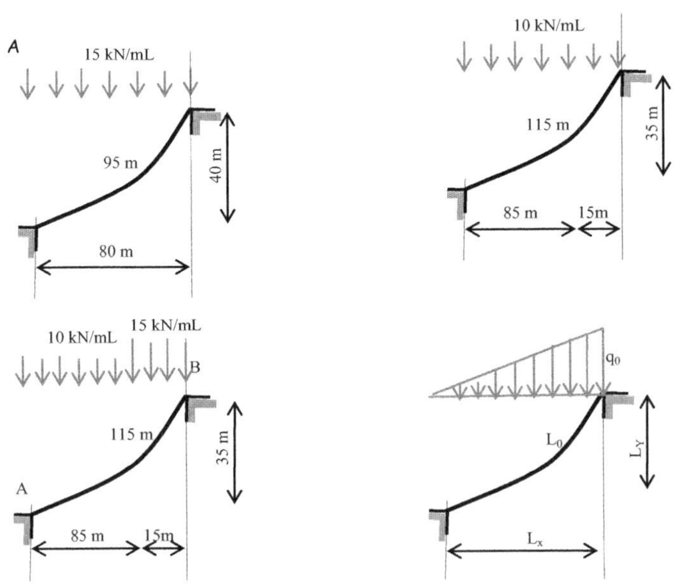

Exercise 7

Analyze the following cable structures. In each case, the cable passes over a frictionless pulley placed at the end of each pylon. Calculate:

- the required diameter of the cable;
- the bending moment in the pylon;
- the normal force at the base of the pylon;
- the weight of the anchor block.

The stress limit of the cables is 540 MPa.

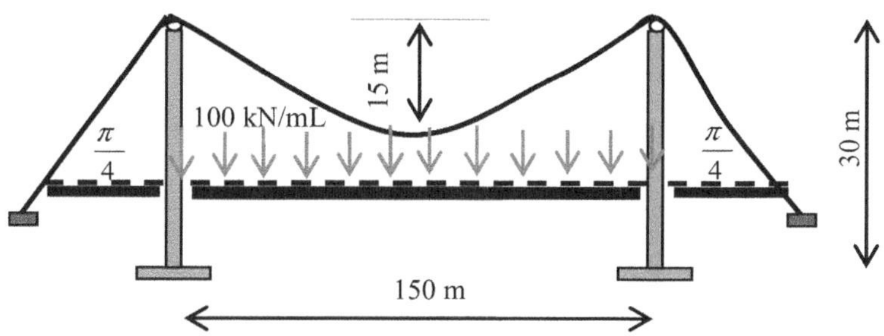

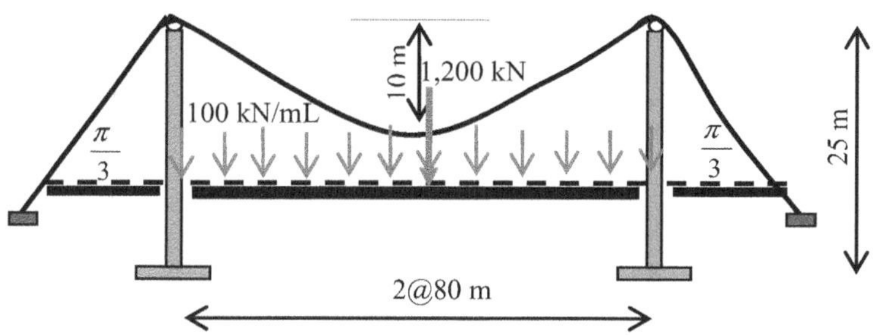

Exercise 8

A cable suspended between two supports with a span of 80 m is stressed by a uniformly distributed load 15 kN/mL. Calculate the minimum deflection at the middle point when the tension in the cable is 900 kN.

9

Analysis of Arches

After studying this chapter, the student will be able to:

– distinguish between different arch structures;

– differentiate between arches with one, two and three hinges;

– be familiar with the advantages of arch structures and their areas of application;

– master the methods of analyzing arches.

In this chapter, we present the essentials of statically determinate arch analysis. In this context, two categories of arch structures have been selected: (1) semicircular arches and (2) parabolic arches. We will also look at two types of loading: concentrated loads and uniformly distributed loads. The purpose of each analysis is to evaluate and draw the diagrams of support reactions and internal actions.

9.1. Introduction

Arch structures are used to reduce the bending effect and their role is absolutely identical to that of reversed cables. They are highly resistant to bending and shear stress and their geometric concavity plays an important role in reducing the effect of external actions.

Arch structures come in many forms and their classification depends on their connections to the external environment. They are bi-fixed, simply supported or fixed with hinges and can be equipped with ties (Figure 9.1).

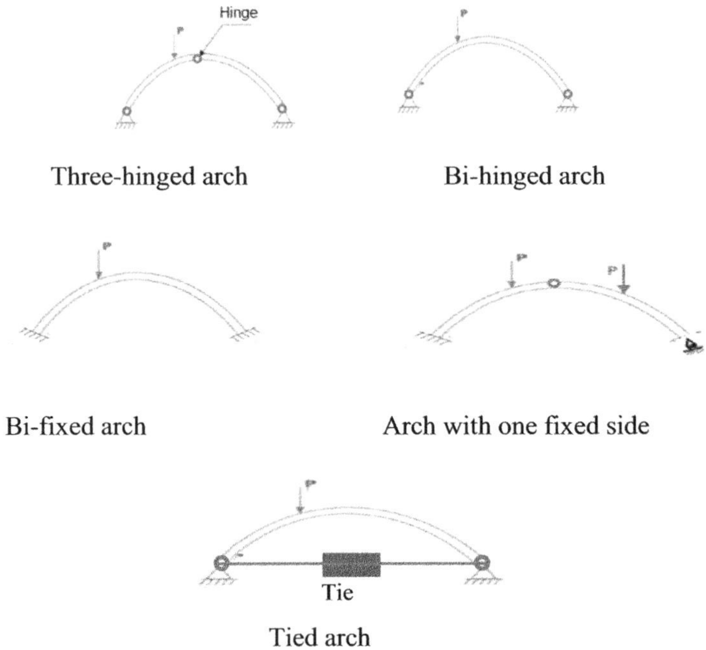

Figure 9.1. *Types of arch structure*[1]

Bi-fixed arches are highly rigid and therefore require less material. They are three times statically indeterminate, while bi-hinged arches are once statically indeterminate. To analyze this type of arch, it is necessary to add a joint to remove the degree of static indeterminacy.

9.2. Three-hinged arch

The three hinges on the arch (Figure 9.2) are found at the top and at the supports. In order to calculate the support reactions, the structure is separated into two parts (Figure 9.3).

The given problem has six unknowns V_A, H_A, V_B, H_B, N_C and T_C. The equations provided are:

$$\sum \overline{F} = \overline{0} \qquad \text{(2 equations)} \tag{9.1}$$

1 All of the figures in this chapter are available to view in full color at www.iste.co.uk/khalfallah/analysis1.zip.

$$\sum \vec{M}_A = \vec{0} \quad \text{(1 equation)} \tag{9.2}$$

$$\sum \vec{M}_B = \vec{0} \quad \text{(1 equation)} \tag{9.3}$$

$$\sum \vec{M}_C = \vec{0} \quad \text{(2 equations)} \tag{9.4}$$

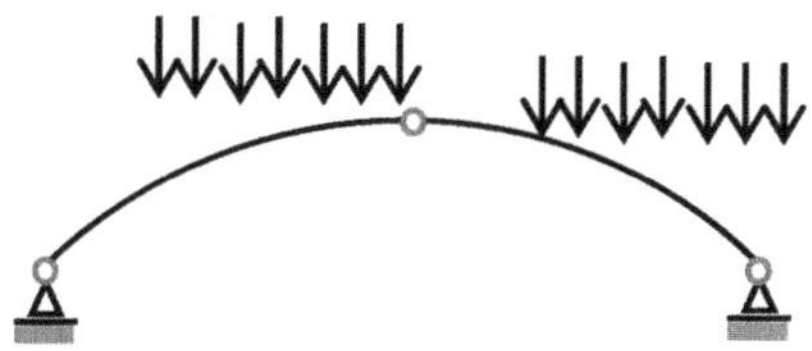

Figure 9.2. *Three-hinged arch*

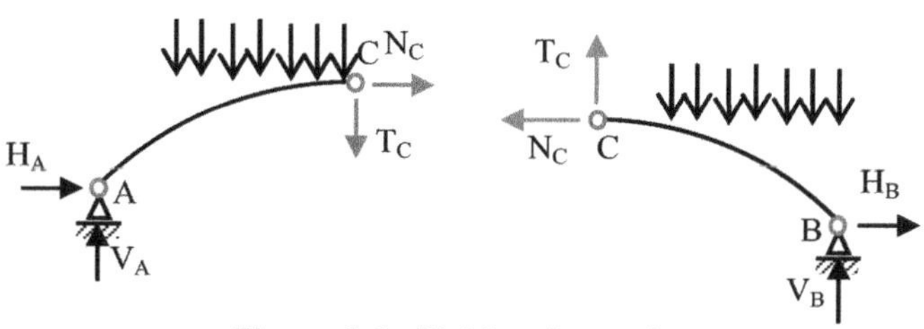

Figure 9.3. *Dividing the arch*

We can construct six equations with six unknowns, therefore the problem is algebraically solvable. Determining the support reactions then leads to calculating the internal actions: the bending moment, the shear force and the normal force.

9.2.1. *Semicircular arch under concentrated load*

We analyze the semicircular arch, as shown in Figure 9.4.

The arch structure is associated with reference (Axy) (Figure 9.5) and then the static equations [9-1]–[9.4] are applied.

$$\sum \vec{M}_B = \vec{0} : V_A.2R - PR(1+\cos\theta) = 0$$

$$V_A = \frac{P}{2}(1+\cos\theta)$$

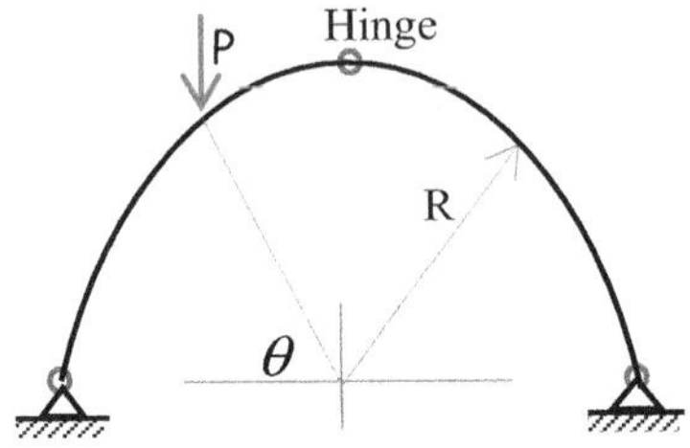

Figure 9.4. *Given arch*

$$\sum \vec{F}_y = \vec{0} : \; V_A + V_B = P$$

$$V_B = \frac{P}{2}(1-\cos\theta)$$

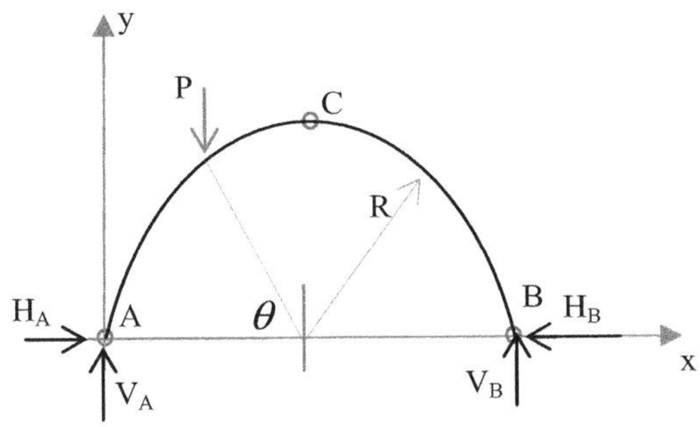

Figure 9.5. *Free-body arch*

$$\sum \vec{M}_C = \vec{0} : \; V_A R(1-\cos\theta) - H_A . R\sin\theta = 0$$

$$H_A = \frac{P}{2}(1-\cos\theta)$$

$$\sum \vec{F}_x = \vec{0} : \; H_A + H_B = 0$$

$$H_B = -H_A = -\frac{P}{2}(1-\cos\theta)$$

The variations of the support reactions depending on the force's application point are shown in Figures 9.6 and 9.7.

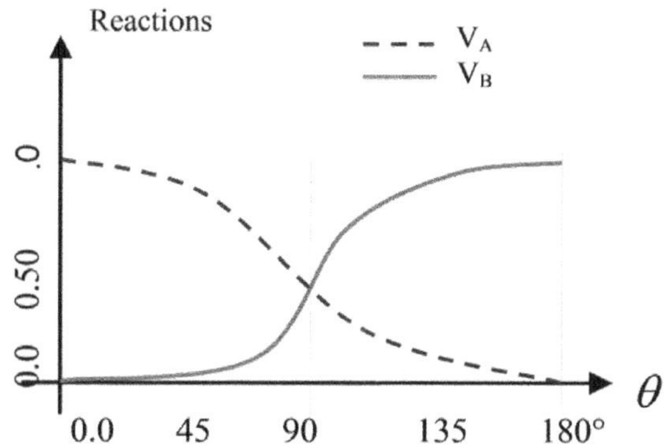

Figure 9.6. *Variation of V_A and V_B*

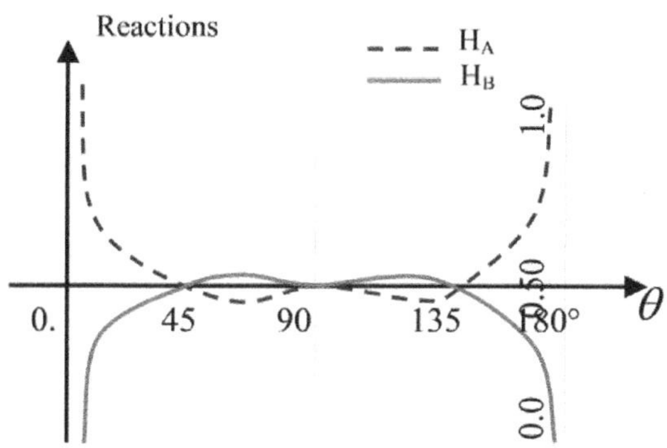

Figure 9.7. *Variation of H_A and H_B*

EXAMPLE 9.1.–

Calculate the internal actions of the arch (9.4).

We divide the part (AD) by section θ_1 (Figure 9.8).

$$\theta_1 \leq \theta$$

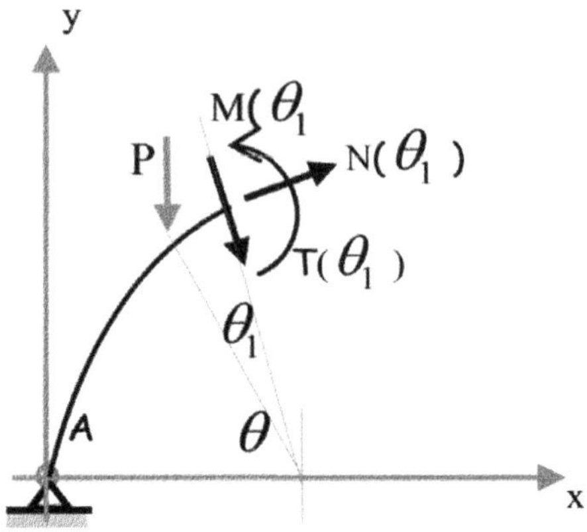

Figure 9.8. *Section* $\theta_1 \leq \theta$

The equilibrium of the right side in relation to the force (Figure 9.9) is:

$$T(\theta_1) = V_A \sin\theta_1 - H_A \cos\theta_1 = \frac{P}{2}(1+\cos\theta)\sin\theta_1 - \frac{P}{2}(1-\cos\theta)\cos\theta_1$$

$$N(\theta_1) = -(H_A \sin\theta_1 + V_A \cos\theta_1) = -\left[\frac{P}{2}(1-\cos\theta)\sin\theta_1 + \frac{P}{2}(1+\cos\theta)\cos\theta_1\right]$$

$$M(\theta_1) = V_A\, R.(1-\cos\theta_1) - H_A R\sin\theta_1 = \frac{PR}{2}(1+\cos\theta)(1-\cos\theta_1) - \frac{PR}{2}(1-\cos\theta)\sin\theta_1$$

We divide the arch at $\theta \leq \theta_1 \leq \frac{\pi}{2}$ (Figure 9.9).

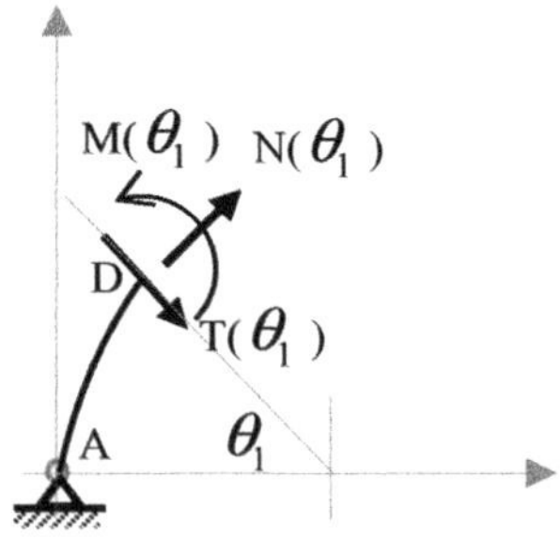

Figure 9.9. *Section* $\theta \leq \theta_1 \leq \frac{\pi}{2}$

The equilibrium of the element (Figure 9.9) allows us to write:

$$\mathrm{T}(\theta_1) = V_A \sin\theta_1 - H_A \cos\theta_1 - P\sin\theta_1 =$$
$$\frac{P}{2}(1+\cos\theta)\sin\theta_1 - \frac{P}{2}(1-\cos\theta)\cos\theta_1 - P\sin\theta_1$$

$$\mathrm{N}(\theta_1) = -(\mathrm{H}_A \sin\theta_1 + V_A \cos\theta_1) + \mathrm{P}\cos\theta =$$
$$-(\frac{P}{2}(1-\cos\theta)\sin\theta_1 + \frac{P}{2}(1+\cos\theta)\cos\theta_1) + \mathrm{P}\cos\theta_1$$

$$\mathrm{M}(\theta_1) = V_A\, \mathrm{R}.(1-\cos\theta_1) - H_A\, \mathrm{R}\sin\theta_1 - PR(\cos\theta_1 - \cos\theta) =$$
$$\frac{PR}{2}(1+\cos\theta)(1-\cos\theta_1) - \frac{PR}{2}(1-\cos\theta)\sin\theta_1 - PR(\cos\theta_1 - \cos\theta)$$

Finally, for $\frac{\pi}{2} \le \theta_1 \le \pi$ (Figure 9.10), we can write

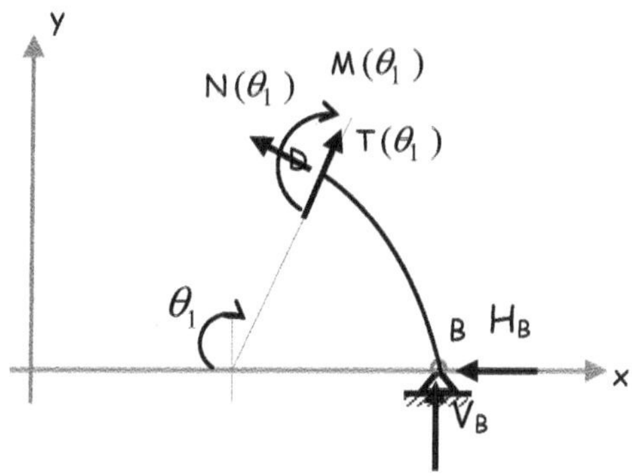

Figure 9.10. *Section* $\frac{\pi}{2} \le \theta_1 \le \pi$

Internal forces can be deduced using equilibrium equations of element (AD) (Figure 9.10).

$$\mathrm{T}(\theta_1) = H_B \sin(\theta_1 - \frac{\pi}{2}) - V_B \cos(\theta_1 - \frac{\pi}{2}) =$$
$$-\frac{P}{2}(1-\cos\theta)\cos\theta_1 - \frac{P}{2}(1-\cos\theta)\sin\theta_1$$

$$N(\theta_1) = -H_B \cos(\theta_1 - \frac{\pi}{2}) - V_B (\sin\theta_1 - \frac{\pi}{2}) =$$

$$-\frac{P}{2}(1-\cos\theta)\sin\theta_1 + \frac{P}{2}(1-\cos\theta)\cos\theta_1$$

$$M(\theta_1) = V_B R.(1-\cos(\pi-\theta_1)) - H_B R\sin(\pi-\theta_1) =$$

$$\frac{PR}{2}(1-\cos\theta)(1+\cos\theta_1) - \frac{PR}{2}(1-\cos\theta)\sin\theta_1$$

To simplify the calculations, $\theta = \frac{\pi}{4}$. The support reactions become:

$$V_A = 0.853P$$

$$V_B = 0.147P$$

$$H_A = -H_B = 0.146P$$

$$\theta_1 \leq \frac{\pi}{4}$$

$$T(\theta_1) = 0.853P\sin\theta_1 - 0.146P\cos\theta_1$$

$$T(0) = -0.146P$$

$$T(\frac{\pi}{4}) = 0.50P$$

$$N(\theta_1) = -(0.146P.\sin\theta_1 + 0.853P.\cos\theta_1)$$

$$N(0) = -0.853P$$

$$N(\frac{\pi}{4}) = -0.706P$$

$$M(\theta_1) = 0.853PR(1-\cos\theta_1) - 0.146PR\sin\theta_1$$

$$M(0) = 0.$$

$$M(\frac{\pi}{4}) = 0.147PR$$

$$\frac{\pi}{4} \le \theta_1 \le \frac{\pi}{2}$$

$$T(\theta_1) = 0.853P\sin\theta_1 - 0.146P\cos\theta_1 - P\sin\theta_1$$

$$T(\frac{\pi}{4}) = -0.207P$$

$$T(\frac{\pi}{2}) = -0.147P$$

$$N(\theta_1) = -(0.146P\sin\theta_1 + 0.146P\cos\theta_1) + \mathrm{P}\cos\theta_1$$

$$N(\frac{\pi}{4}) = 0.50P$$

$$N(\frac{\pi}{2}) = -0.146P$$

$$M(\theta_1) = 0.853PR(1-\cos\theta_1) - 0.146PR\sin\theta_1 + PR(\cos\theta_1 - \cos\theta)$$

$$M(\frac{\pi}{4}) = 0.146PR$$

$$M(\frac{\pi}{2}) = 0.0$$

$$\frac{\pi}{2} \le \theta_1 \le \pi$$

$$T(\theta_1) = 0.147P\cos\theta_1 - 0.146P\sin\theta_1$$

$$T(\frac{\pi}{2}) = -0.103P$$

$$T(\pi) = -0.147P$$

$$N(\theta_1) = -0.146P\sin\theta_1 + 0.146P\cos\theta_1$$

$$N(\frac{\pi}{2}) = -0.146P$$

$$N(\pi) = -0.146P$$

$$M(\theta_1) = 0.146PR(1+\cos\theta_1) - 0.146PR\sin\theta_1$$

$$M(\frac{\pi}{2}) = 0.0$$

$$M(\pi) = 0.0$$

The diagrams of internal actions are given by Figure 9.11.

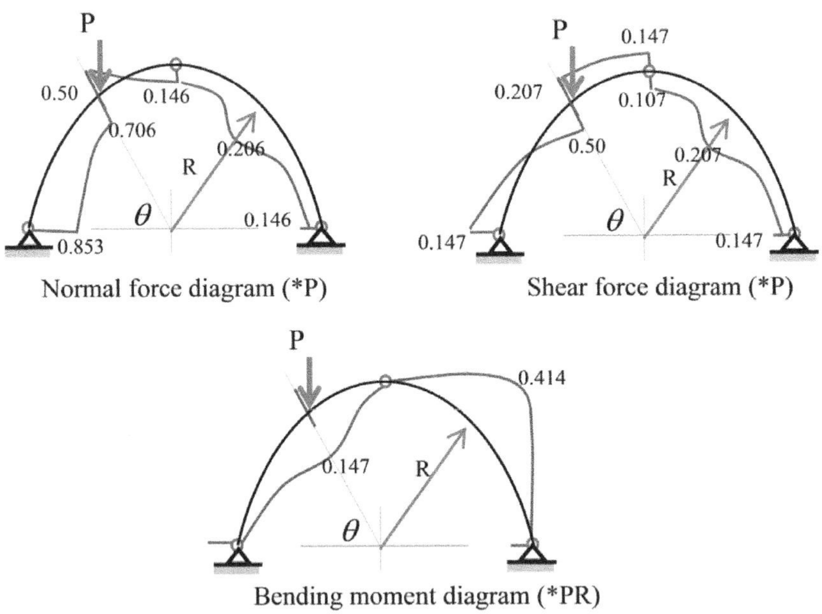

Figure 9.11. *Diagrams of internal actions*

9.2.2. *Semicircular arch under uniformly distributed load*

We analyze the arch given in Figure 9.12.

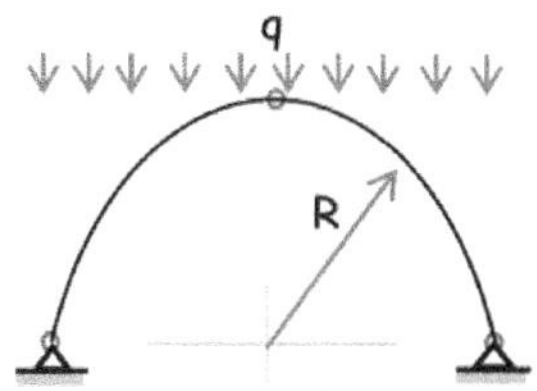

Figure 9.12. *Given arch*

The symmetry of the arch makes it possible to calculate the vertical reactions (Figure 9.13).

$$V_A = V_B = qR$$

The sum of the moments relating to point C is written as:

$$\sum \overrightarrow{M}_C = \vec{0}$$

$$V_A R - H_A R - \frac{1}{2} q R^2 = 0$$

$$H_A = \frac{qR}{2}$$

$$H_B = -H_A = -\frac{qR}{2}$$

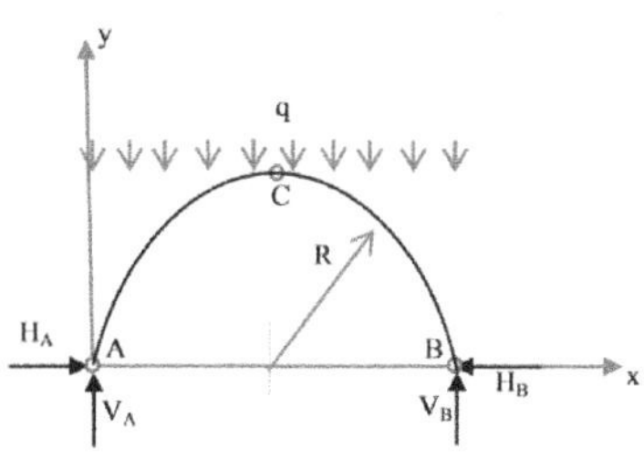

Figure 9.13. *Free-body arch*

Calculation of internal actions

We cut the section (AC) at an angle θ (Figure 9.14).

$\theta \leq \frac{\pi}{2}$

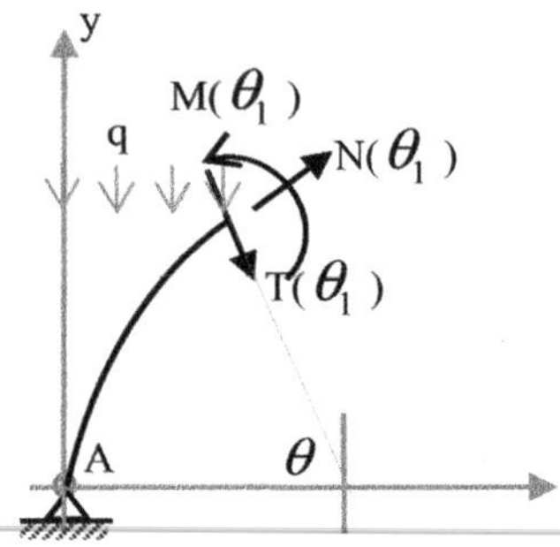

Figure 9.14. *Section* $\theta \leq \frac{\pi}{2}$

The equilibrium of section (Figure 9.14) allows us to write:

$$\mathrm{N}(\theta) = -V_A \cos\theta - H_A \sin\theta + \int_0^{R\theta} q \cos\theta.dR = -qR\cos\theta - \frac{1}{2}qR\sin\theta + qR\theta.\cos\theta$$

$$\theta = 0 \quad \mathrm{N}(0) = -qR$$

$$\theta = \frac{\pi}{2} \quad \mathrm{N}(\frac{\pi}{2}) = -\frac{1}{2}qR$$

$$\mathrm{T}(\theta) = V_A \sin\theta - H_A \cos\theta - \int_0^{R\theta} q \sin\theta.dR = -qR\sin\theta - \frac{1}{2}qR\cos\theta + qR\theta.\sin\theta$$

$$\theta = 0 \quad \mathrm{T}(0) = -\frac{1}{2}qR$$

$$\theta = \frac{\pi}{2} \quad \mathrm{T}(\frac{\pi}{2}) = \frac{1}{2}qR(\pi - 2)$$

$$M(\theta) = V_A\,R.(1-\cos\theta) - H_A\,R\sin\theta - \frac{qR^2}{2}(1-\cos\theta)^2 =$$

$$qR^2.(1-\cos\theta) - \frac{1}{2}qR^2\sin\theta - \frac{qR^2}{2}(1-\cos\theta)^2$$

$$\theta = 0 \quad M(0) = 0.$$

$$\theta = \frac{\pi}{4} \quad M(\frac{\pi}{4}) = -0.104\,qR^2$$

$$\theta = \frac{\pi}{2} \quad M(\frac{\pi}{2}) = 0.$$

The symmetry of the arch makes it possible to draw the internal actions diagrams (Figure 9.13).

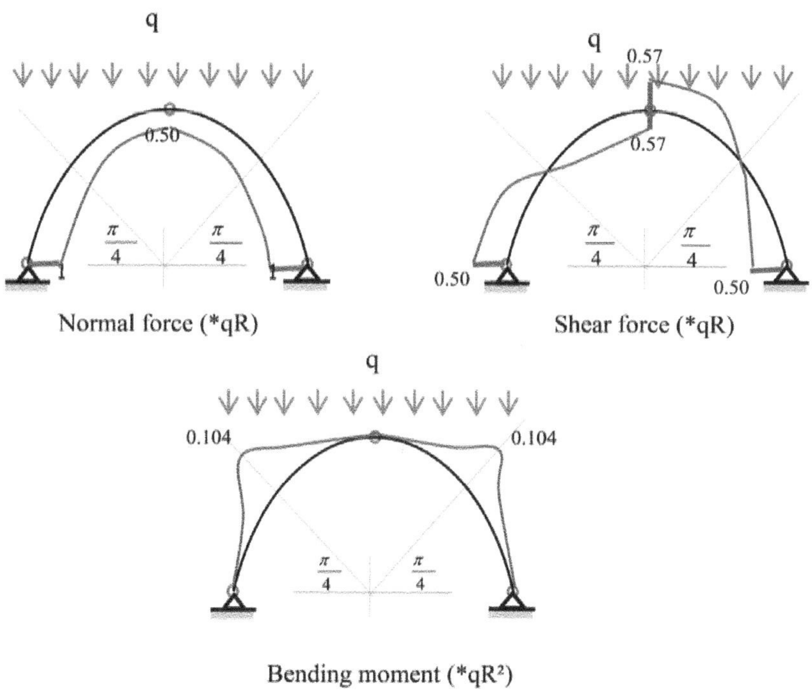

Figure 9.15. *Diagrams of internal actions*

9.2.3. *Parabolic arch under concentrated load*

The arch (Figure 9.16) has a curve whose equation is defined by $y = 2\frac{h}{L}x - \frac{h}{L^2}x^2$

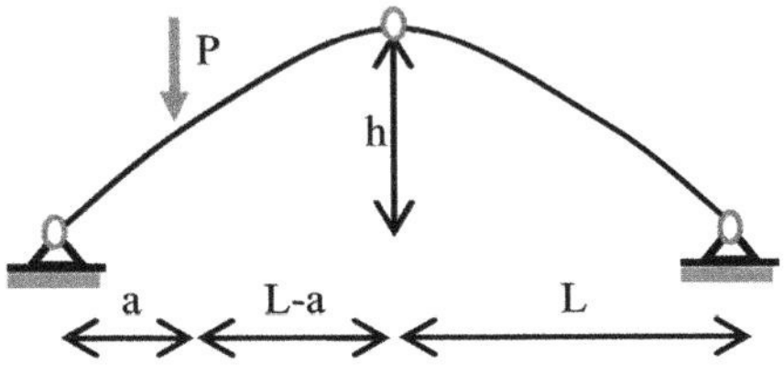

Figure 9.16. *Parabolic arch*

Consider the sum of the moments relating to point B (Figure 9.17).

The vertical reaction in A is:

$$V_A = P(1 - \frac{a}{2L})$$

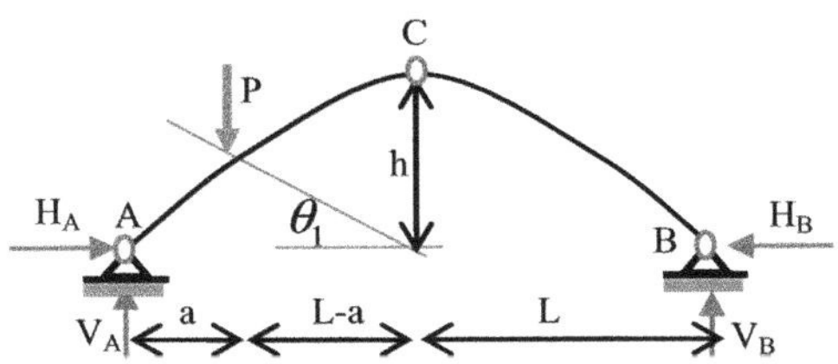

Figure 9.17. *Support reactions*

$$V_B = P - V_A = P\frac{a}{2L}$$

$$\sum \overline{M}_C = \vec{0} : \ V_A.L - H_A.h - P(L-a) = 0$$

$$H_A = \frac{Pa}{2h}$$

The horizontal reaction is:

$$H_B = -H_A = -\frac{Pa}{2h}$$

The internal actions expressions (Figure 9.18) are given as:

$0 \le x \le a$

$$N(\mathrm{x}) = -V_A \cos\theta - H_A \sin\theta = -P(1-\frac{a}{2L})\cos\theta - \frac{1}{2}\frac{Pa}{h}\sin\theta$$

$$N(0) = -P(1-\frac{a}{2L})$$

$$N(\mathrm{a}) = -P(1-\frac{a}{2L})\cos\theta_1 - \frac{1}{2}\frac{Pa}{h}\sin\theta_1$$

Knowing that $tg\theta_1 = \frac{a}{L^2}\frac{2h-\frac{ha}{L}}{1-\frac{a}{L}}$

$$M(x) = V_A x - H_A y = P(1-\frac{3a}{2L})x + \frac{Pa}{2L^2}x^2$$

$$M(0) = 0.$$

$$M(a) = Pa(1-\frac{3}{2}\frac{a}{L}+\frac{1}{2}\frac{a^2}{L^2})$$

$$\mathrm{T}(x) = \frac{dM(x)}{dx} = P(1-\frac{3a}{2L}) + \frac{Pa}{L^2}x$$

$$\mathrm{T}(0) = P(1-\frac{3a}{2L})$$

$$\mathrm{T}(\mathrm{a}) = P(1-\frac{3a}{2L}) + \frac{Pa}{L^2}a = P(1-\frac{3a}{2L}+\frac{a^2}{L^2})$$

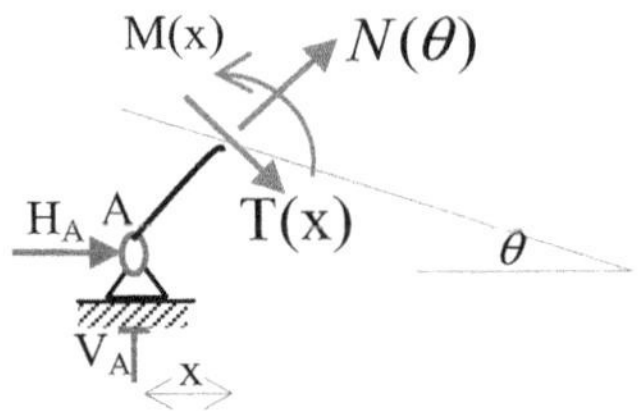

Figure 9.18. *Section* $0 \leq x \leq a$

$a \leq x \leq L$ (Figure 9.19)

$$N(\theta) = -V_A \cos\theta - H_A \sin\theta + P\sin\theta = -P(1-\frac{a}{2L})\cos\theta - \frac{1}{2}\frac{Pa}{h}\sin\theta + P\sin\theta$$

$$N(\theta_1) = -P(1-\frac{a}{2L})\cos\theta_1 - \frac{1}{2}\frac{Pa}{h}\sin\theta_1$$

$$N(\frac{\pi}{2}) = P(1-\frac{a}{2h})$$

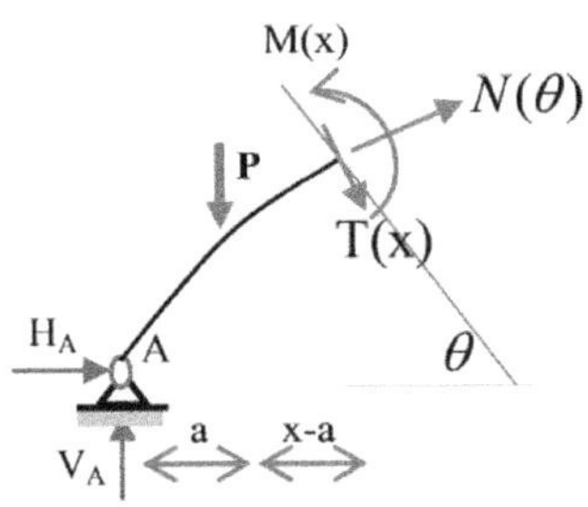

Figure 9.19. *Section* $a \leq x \leq L$

$$M(x) = V_A x - H_A y - P(x-a) = Pa\left[1 - \frac{3x}{2L} + \frac{x^2}{2L^2}\right]$$

$$M(x=a) = Pa(1-\frac{3}{2}\frac{a}{L}+\frac{1}{2}\frac{a^2}{L^2})$$

$$M(x=L) = 0.$$

$$T(x) = \frac{dM(x)}{dx} = Pa\left[-\frac{3}{2L} + \frac{x}{L^2}\right]$$

$$T(x = a) = Pa(-\frac{3}{2L} + \frac{a}{L^2})$$

$$T(x = L) = -\frac{Pa}{2L}$$

$L \le x \le 2L$ (Figure 9.20)

$$N(\theta) = -V_B \cos(\pi - \theta) - H_B \sin(\pi - \theta) = P\frac{a}{2L}\cos\theta - \frac{1}{2}\frac{Pa}{h}\sin\theta$$

$$N(\frac{\pi}{2}) = -\frac{1}{2}\frac{Pa}{h}$$

$$N(\pi) = -P\frac{a}{2L}$$

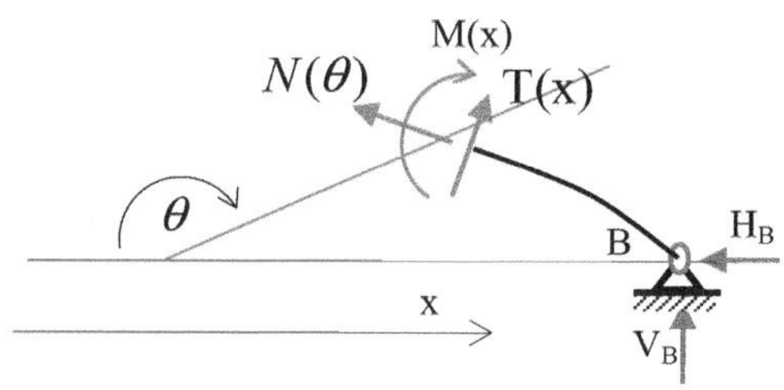

Figure 9.20. *Section* $L \le x \le 2L$

$$M(x) = V_B(2L - x) - H_A y(x) = Pa(1 - \frac{3x}{2L} + \frac{x^2}{2L^2})$$

$$M(x = L) = 0.$$

$$M(x = 2L) = 0.$$

$$T(x) = \frac{dM(x)}{dx} = Pa\left[-\frac{3}{2L} + \frac{x}{L^2}\right]$$

$$\mathrm{T}(x=L)=-\frac{Pa}{2L}$$

$$\mathrm{T}(x=2L)=\frac{Pa}{2L}$$

Numerical application

Either $a=\frac{L}{2}$ or $h=\frac{L}{10}$, the support reactions become

$$V_A=\frac{3}{4}P$$

$$V_B=\frac{1}{4}P$$

$$H_A=-H_B=\frac{5}{2}P$$

$$0\leq x\leq\frac{L}{2}$$

$$N(\mathrm{x})=-\frac{3}{4}P\cos\theta-\frac{5}{2}P\sin\theta$$

Knowing that $tg\theta_1=\frac{3}{20}$, $\theta_1=8.53°$, $\sin\theta_1=0.148$ and $\cos\theta_1=0.988$

$$N(0)=-\frac{3}{4}P$$

$$N(\frac{L}{2})=-1.11P$$

$$M(x)=\frac{1}{4}Px+\frac{1}{4}P\frac{x^2}{L}$$

$$M(0)=0.$$

$$M(\frac{L}{2}) = \frac{3}{16}PL$$

$$\text{T}(x) = \frac{dM(x)}{dx} = \frac{1}{4}P + \frac{1}{2}P\frac{x}{L}$$

$$\text{T}(0) = \frac{1}{4}P$$

$$\text{T}(\frac{L}{2}) = \frac{1}{2}P$$

$$\frac{L}{2} \leq x \leq L$$

$$N(\theta) = -\frac{3}{4}P\cos\theta - \frac{5}{2}P\sin\theta + P\sin\theta$$

$$N(\theta_1) = -0.962P$$

$$N(\frac{\pi}{2}) = -1.5P$$

$$M(x) = -\frac{3}{4}Px + \frac{1}{4}P\frac{x^2}{L} + \frac{PL}{2}$$

$$M(x = \frac{L}{2}) = \frac{3}{16}PL$$

$$M(x = L) = 0.$$

$$\text{T}(x) = \frac{dM(x)}{dx} = -\frac{3}{4}P + \frac{1}{2}\frac{Px}{L}$$

$$\text{T}(x = \frac{L}{2}) = -\frac{1}{2}P$$

$$\text{T}(x = L) = -\frac{P}{4}$$

$L \leq x \leq 2L$

$$N(\theta) = V_B \cos\theta - H_B \sin\theta = -0.123P$$

$$M(x) = V_B(2L - x) - H_A y(x) = \frac{1}{2}PL - \frac{3}{4}Px + \frac{P}{4}\frac{x^2}{L}$$

$$M(x = L) = 0.$$

$$M(x = 2L) = 0.$$

$$\mathrm{T}(x) = \frac{dM(x)}{dx} = -\frac{3}{4}P + \frac{Px}{2L}$$

$$\mathrm{T}(x = L) = -\frac{P}{4}$$

$$\mathrm{T}(x = 2L) = \frac{P}{4}$$

The diagrams of internal actions are given by Figure 9.21.

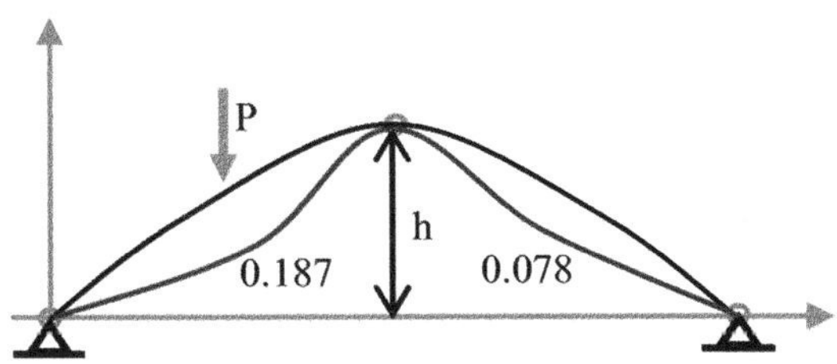

Bending moment diagram *(PL)

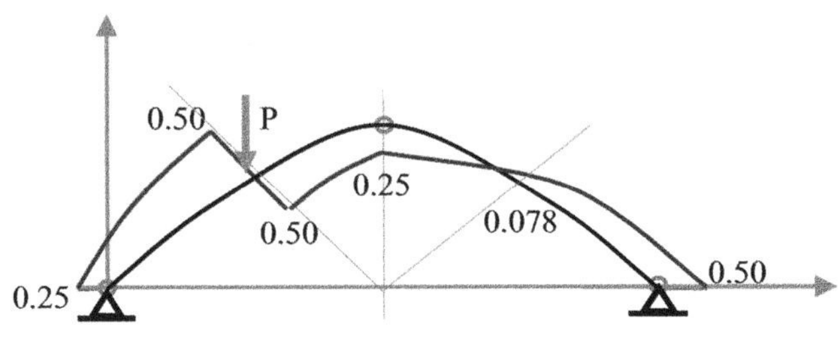

Shear force diagram *(P)

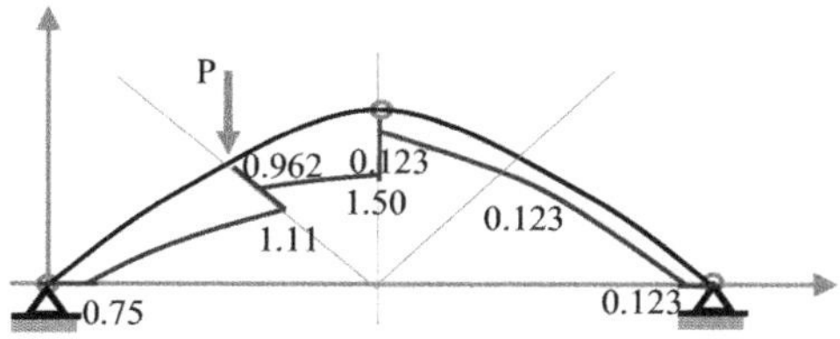

Normal force diagram *(P)

Figure 9.21. *Diagram of internal actions*

9.2.4. *Parabolic arch under uniformly distributed load*

We analyze the parabolic arch stressed by a uniformly distributed load (Figure 9.22). The curve of the arch is $y = 2\dfrac{h}{L}x - \dfrac{h}{L^2}x^2$.

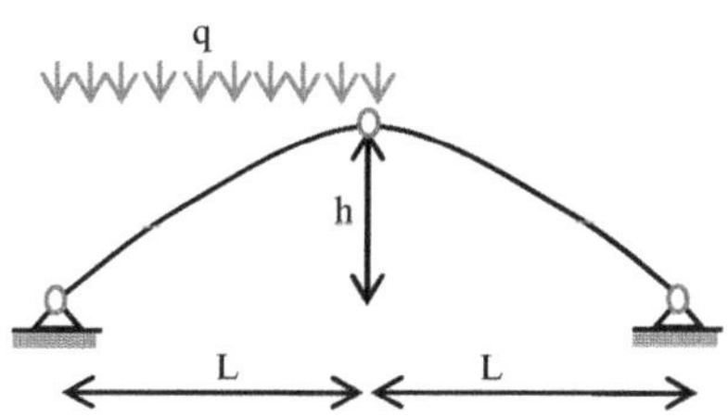

Figure 9.22. *Parabolic arch*

Support reactions can be calculated using the static equations, which are shown in Figure 9.23.

$$V_A.2L - qL(L + \frac{L}{2}) = 0 : V_A = \frac{3}{4}qL$$

$$V_B = qL - \frac{3}{4}qL = \frac{1}{4}qL$$

$$\sum \overrightarrow{M}_C = \vec{0} : \text{V}_A.L - H_A.h - qL.\frac{L}{2} = 0$$

$$H_A = -H_B = \frac{qL^2}{4h}$$

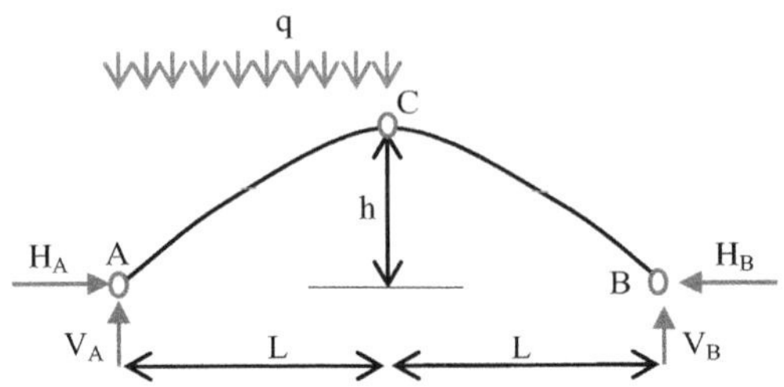

Figure 9.23. *Free-body arch*

The internal action expressions are written as (Figure 9.24):

$$0 \le x \le L$$

$$N(\mathrm{x}) = -V_A \cos\theta - H_A \sin\theta = -\frac{3}{4}qL\cos\theta - \frac{1}{4}\frac{qL^2}{h}\sin\theta$$

$$N(0) = -\frac{3}{4}qL$$

$$N(\frac{\pi}{2}) = -\frac{1}{4}\frac{qL^2}{h}$$

$$M(x) = V_A x - H_A y(x) - \frac{qx^2}{2} = \frac{1}{4}qLx - \frac{1}{4}qx^2$$

$$M(0) = 0$$

$$M(\frac{L}{2}) = \frac{1}{16}qL^2$$

$$M(L) = 0$$

$$\mathrm{T}(x) = \frac{dM(x)}{dx} = \frac{1}{4}qL - \frac{1}{2}qx$$

$$\mathrm{T}(0) = \frac{1}{4}qL$$

$$\mathrm{T(L)} = -\frac{1}{4}qL$$

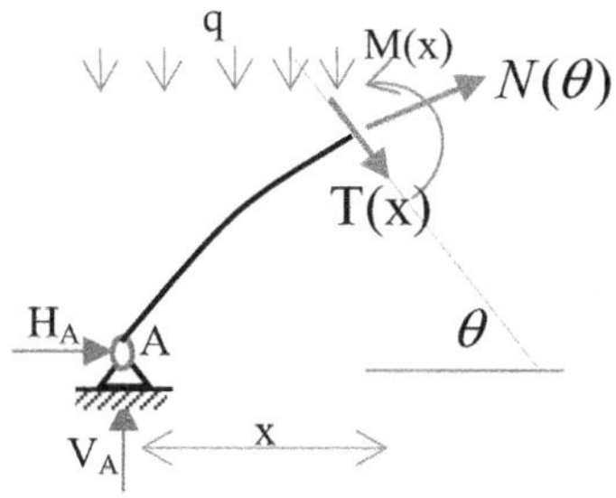

Figure 9.24. *Section* $0 \le x \le L$

$L \le x \le 2L$ (Figure 9.25)

$$N(\theta) = -V_B \cos(\pi - \theta) - H_B \sin(\pi - \theta) = \frac{1}{4} qL \cos\theta - \frac{1}{4}\frac{qL^2}{h} \sin\theta$$

$$N(\frac{\pi}{2}) = -\frac{1}{4}\frac{qL^2}{h}$$

$$N(\pi) = -\frac{1}{4} qL$$

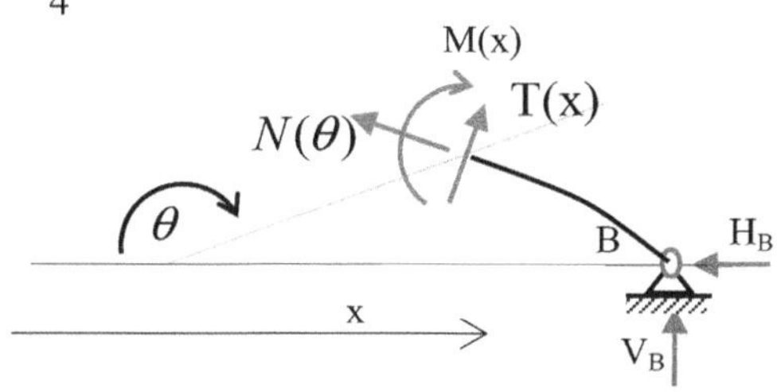

Figure 9.25. *Section* $L \le x \le 2L$

$$M(x) = V_B (2L - x) - H_A y(x) = \frac{1}{4} qL^2 (2 - 3\frac{x}{L} + \frac{x^2}{L^2})$$

$$M(x = L) = 0$$

$$M(\frac{3L}{2}) = -\frac{1}{16} qL^2$$

$$M(x = 2L) = 0$$

$$T(x) = \frac{dM(x)}{dx} = \frac{1}{4}qL^2(-\frac{3}{L} + \frac{2x}{L^2})$$

$$T(x = L) = -\frac{1}{4}qL$$

$$T(x = 2L) = \frac{1}{4}qL$$

Numerical application for $h = \frac{L}{10}$

$$0 \le x \le L$$

$$N(\mathrm{x}) = -\frac{3}{4}qL\cos\theta - \frac{5}{2}qL\sin\theta$$

$$N(0) = -\frac{3}{4}qL$$

$$N(\frac{\pi}{2}) = -\frac{5}{2}qL$$

$$M(x) = V_A x - H_A y(x) - \frac{qx^2}{2} = \frac{1}{4}qLx - \frac{1}{4}qx^2$$

$$M(0) = 0.$$

$$M(\frac{L}{2}) = \frac{1}{16}qL^2$$

$$M(L) = 0.$$

$$T(x) = \frac{dM(x)}{dx} = \frac{1}{4}qL - \frac{1}{2}qx$$

$$T(0) = \frac{1}{4}qL$$

$$T(L) = -\frac{1}{4}qL$$

$$L \le x \le 2L$$

$$N(\theta) = \frac{1}{4}qL\cos\theta - \frac{5}{2}qL\sin\theta$$

$$N(\frac{\pi}{2}) = -\frac{5}{2}qL$$

$$N(\pi) = -\frac{1}{4}qL$$

$$M(x) = \frac{1}{4}qL^2(2 - \frac{3x}{L} + \frac{x^2}{L^2})$$

$$M(x = L) = 0.$$

$$M(\frac{3L}{2}) = -\frac{1}{16}qL^2$$

$$M(x = 2L) = 0.$$

$$\mathrm{T}(x) = \frac{dM(x)}{dx} = \frac{1}{4}qL^2(-\frac{3}{L} + \frac{2x}{L^2})$$

$$\mathrm{T}(x = L) = -\frac{1}{4}qL$$

$$\mathrm{T}(x = 2L) = \frac{1}{4}qL$$

The diagrams of internal actions are given by Figure 9.26.

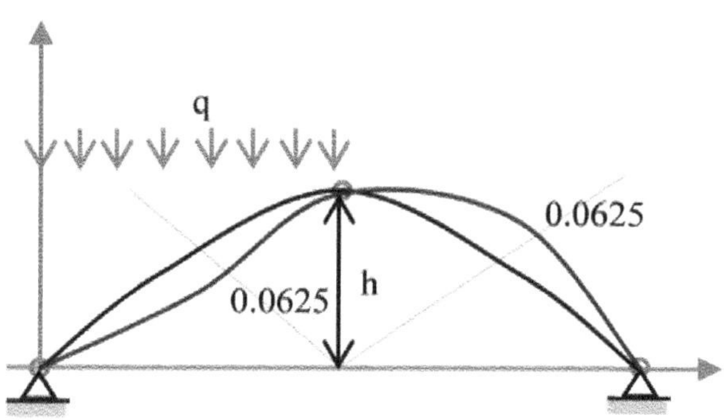

Bending moment diagram *(qL^2)

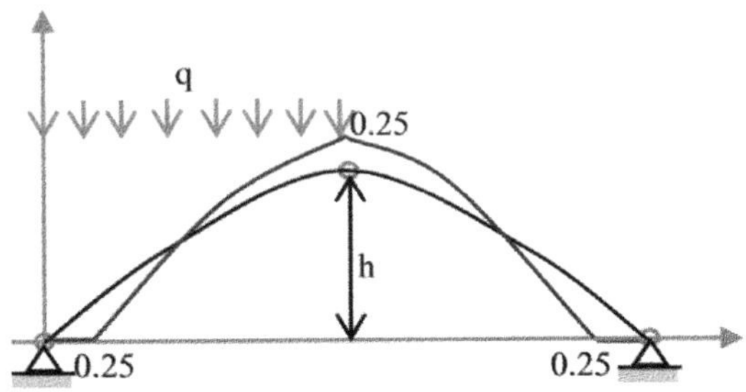

Shear force diagram *(qL)

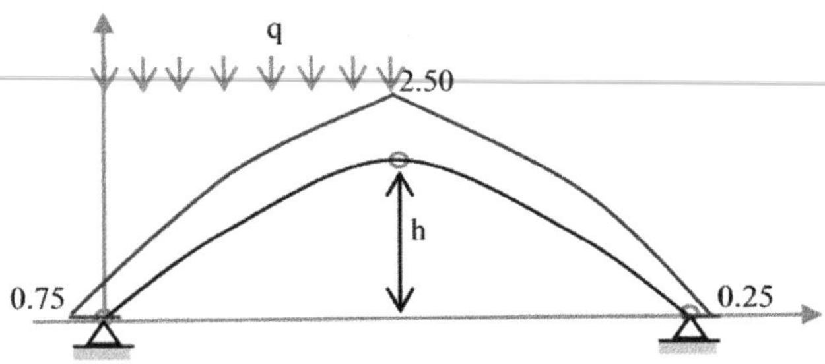

Normal force diagram *(qL)

Figure 9.26. *Diagram of internal actions*

9.2.5. *Semicircular arch with support settlements*

Figure 9.27 shows a three-hinged arch with support settlements and a bend radius R = 10 m. Draw the diagrams of internal actions.

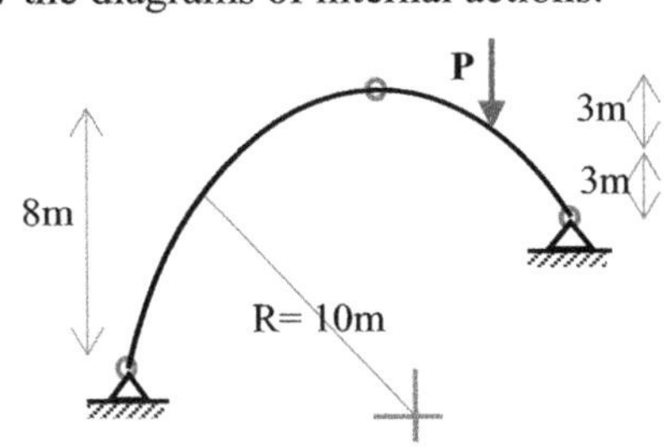

Figure 9.27. *Arch with level supports*

The support reactions are shown in Figure 9.28 and static equations are used to determine them.

The equation of the arch is $x^2 + y^2 = R^2$.

We apply static equations [9.1]–[9.4].

$$\sum \overrightarrow{M}_B = \vec{0} : V_A.(9.80 + 9.165) + \mathrm{H}_A.2 - P(9.165 - 7.14) = 0$$

Hence:

$$18.965.V_A + 2H_A = 2.025P$$

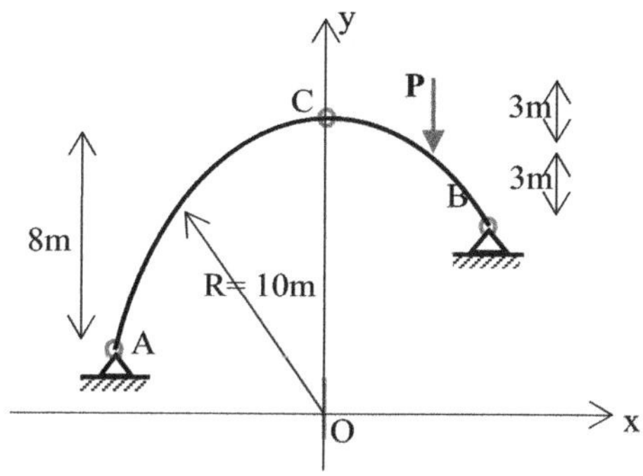

Figure 9.28. *Identifying the arch*

$$\sum \overrightarrow{M}_C = \vec{0} : 9.80.V_A - 8.H_A = 0$$

The reactions at support A are given as:

$$H_A = 0.15P$$

$$V_A = 0.123P$$

$$\sum \overrightarrow{F}_y = \vec{0}$$

$$V_A + V_B = P$$

$$V_B = 0.877P$$

$$\sum \vec{F}_x = \vec{0}$$

$$H_A + H_B = 0$$

$$H_B = -H_A = -0.15P$$

$11.53° \leq \theta \leq 90°$ (Figure 9.29)

The equilibrium of the right side (Figure 9.29) is given as

$$\text{N}(\theta) = -(\text{H}_A \sin\theta + V_A \cos\theta) = -(0.15P\sin\theta + 0.123P\cos\theta)$$

$$\text{N}(11.53) = -0.466P$$

$$\text{N}(90) = -0.15P$$

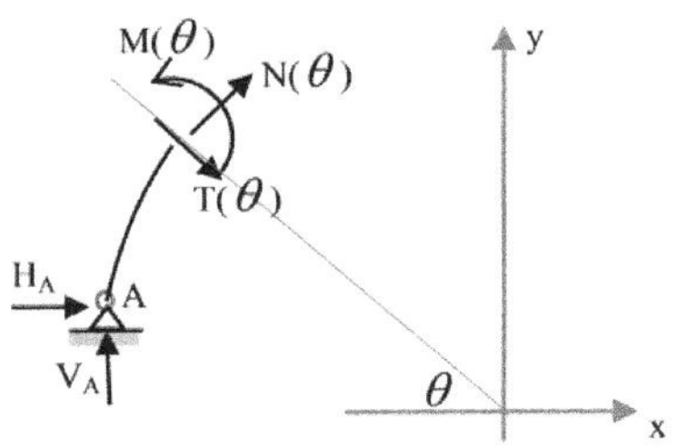

Figure 9.29. *Section* $11.53° \leq \theta \leq 90°$

$$\text{T}(\theta) = V_A \sin\theta - H_A \cos\theta = 0.123P\sin\theta - 0.15P\cos\theta$$

$$\text{T}(11.53) = -0.122P$$

$$\text{T}(90) = 0.123P$$

$$\text{M}(\theta) = V_A\,\text{R}.(1 - \cos\theta) - H_A(\text{R}\sin\theta - 2) - 1.23P(1-\cos\theta) - 0.15P(10\sin\theta - 2)$$

$$M(11.53) \approx 0$$

$$M(56.63) \approx -0.40P$$

$$M(90) \approx 0$$

$90° \leq \theta \leq 135.56°$ (Figure 9.30)

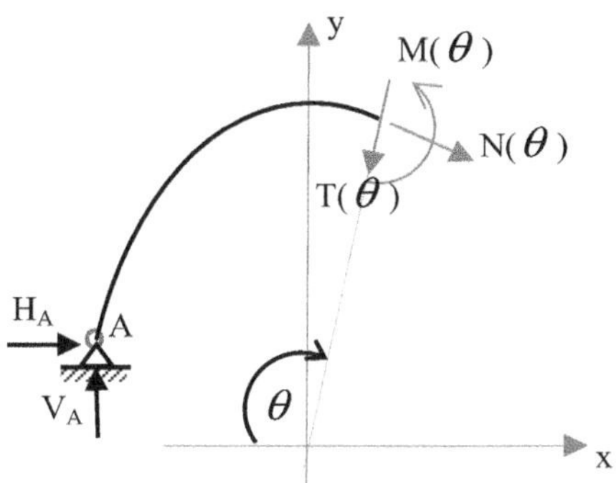

Figure 9.30. *Section* $90° \leq \theta \leq 135.56°$

The equilibrium of the considered side (Figure 9.30) is

$$N(\theta) = -H_A \sin(\pi - \theta) + V_A \cos(\pi - \theta) = -H_A \sin\theta - V_A \cos\theta = -0.15P\sin\theta - 0.123P\cos\theta$$
$$N(90) = -0.15P$$

$$N(135.56) = -0.017P$$

$$T(\theta) = H_A \cos(\pi - \theta) + V_A \sin(\pi - \theta) = -H_A \cos\theta + V_A \sin\theta = -0.15P\cos\theta + 0.123P\sin\theta$$
$$T(90) = 0.123P$$

$$T(135.56) = 0.193P$$

$$M(\theta) = V_A(9.80 + R\cos(\pi - \theta)) - H_A(R\sin(\pi - \theta) - 2) = V_A(9.80 - R\cos\theta) - H_A(R\sin\theta - 2)$$

$$M(90) \approx 0$$

$$M(135.65) \approx 1.33P$$

$135.56° \leq \theta \leq 156.42°$ (Figure 9.31)

$$N(\theta) = -H_B \sin\theta + V_B \cos\theta = -0.15P\sin\theta + 0.877P\cos\theta$$

$$N(135.56) = -0.43P$$

$$N(156.42) = -0.56P$$

$$T(\theta) = -H_B \cos\theta - V_B \sin\theta = -0.15P\cos\theta - 0.877P\sin\theta$$

$$T(135.56) = -0.207P$$

$$T(156.42) = -0.085P$$

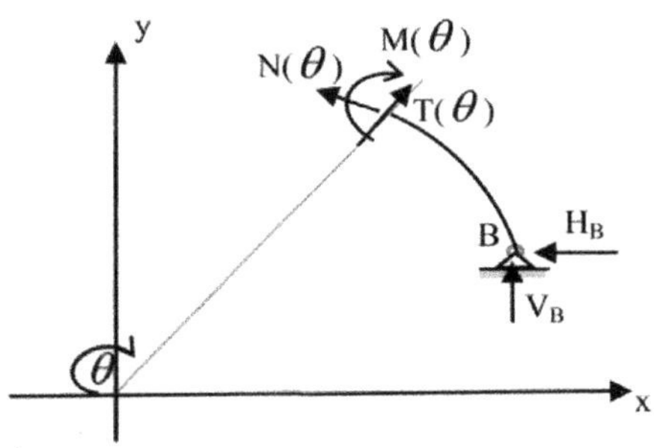

Figure 9.31. *Section* $135.56° \leq \theta \leq 156.42°$

$$M(\theta) = V_B(9.165 - R\cos(\pi - \theta)) - H_B(R\sin(\pi - \theta) - 4) = 0.877.(9.165 + 10.\cos\theta) - 0.15.(10.\sin\theta - 4)$$

$$M(135.56) = 1.33P$$

$$M(156.42) = 0.0$$

The diagrams of internal actions are given by Figure 9.32.

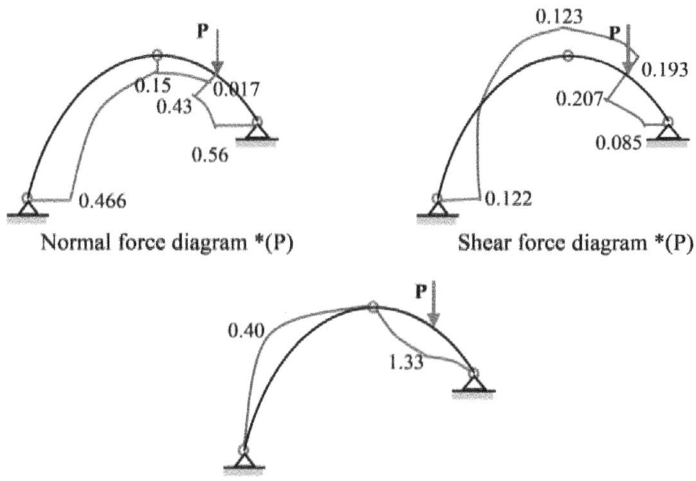

Figure 9.32. *Diagram of internal actions*

9.3. Bi-hinged arch

We analyze the bi-hinged arch stressed by concentrated load P (Figure 9.33).

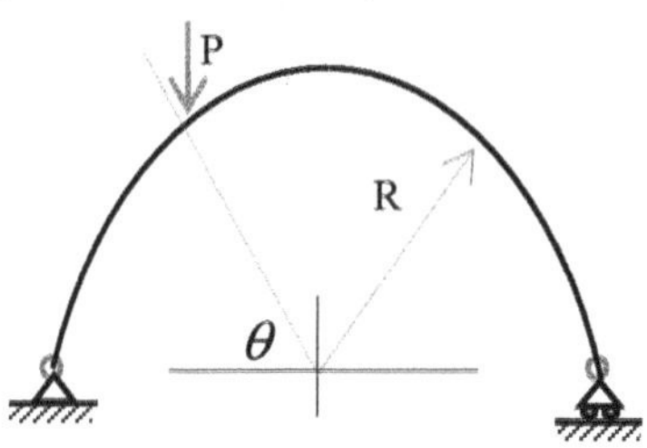

Figure 9.33. *Bi-hinged arch*

The given arch is statically determinate. Calculating the support reactions is done using the static equations (Figure 9.34).

$$\sum \overrightarrow{M}_B = \vec{0}: \; V_A.2R - PR(1+\cos\theta) = 0$$

$$V_A = \frac{P}{2}(1+\cos\theta)$$

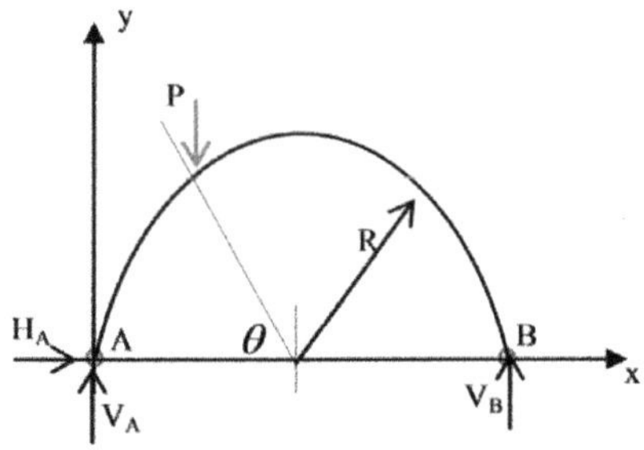

Figure 9.34. *Free-body arch*

$$\sum \vec{F}_y = \vec{0} : V_A + V_B = P$$

$$V_B = \frac{P}{2}(1 - \cos\theta)$$

$$\sum \vec{F}_x = \vec{0} : H_A = 0.0$$

$\theta_1 \leq \theta$ (Figure 9.35)

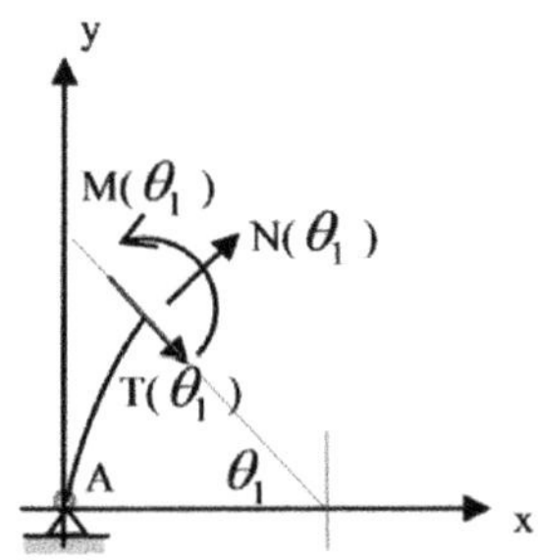

Figure 9.35. *Section* $\theta_1 \leq \theta$

$$\mathrm{T}(\theta_1) = V_A \cos\theta_1 = \frac{P}{2}(1 + \cos\theta)\cos\theta_1$$

$$\mathrm{T}(0) = V_A \cos\theta_1 = \frac{P}{2}(1 + \cos\theta)$$

$$T(\theta) = V_A \cos\theta_1 = \frac{P}{2}(1+\cos\theta)\cos\theta$$

$$N(\theta_1) = V_A \sin\theta_1 = \frac{P}{2}(1+\cos\theta)\sin\theta_1$$

$$N(0) = \frac{P}{2}(1+\cos\theta)\sin\theta_1 = 0.$$

$$N(\theta) = \frac{P}{2}(1+\cos\theta)\sin\theta$$

$$M(\theta_1) = V_A\, R.(1-\cos\theta_1) = \frac{PR}{2}(1+\cos\theta)(1-\cos\theta_1)$$

$$M(0) = 0.$$

$$M(\theta) = \frac{PR}{2}(1-\cos^2\theta)$$

$\theta \le \theta_1 \le \pi$ (Figure 9.36)

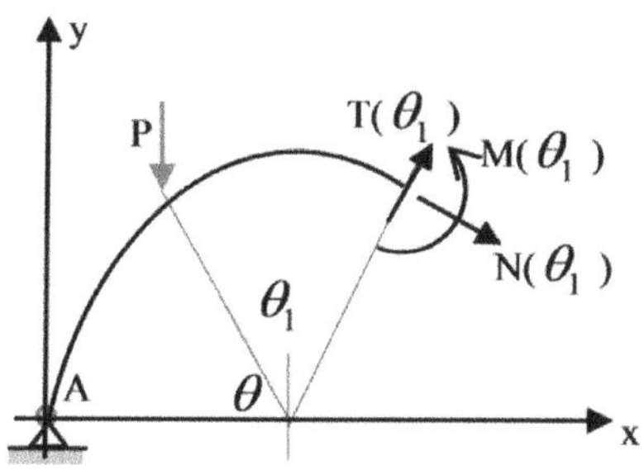

Figure 9.36. *Section* $\theta \le \theta_1 \le \pi$

$$T(\theta_1) = -V_A \cos\theta_1 - P\sin\theta_1 = \frac{P}{2}(1+\cos\theta)\cos\theta_1 - P\sin\theta_1$$

$$T(\theta) = \frac{P}{2}(1+\cos\theta)\cos\theta - P\sin\theta$$

$$\mathrm{T}(\pi) = -\frac{P}{2}(1+\cos\theta)$$

$$\mathrm{N}(\theta_1) = -V_A \cos\theta_1 - \mathrm{P}\sin\theta_1 = -\frac{P}{2}(1+\cos\theta)\cos\theta_1 - \mathrm{P}\sin\theta_1$$

$$\mathrm{N}(\theta) = -\frac{P}{2}(1+\cos\theta)\cos\theta - \mathrm{P}\sin\theta$$

$$\mathrm{N}(\pi) = \frac{P}{2}(1+\cos\theta)$$

$$\mathrm{M}(\theta_1) = V_A\,\mathrm{R}.(1-\cos\theta_1) - PR(\cos\theta - \cos\theta_1) =$$
$$\frac{PR}{2}(1+\cos\theta)(1-\cos\theta_1) - PR(\cos\theta - \cos\theta_1)$$

$$M(\theta) = \frac{PR}{2}(1-\cos^2\theta) = \frac{PR}{2}\sin^2\theta$$

$$M(\pi) = 0.$$

The internal actions diagrams are given by Figure 9.37.

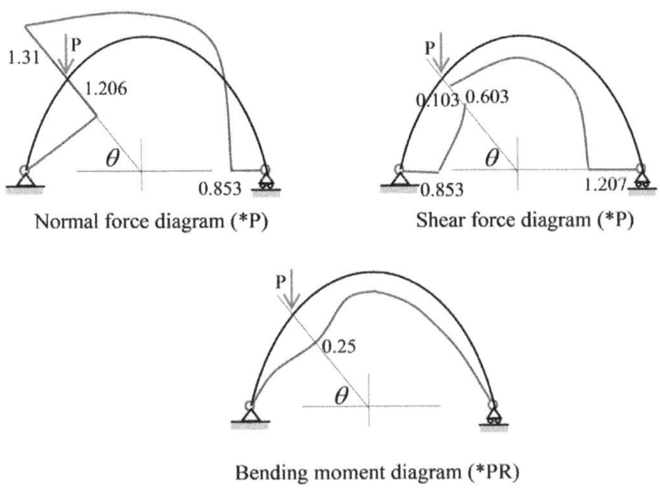

Figure 9.37. *Diagrams of internal actions*

9.4. Conclusion

In this chapter, we have presented the methodology for analyzing statically determinate arches. Three-hinged arches and two-hinged arches under a concentrated load or a uniformly distributed load were discussed in this chapter. The focus of this chapter has been on:

– a general description of arches and the distinction between different types of arches;

– the analysis of arches with three hinges and two hinges. We have looked at two types of loading: concentrated loads and uniformly distributed loads.

The internal action diagrams were constructed to distinguish the most unfavorable section and to take into account the design requirements.

Some examples have been given to show how to use the analysis method for statically determinate arches.

9.5. Problems

Exercise 1

Determine the support reactions of the following semicircular arches:

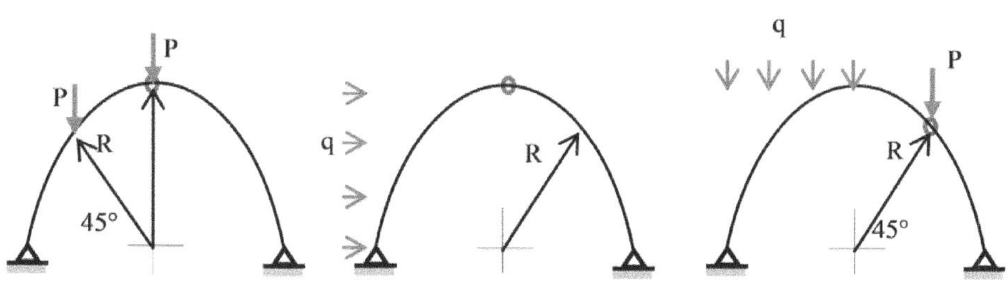

Exercise 2

Determine the support reactions of parabolic arches of equation $y(x) = \dfrac{4h}{L^2}x(L-x)$, where L and h are, respectively, the length and height of the arch.

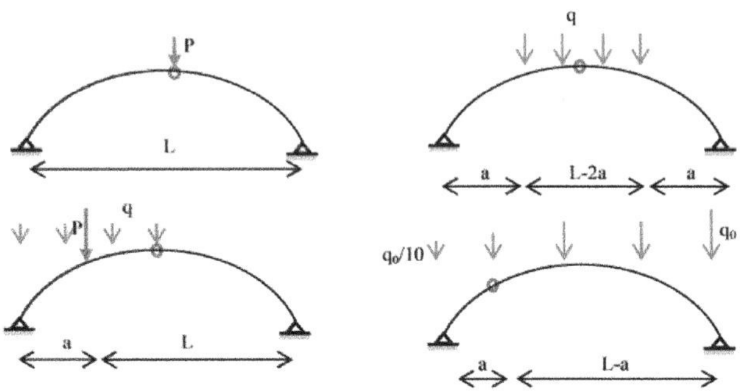

Exercise 3

Calculate the bending moment, the shear force and the normal force at the top of each arch. Parabolic arches are defined by their curve: $y(x) = \dfrac{4h}{L^2}x(L-x)$.

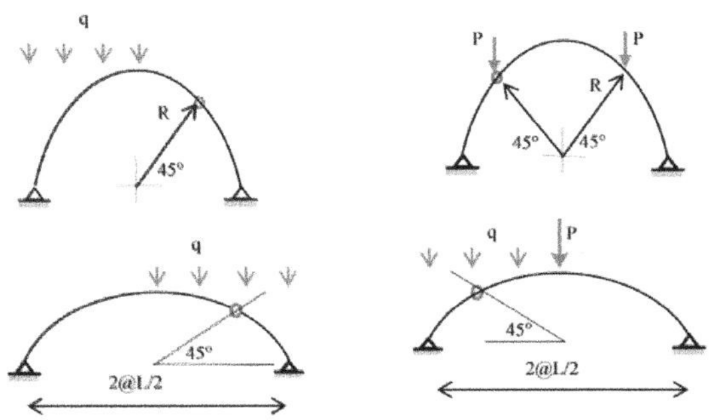

Exercise 4

Determine the variations in internal actions of the following arches.

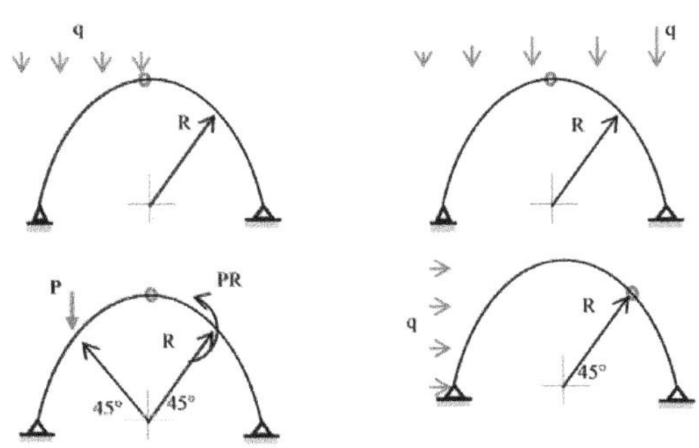

Exercise 5

Determine the internal actions of the following arches. EI is a constant and $y(x)=\dfrac{4h}{L^2}x(L-x)$.

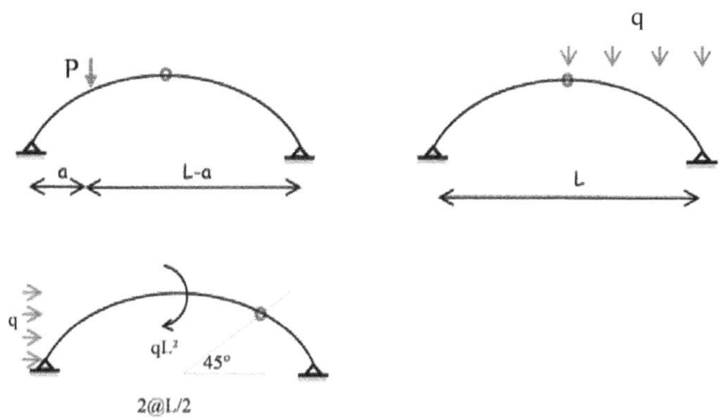

Exercise 6

Determine the internal actions and draw their diagrams. EI is a constant. To remove the static indeterminacy of the arch, use joints.

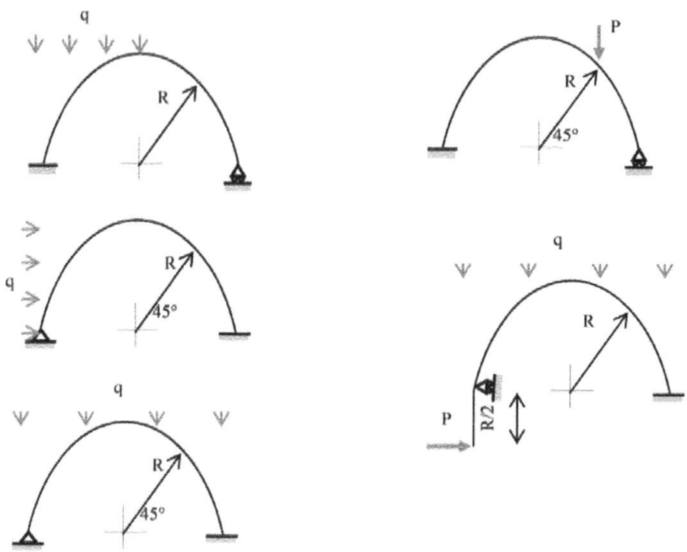

Exercise 7

Determine the internal actions and draw their diagrams. EI is a constant and $y(x) = \dfrac{4h}{L^2} x(L-x)$.

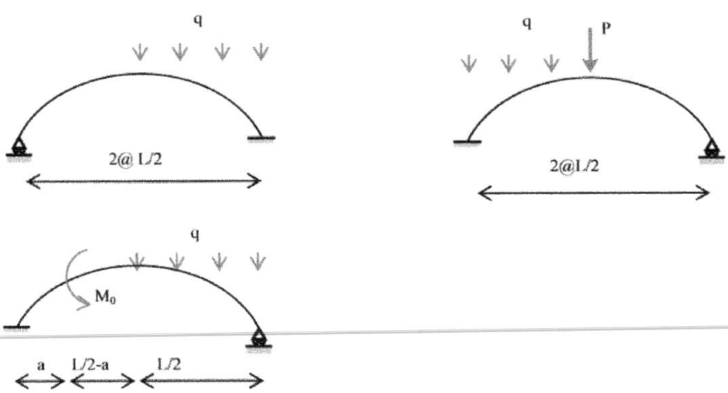

Exercise 8

Determine the internal actions and draw their diagrams. EI is a constant.

Calculate the tension in the spring and deduce the displacement at its end.

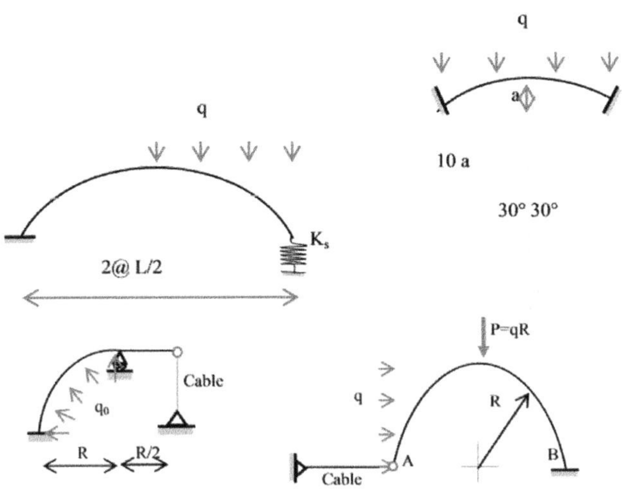

Exercise 9

1) Draw the influence lines of external actions for each arch stressed by a moving unit load. Use the necessary joints.

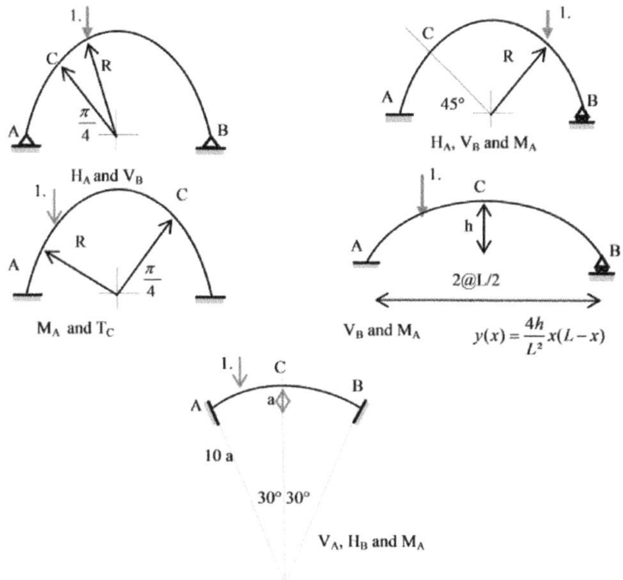

2) Draw the influence lines of internal actions relating to point C for each example.

10

Influence Lines

The teaching objectives for this chapter on influence lines are as follows:

– understanding the variation of support reactions and internal actions of a statically determinate structure under the effect of a moving force;

– establishing influence lines for statically determinate structures;

– understanding the shape of the influence lines of internal and external actions;

– locating the load corresponding to maximum stresses that leads to dimensioning the sections.

Influence lines are very important when designing structures acted on by moving loads. In this chapter, we show the variation of an internal or external action depending on the position of the moving force along a statically determinate structure. This structure may be a beam, a frame or a truss. The second part focuses on constructing influence lines using the Muller–Breslau principle, while the last part describes the deflections of influence lines.

10.1. Introduction

In the previous chapters, we assumed that the loads were constant in values and positions, such as the external actions, the dead weight of both the structure itself and the equipment permanently attached to it. Nevertheless, structures are also subjected to constant and moving forces.

In Chapters 3–9, we did not distinguished between permanent loads with a fixed position and live loads that may change position. We underlined that operating or environmental loads act on a structure for a definite time.

This chapter shows how statically determinate structures respond when subjected to moving loads. The response of the structure to this type of stress is called the influence line. Figure 10.1 represents this phenomenon; it is a vehicle that crosses the bridge (AE). Therefore, for each position of the vehicle over the distance (AE), there is an internal force in the bars of the truss.

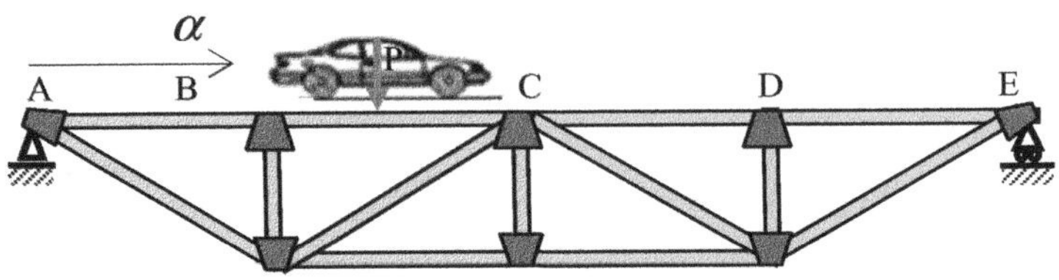

Figure 10.1. *Moving mass on a bridge*

Starting from this concept, analyzing structures subjected to moving loads consists of determining (1) the position of the moving force that corresponds to extreme stresses and (2) the magnitude of the action. This action can be a support reaction, a shear force or a bending moment.

10.2. Influence line definition

When a force is moving on a statically determinate structure, the internal actions vary in relation to the position of the moving force. In the case of a statically determinate symmetrical beam, the stresses are extreme when the force is at the middle of the beam. To design a statically determinate structure solicited by a moving load, we must look at how it responds at the action points. The curve that is obtained is called the influence line of the studied action.

The influence line of an action is a diagram where the action's value is expressed depending on the abscissa of the moving force. In general, the moving force is a unit force. Once the influence line is constructed, we are able to specify the position of the force and the relevant studied action, which has been adopted as a structural design value.

10.3. Influence lines of a beam using the equilibrium method

To establish an action's influence line, we consider a statically determinate beam stressed by a unitary load moving along the beam (Figure 10.2).

10.3.1. *Influence lines of a support reaction*

To develop the influence line of support reaction V_A, for example, it is necessary to formulate the support reaction expression according to the position of the moving load along the beam (AB).

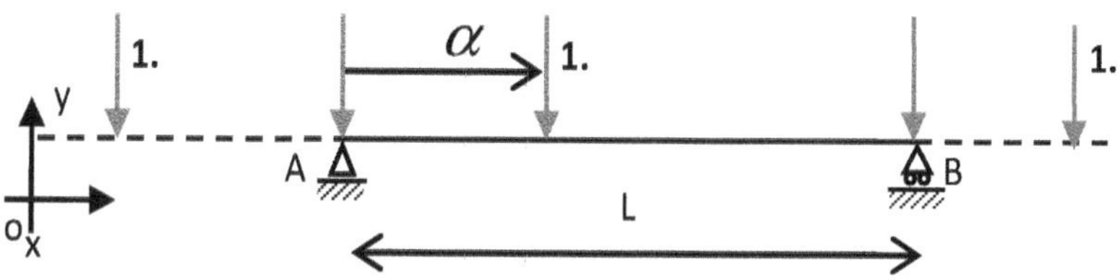

Figure 10.2. *Moving unit force on a beam*[1]

$$\sum \vec{M}_B = \vec{0} \; : \; V_A \vec{j} \wedge (L.\vec{i}) + (-1)\vec{j} \wedge (L-\alpha).\vec{i} = \vec{0}$$

$$V_A(\alpha) = (1 - \frac{\alpha}{L}) \qquad [10.1]$$

$$\sum \vec{F}_y = \vec{0} : \; V_B(\alpha) = \frac{\alpha}{L} \qquad [10.2]$$

$V_A(\alpha = 0) = 1$ and $V_A(\alpha = L) = 0$.

$V_B(\alpha = 0) = 1$ and $V_B(\alpha = L) = 1$.

Expressions [10.1] and [10.2] describe the variation in support reactions V_A and V_B according to the position of the moving force.

The influence line of support reaction V_A (I.L.V_A) is obtained by evaluating the expression of V_A according to the successive position of the unit force along the length of the beam (AB). When the unit force is in A, the reaction V_A is equal to the unit. As the unit force moves away from point A, the value of reaction V_A decreases, disappearing when the unit force reaches point B. The influence line of reaction V_B (I.L.V_B) is found in the same way (Figure 10.3).

1 All of the figures in this chapter are available to view in full color at www.iste.co.uk/khalfallah/analysis1.zip.

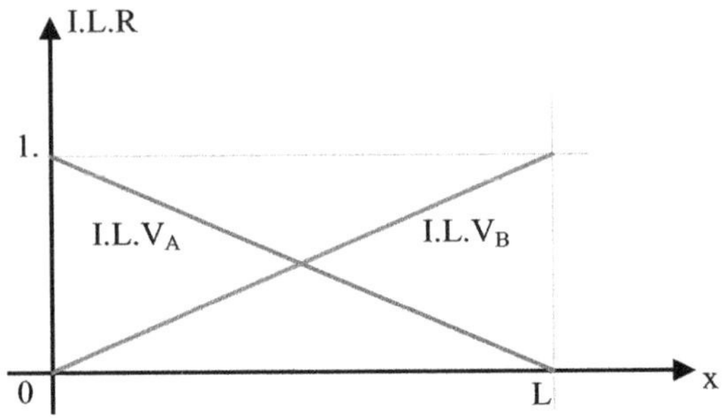

Figure 10.3. *Influence lines of support reactions*

10.3.2. *Influence line of a shear force*

The influence line of a shear force can be developed by constructing the influence line of reaction V_A. To establish the influence line of a shear force in relation to point C (I.L.T_C) of the *x*-axis (Figure 10.4), we calculate the shear force expressions $T_C(x,\alpha)$ for the different positions of the unit force.

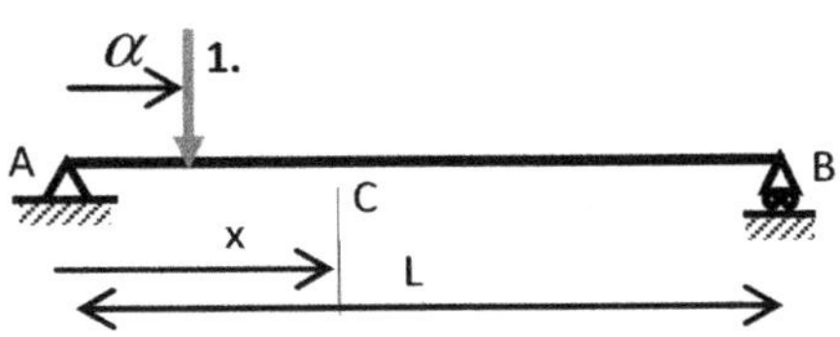

Figure 10.4. *Beam studied*

In this case, we identify:

$0 \leq \alpha \leq x$

$$T_C(x,\alpha) = -\frac{\alpha}{L} \quad [10.3]$$

$x \leq \alpha \leq L$

$$T_C(x,\alpha) = 1 - \frac{\alpha}{L} \quad [10.4]$$

Figure 10.5 shows the influence line of a shear force relating to section C.

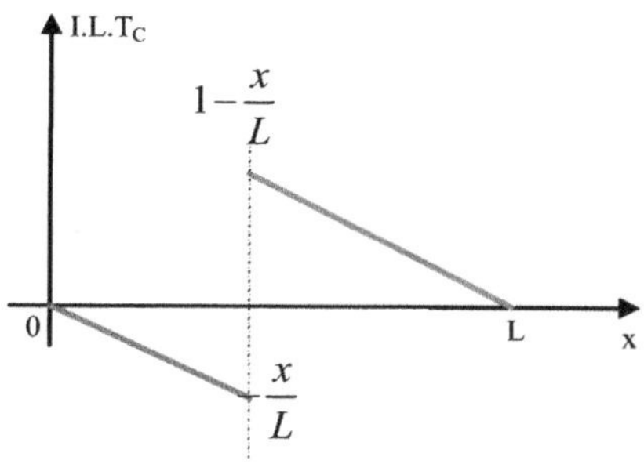

Figure 10.5. *Influence line of a shear force*

10.3.3. *Influence line of a bending moment*

The influence line of the bending moment in relation to point C (I.L.M_C) is obtained by dismissing the positions of the unit force. The bending moment expressions can be deduced by:

$$0 \le \alpha \le x$$

$$M_C(x,\alpha) = \alpha - \frac{\alpha}{L}x \qquad [10.5]$$

$$M_C(x,0) = 0$$

$$M_C(x,x) = x(1-\frac{x}{L})$$

When the unit force is to the right of section C, the bending moment expression is:

$$x \le \alpha \le L$$

$$M_c(x,\alpha) = (1-\frac{\alpha}{L})x \qquad [10.6]$$

$$M_C(x,x) = x(1-\frac{x}{L})$$

$$M_C(x,L) = 0$$

The influence line of the bending moment at section C is shown in Figure 10.6.

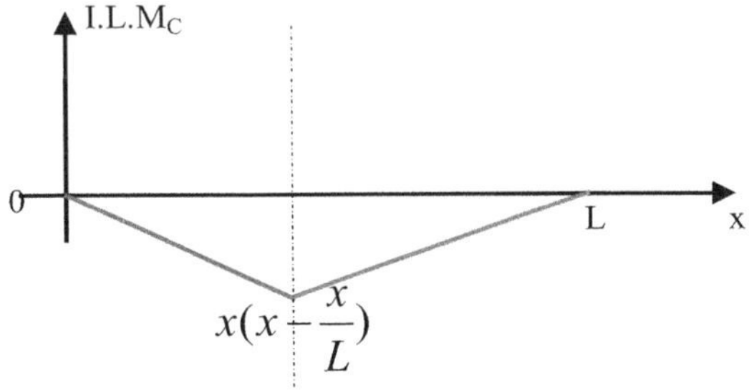

Figure 10.6. *Influence line of a bending moment*

EXAMPLE 10.1.–

Draw the influence lines of the support reaction V_B, the bending moment and the shear force relating to point C of the beam (Figure 10.7).

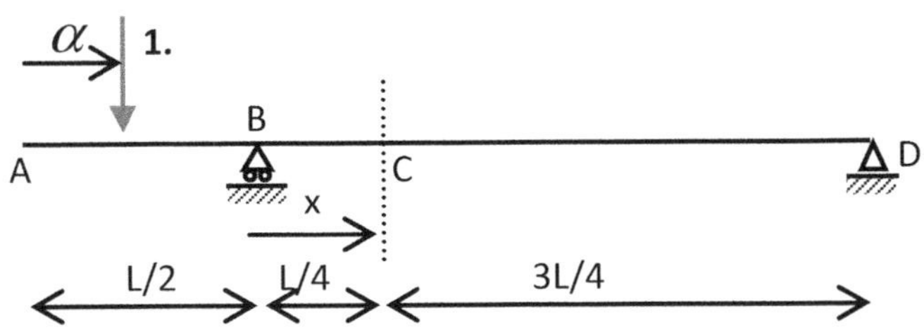

Figure 10.7. *Given beam*

Influence line of support reaction V_B

$$0 \leq \alpha \leq \frac{L}{2}$$

$$\sum \vec{M}_D = \vec{0}: \; V_B.L - 1.(\frac{3L}{2} - \alpha) = 0$$

$$V_B(\alpha) = (\frac{3}{2} - \frac{\alpha}{L})$$

$$V_B(\alpha = 0) = \frac{3}{2}$$

$$V_B(\alpha = \frac{L}{2}) = 1.$$

$$\frac{L}{2} \le \alpha \le \frac{3L}{2}$$

$$V_B(\alpha) = (\frac{3}{2} - \frac{\alpha}{L})$$

$$V_B(\alpha = \frac{L}{2}) = 1.$$

$$V_B(\alpha = \frac{3L}{2}) = 0.$$

The influence line of support reaction V_B is given in Figure 10.8.

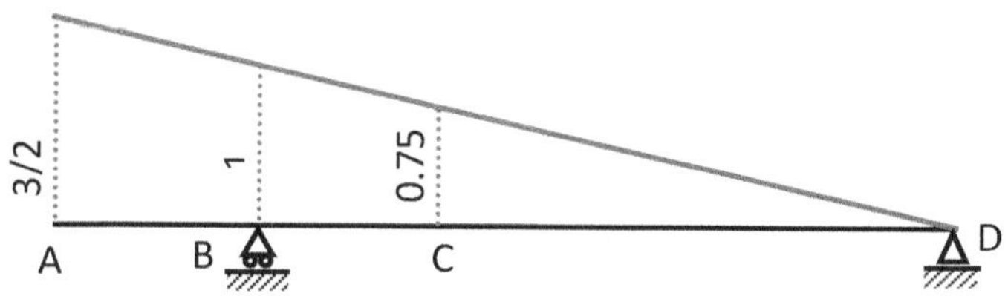

Figure 10.8. *Influence line of support reaction V_B*

We formulate the expression of the bending moment relating to point C.

$$0 \le \alpha \le \frac{L}{2}$$

$$M_C(\mathrm{x}, \alpha) = V_B x - 1(\frac{L}{2} + \mathrm{x} - \alpha)$$

Substituting the expression V_B, we obtain:

$$M_C(\mathrm{x}, \alpha) = \frac{1}{2}(x - L) + \alpha(1 - \frac{x}{L})$$

$$M_C(\mathrm{x},0) = \frac{1}{2}(x-L) = -\frac{L}{2}(1-\frac{x}{L})$$

$$M_C(\mathrm{x},\frac{L}{2}) = 0.$$

$$\frac{L}{2} \leq \alpha \leq \frac{L}{2} + x$$

$$M_C(\mathrm{x},\alpha) = V_B x - 1.(\frac{L}{2} + \mathrm{x} - \alpha)$$

Substituting the expression V_B, the term of the bending moment M_C is:

$$M_C(\mathrm{x},\alpha) = \frac{1}{2}(x-L) + \alpha(1-\frac{x}{L})$$

$$M_C(\mathrm{x},\frac{L}{2}) = 0.$$

$$M_C(\mathrm{x},x+\frac{L}{2}) = x(1-\frac{x}{L})$$

$$\frac{L}{2} + x \leq \alpha \leq \frac{3L}{2}$$

$$M_C(\mathrm{x},\alpha) = V_B x$$

$$M_C(\mathrm{x},\alpha) = (\frac{3}{2} - \frac{\alpha}{L}).x$$

$$M_C(\mathrm{x},x+\frac{L}{2}) = x(1-\frac{x}{L})$$

$$M_C(\mathrm{x},\frac{3L}{2}) = 0.$$

The influence line of the bending moment at point C is shown in Figure 10.9.

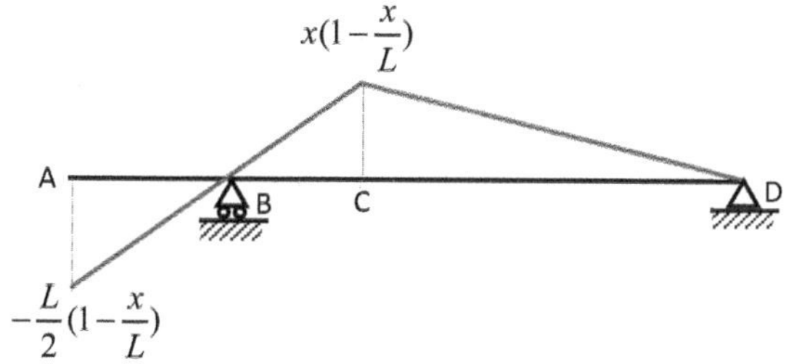

Figure 10.9. *Influence line of moment M_C*

Finally, the expressions of the shear force can be deduced by:

$$0 \leq \alpha \leq \frac{L}{2}$$

$$T_C(x,\alpha) = V_B - 1 = \frac{1}{2} - \frac{\alpha}{L}$$

$$T_C(x,0) = \frac{1}{2}$$

$$T_C(x,\frac{L}{2}) = 0.$$

$$\frac{L}{2} \leq \alpha \leq \frac{L}{2} + x$$

$$T_C(x,\alpha) = V_B - 1 = \frac{1}{2} - \frac{\alpha}{L}$$

$$T_C(x,\frac{\mathrm{L}}{2}) = 0.$$

$$T_C(x,x+\frac{L}{2}) = -\frac{x}{L}$$

$$\frac{L}{2} + x \leq \alpha \leq \frac{3L}{2}$$

$$T_C(x, \frac{L}{2} + x) = 1 - \frac{x}{L}$$

$$T_C(x, \frac{3L}{2}) = 0.$$

The influence line of the shear force at point C is represented in Figure 10.10.

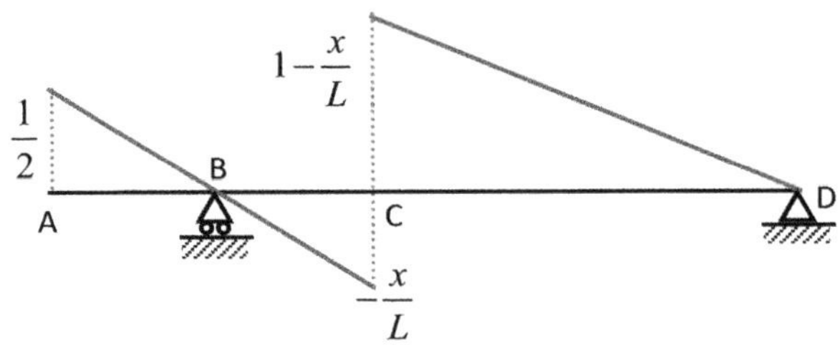

Figure 10.10. *Influence line of T_C*

10.4. Influence lines of a frame using the equilibrium method

In this section, we construct the influence lines of support reaction V_A, the shear force and the bending moment relating to point Q at a distance x from support B (Figure 10.11).

The moving force passes through the beam (BQCE) of the frame (Figure 10.11).

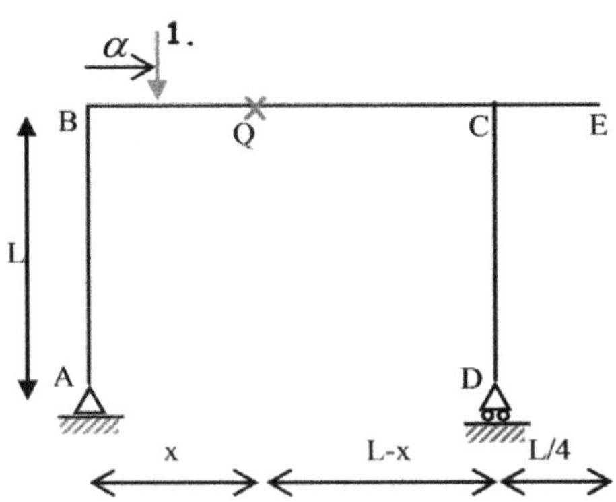

Figure 10.11. *Given frame*

10.4.1. *Influence line of support reaction V_A*

$$0 \le \alpha \le L$$

$$\sum \overrightarrow{M}_D = \vec{0} : \mathrm{V_A}(\alpha) = 1 - \frac{\alpha}{L} \qquad [10.7]$$

$$\mathrm{V_A}(0) = 1.$$

$$\mathrm{V_A}(\mathrm{L}) = 0.$$

$$L \le \alpha \le \frac{5L}{4}$$

$$\sum \overrightarrow{M}_D = \vec{0} : \mathrm{V_A}(\alpha) = 1 - \frac{\alpha}{L} \qquad [10.8]$$

$$\mathrm{V_A}(\mathrm{L}) = 0.$$

$$\mathrm{V_A}\left(\frac{5\mathrm{L}}{4}\right) = -\frac{1}{4}$$

The influence line of support reaction V_A is given in Figure 10.12.

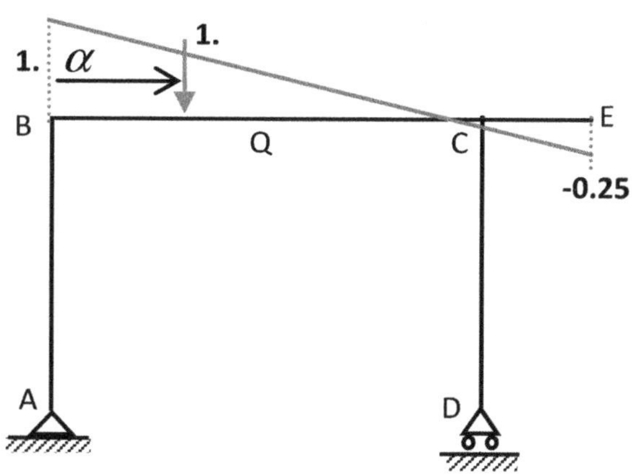

Figure 10.12. *Influence line of V_A*

In the same way, the influence lines of the bending moment at point Q can be obtained as follows:

$$0 \leq \alpha \leq x$$

Substituting the expression of the reaction V_A, we obtain:

$$M_Q(x,\alpha) = \alpha(1-\frac{x}{L}) \quad [10.9]$$

$$M_Q(x,0) = 0$$

$$M_Q(x,x) = x(1-\frac{x}{L})$$

When the unit force is to the right of section Q, the bending moment expression takes the following form:

$$x \leq \alpha \leq L$$

$$M_Q(x,\alpha) = (1-\frac{\alpha}{L})x \quad [10.10]$$

$$M_Q(x,x) = x(1-\frac{x}{L})$$

$$M_Q(x,L) = 0$$

Finally, when the unit force is on branch (CE):

$$L \leq \alpha \leq \frac{5L}{4}$$

$$M_Q(x,\alpha) = V_A.\mathrm{x} = (1-\frac{\alpha}{L})x \quad [10.11]$$

$$M_Q(x,\mathrm{L}) = 0.$$

$$M_Q(x,\frac{5L}{4}) = -\frac{x}{4}$$

The expressions [10.9] to [10.11] allow us to draw the influence line of a bending moment relating to section Q (Figure 10.13).

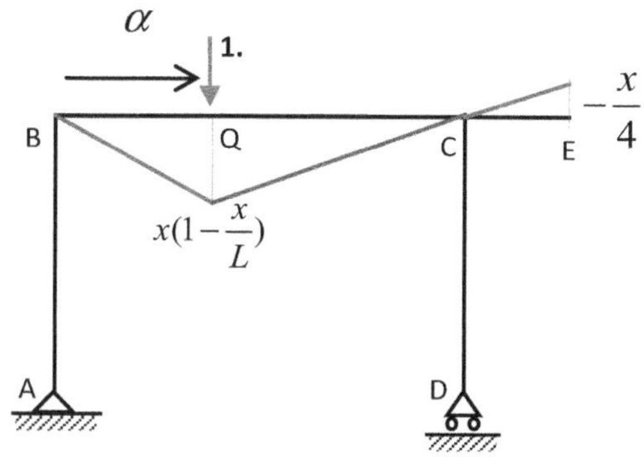

Figure 10.13. *Influence line of M_Q*

The influence line of a shear force relating to section Q can be obtained by applying the above procedure.

$$0 \leq \alpha \leq x$$

$$\mathrm{T_Q}(x, \alpha) = V_A - 1 = -\frac{\alpha}{L} \qquad [10.12]$$

$$\mathrm{T_Q}(x, 0) = 0.$$

$$\mathrm{T_Q}(x, x) = -\frac{x}{L}$$

$$x \leq \alpha \leq L$$

$$\mathrm{T_Q}(x, \alpha) = (1 - \frac{\alpha}{L}) \qquad [10.13]$$

$$\mathrm{T_Q}(x, x) = (1 - \frac{x}{L})$$

$$\mathrm{T_Q}(x, L) = 0$$

Finally, when the unit force is on branch (CE):

$$L \leq \alpha \leq \frac{5L}{4}$$

$$\mathrm{T_Q}(x,\alpha) = V_A = (1 - \frac{\alpha}{L}) \qquad [10.14]$$

$$\mathrm{T_Q}(x, \mathrm{L}) = 0.$$

$$\mathrm{T_Q}(x, \frac{5L}{4}) = -\frac{1}{4}$$

The expressions [10.12] to [10.14] allow us to draw the influence line of a shear force relating to section Q (Figure 10.14).

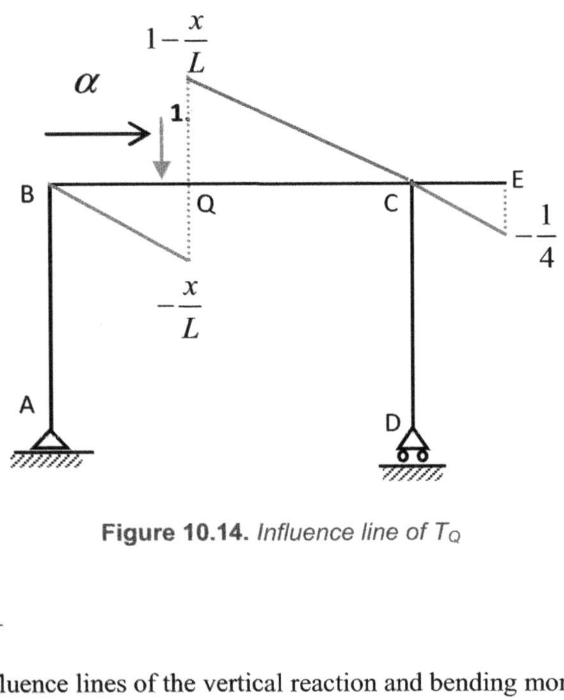

Figure 10.14. *Influence line of* T_Q

EXAMPLE 10.2.–

Draw the influence lines of the vertical reaction and bending moment at the fixed side and the influence lines of the bending moment and the shear force with respect to section D (Figure 10.15).

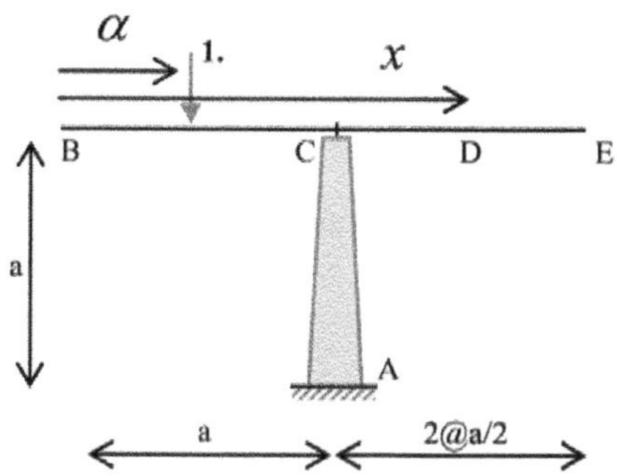

Figure 10.15. *Given structure*

$0 \le \alpha \le a$

$V_A = 1.$

$M_A(\alpha) = -(a - \alpha)$

$T_D(\alpha) = 0.$

$M_D(x, \alpha) = \alpha - a$

$a \le \alpha \le x$

$V_A = 1.$

$M_A(\alpha) = (\alpha - a)$

$T_D(\alpha) = 0.$

$M_D(x, \alpha) = \alpha - a$

$x \le \alpha \le 2a$

$V_A = 1.$

$M_A(\alpha) = (\alpha - a)$

$$T_D(\alpha) = 1.$$

$$M_D(x, \alpha) = x - \alpha$$

The influence lines of the external and internal actions at points A and D are shown in Figure 10.16.

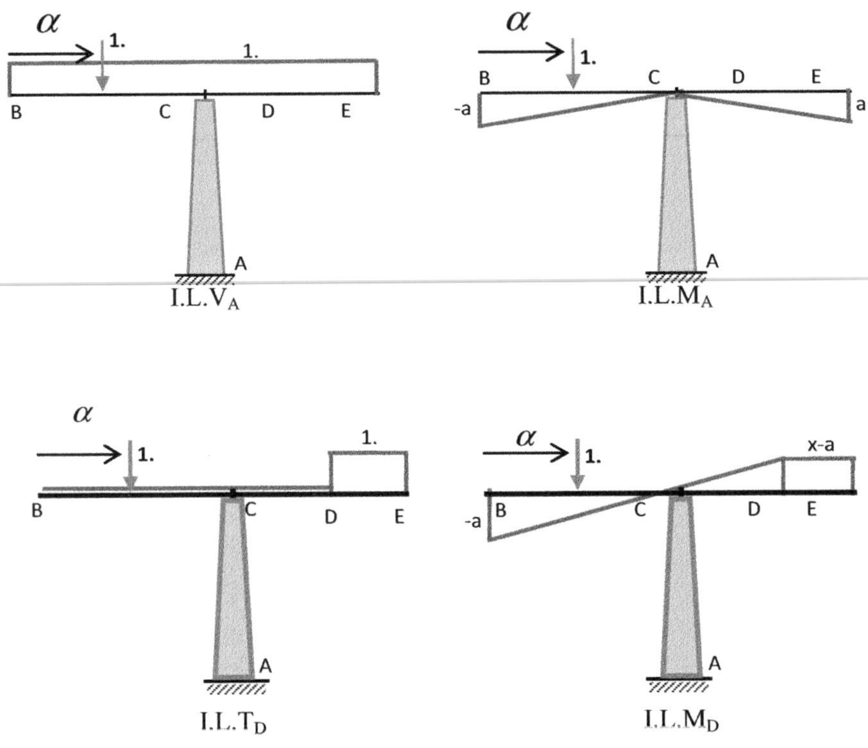

Figure 10.16. *Influence lines at points A and D*

10.5. Analysis steps

In general, to construct the influence line of an internal or external action, it is necessary to adhere to the following steps:

1) We associate a reference to the structure on which we fix the direction of movement of the moving force. In general, it is moving from left to right with a unit value and oriented toward the negative direction of the ordinates.

2) We position the unit force at a distance of α from the left end of the structure and formulate the expression of the action.

3) Once the expressions of the actions have been formulated, it is easy to construct a diagram of the action. The influence line is the magnitude of the action (y-axis) according to the position of the moving force (x-axis).

4) The influence line is used to determine the value and direction of the studied action, the extreme value of which must be taken in the design phase.

EXAMPLE 10.3.–

Draw the influence lines of support reaction V_E, the bending moment relating to point B and the shear force relating to point C of the beam (Figure 10.17).

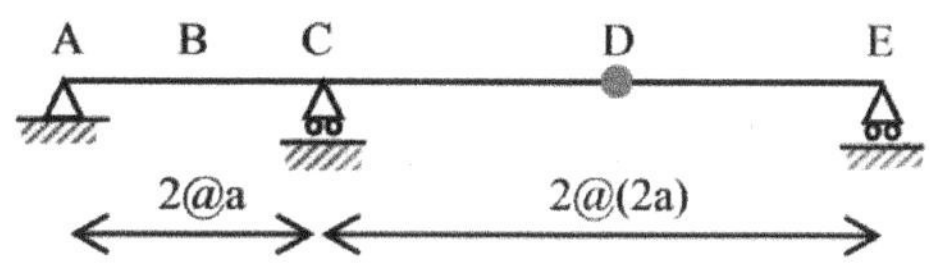

Figure 10.17. *Given beam*

We associate the reference (oxy) and a moving force crosses the beam (AE) (Figure 10.18); a hinge is placed at point D.

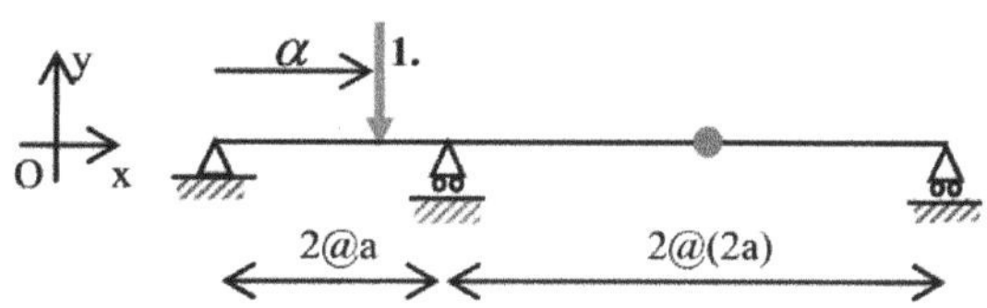

Figure 10.18. *Beam with a hinge*

The beam (AE) is once statically indeterminate and introducing a hinge at point D makes the structure statically determinate.

Influence line of support reaction V_E

The moving force has no effect on reaction V_E when it is in interval [0,4a].

$$4a \leq \alpha \leq 6a$$

$$\sum \overrightarrow{M}_D = \vec{0} : V_E(\alpha) = \frac{\alpha - 4a}{2a}$$

$$V_E(4a) = 0.$$

$$V_E(6a) = a$$

The influence line of support reaction V_E is given in Figure 10.19.

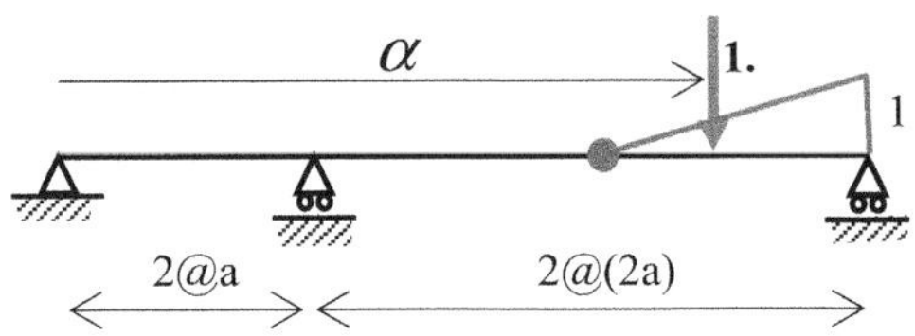

Figure 10.19. *Influence line of V_E*

Influence line of bending moment M_B

$$0 \le \alpha \le a$$

$$V_A(\alpha) = 1 - \frac{\alpha}{2a}$$

$$M_B(\alpha) = (1 - \frac{\alpha}{2a})a - 1.(a - \alpha)$$

$$M_B(0) = 0.$$

$$M_B(\text{a}) = \frac{a}{2}$$

$$a \le \alpha \le 2a$$

$$V_A(\alpha) = 1 - \frac{\alpha}{2a}$$

$$M_B(\alpha) = (1 - \frac{\alpha}{2a})a$$

$$M_B(\text{a}) = \frac{a}{2}$$

$$M_B(2\text{a}) = 0.$$

$$2a \le \alpha \le 4a$$

$$\sum \overrightarrow{M}_D = \vec{0}: \; V_A.4a + V_C.2a = 1.(4a - \alpha)$$

Knowing that $V_A + V_C = 1.$ and V_E=0.

$$V_A(\alpha) = 1 - \frac{\alpha}{2a}$$

$$V_C(\alpha) = \frac{\alpha}{2a}$$

$$M_B(\alpha) = (1 - \frac{\alpha}{2a})a$$

$$M_B(2\text{a}) = 0.$$

$$M_B(4\text{a}) = -a$$

$$4a \le \alpha \le 6a$$

$$\sum \overrightarrow{M}_D = \vec{0}: \quad V_E.\text{a} = 1.(\alpha - 4a)$$

$$V_E(\alpha) = \frac{\alpha}{a} - 4$$

$$V_A(\alpha) = \frac{\alpha}{2a} - 3$$

$$V_C(\alpha) = 6 - \frac{\alpha}{2a}$$

$$M_B(\alpha) = (\frac{\alpha}{2a} - 3).a$$

$$M_B(4\text{a}) = -a$$

$$M_B(6\text{a}) = 0.$$

The influence line is represented in Figure 10.20.

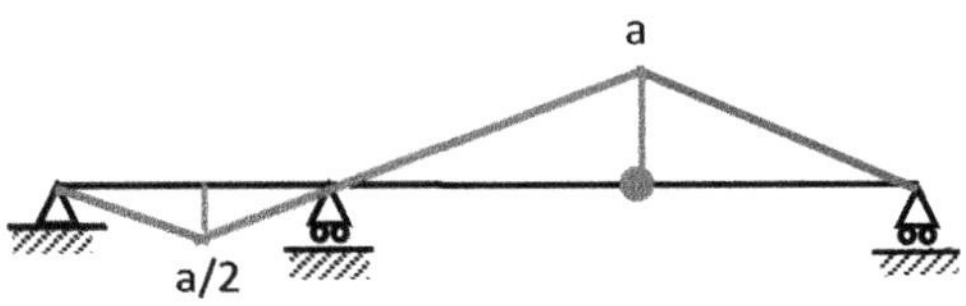

Figure 10.20. *Influence line of moment M_B*

Influence line of shear force T_C

$$0 \le \alpha \le 2a$$

$$V_A(\alpha) = 1 - \frac{\alpha}{2a}$$

$$T_C(\alpha) = V_A(\alpha) - 1 = -\frac{\alpha}{2a}$$

$$T_C(0) = 0.$$

$$T_C(2\text{a}) = -1.$$

$$2a \le \alpha \le 4a$$

$$V_A(\alpha) = 1 - \frac{\alpha}{2a}$$

$$V_C(\alpha) = \frac{\alpha}{2a}$$

$$T_C(\alpha) = V_A(\alpha) = (1 - \frac{\alpha}{2a})$$

$$T_C(2\text{a}) = 0.$$

$T_C(4\text{a}) = -1.$

$4a \le \alpha \le 6a$

$$V_A(\alpha) = \frac{\alpha}{2a} - 3$$

$$T_C(\alpha) = V_A(\alpha) = (\frac{\alpha}{2a} - 3)$$

$T_C(4\text{a}) = 1.$

$T_C(4\text{a}) = 0.$

The influence line is represented in Figure 10.21.

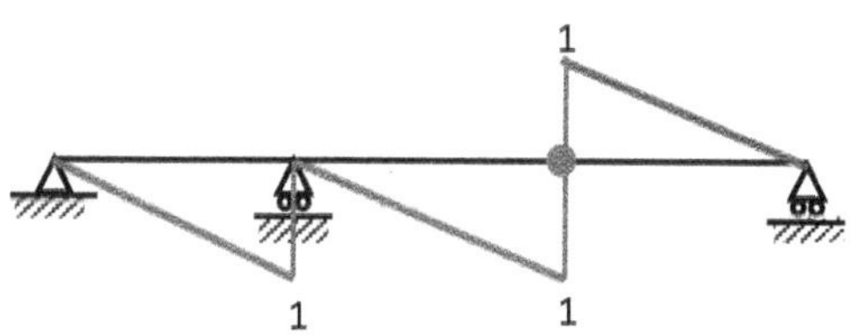

Figure 10.21. *Influence line of Tc*

EXAMPLE 10.4.–

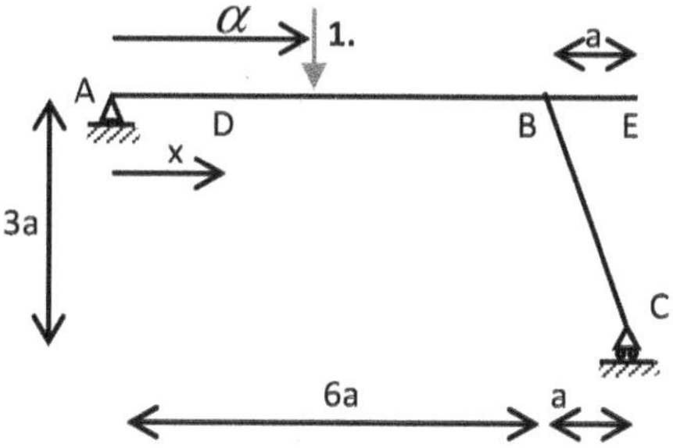

Figure 10.22. *Given frame*

Draw the influence lines of support reaction H_A, the shear force and the bending moment relating to point D (Figure 10.22).

The bar (BC) generates a force R_B at point B (Figure 10.23).

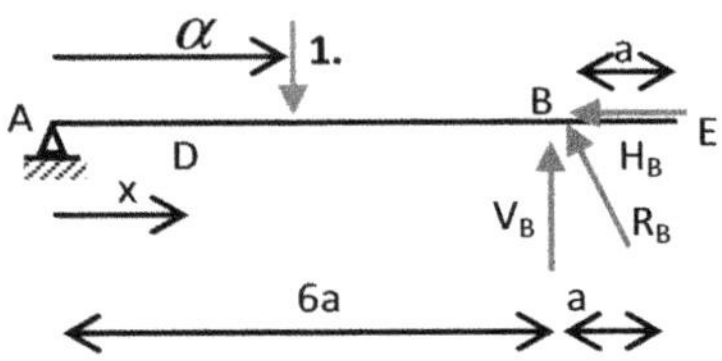

Figure 10.23. *Bar (AB)*

$$\frac{V_B}{H_B} = \frac{1}{3}$$

$$H_B = \frac{V_B}{3}$$

$$\sum \vec{M}_A = \vec{0}$$

$$V_B = \frac{\alpha}{6a}$$

$$H_B = \frac{\alpha}{18a}$$

$$\sum \vec{F}_x = \vec{0}$$

$$H_A = \frac{\alpha}{18a}$$

Draw the influence lines of the support reaction H_A; the bending moment and the shear force relating to point D of the different intervals are given as:

$$0 \leq \alpha \leq x$$

$$H_A = \frac{\alpha}{18a}$$

$$T_D(\mathrm{x},\alpha) = V_A - 1 = -\frac{\alpha}{6a}$$

$$T_D(\mathrm{x},0) = 0.$$

$$T_D(\mathrm{x},\mathrm{x}) = -\frac{x}{6a}$$

$$M_D(\mathrm{x},\alpha) = V_A x - 1.(x-\alpha) = \alpha - \frac{\alpha x}{6a}$$

$$M_D(\mathrm{x},0) = V_A x - 1.(x-\alpha) = 0.$$

$$M_D(\mathrm{x},\mathrm{x}) = x(1-\frac{x}{6a})$$

$$x \le \alpha \le 7a$$

$$H_A = \frac{\alpha}{18a}$$

$$T_D(\mathrm{x},\alpha) = V_A = 1 - \frac{\alpha}{6a}$$

$$T_D(\mathrm{x},\mathrm{x}) = 1 - \frac{x}{6a}$$

$$T_D(\mathrm{x},7\,\mathrm{a}) = -\frac{1}{6}$$

$$M_D(\mathrm{x},\alpha) = V_A x = (1-\frac{\alpha}{6a})x$$

$$M_D(\mathrm{x},\mathrm{x}) = (1-\frac{x}{6a})x$$

$$M_D(\mathrm{x},7\,\mathrm{a}) = -\frac{x}{6}$$

The influence lines of reaction H_A, the shear force and the bending moment relating to point D are represented, respectively, in Figures 10.24–10.26.

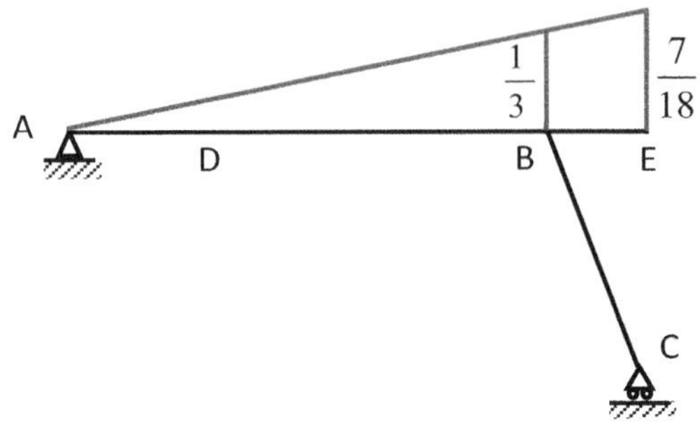

Figure 10.24. *I.L.H*

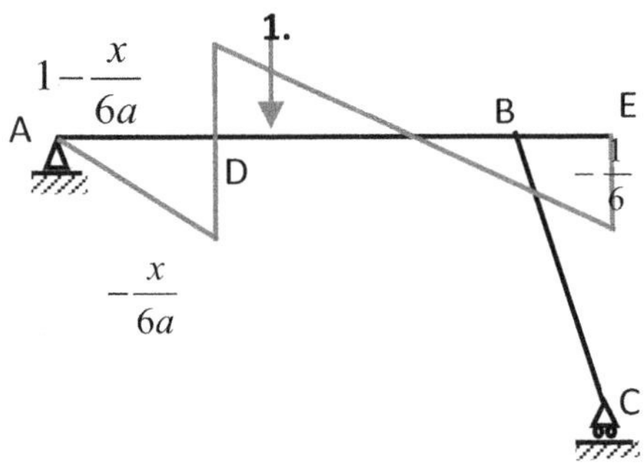

Figure 10.25. *I.L.*T_D

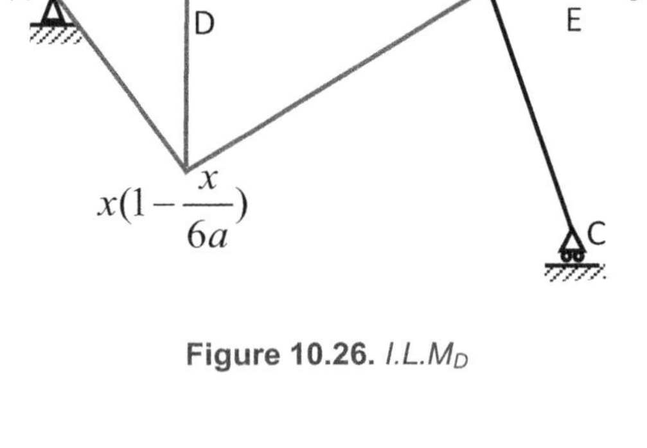

Figure 10.26. *I.L.*M_D

10.6. Influence lines of trusses

Trusses are often used for bridges, sports and conference halls, crane bridges, etc.

During their design, it is important to construct the influence lines of the structure or one of its parts. Figure 10.1 clearly shows this phenomenon; the applied loading is transmitted from the deck to the carrying elements of the bridge.

The influence lines of one or more bars of the truss can be constructed by subjecting the joints to unit joint forces (Figure 10.27). We can determine the internal forces using one of the methods for analyzing trusses (Chapter 4). The results are graphically presented in order to show the design phase.

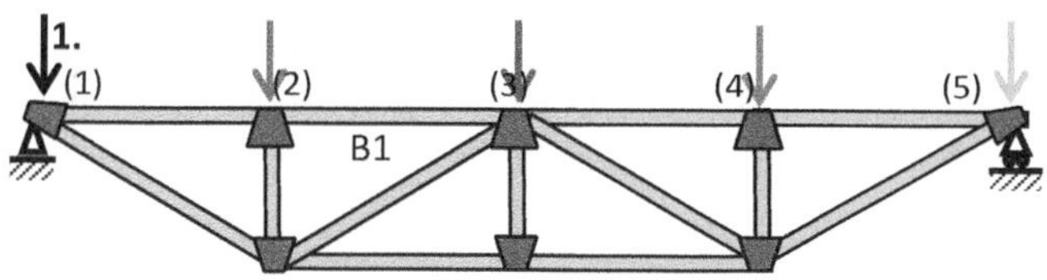

Figure 10.27. *Truss*

Figure 10.27 shows how to construct the influence line of the internal force of bar (B1). The unit force is applied to joint (1) and the force in bar (B1) is calculated. This procedure applies when the moving force is alternately applied to joints (2)–(5). The curve enveloping the variation of the intensity of the normal force in the bar (B1) depending on the unit force's application point is the normal force's influence line.

EXAMPLE 10.5.–

Draw the influence lines of the internal force in bars (B1), (B2), (B3), and the vertical reaction V_A of the structure (Figure 10.28).

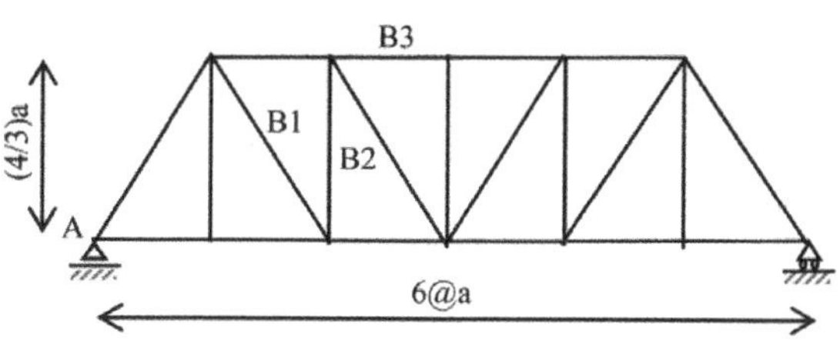

Figure 10.28. *Truss*

To determine the influence line of reaction V_A and the influence line of the normal force on bars (B1), (B2) and (B3), we use a numerical system developed for the analysis of trusses.

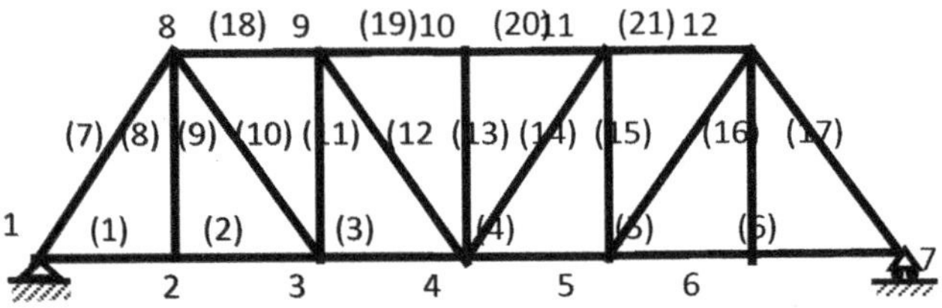

Figure 10.29. *Numbering of bars*

To construct the influence lines of reaction V_A and the internal force in bars (B1≡ (10)), (B2≡ (11)) and (B3≡ (19)), the procedure described in the previous sections is applied.

Using the system gives us the results shown in Table 10.1.

Joint	1	2	3	4	5	6	7
VA	1	0.933	0.667	0.50	0.334	0.167	0
B1	0	–0.209	0.933	O.625	O.416	0.209	0
B2	0	0.166	0.344	–0.50	–0.334	–0.166	0
B3	0	–0.375	–0.750	–1.125	–0.750	–0.375	0

Table 10.1. *Forces in bars B1, B2, B3*

The influence lines are given in Figures 10.30–10.33.

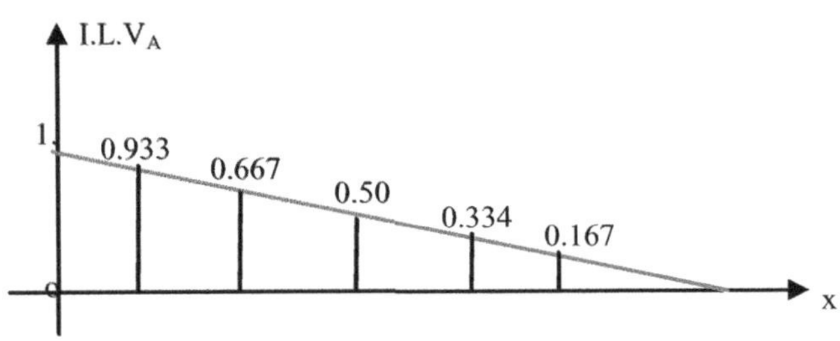

Figure 10.30. *I.L.* V_A

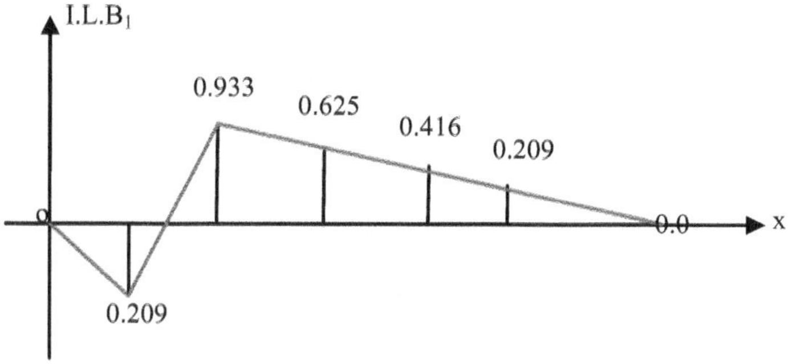

Figure 10.31. *I.L.B*$_1$

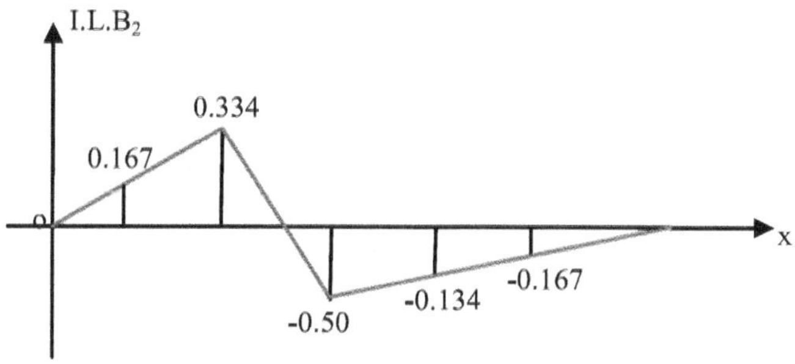

Figure 10.32. *I.L.B*$_2$

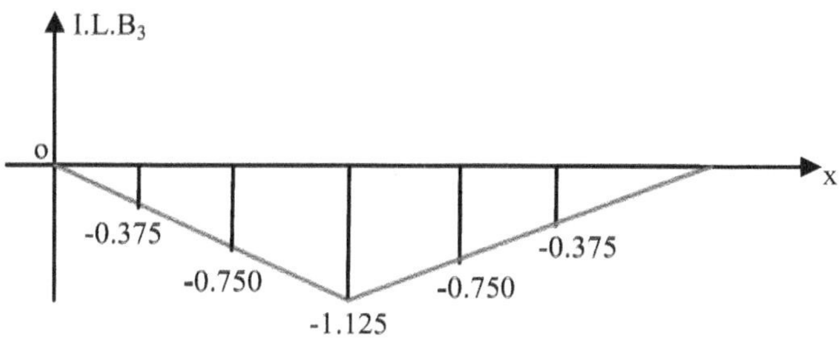

Figure 10.33. *I.L.B*$_3$

10.7. Influence lines using the Muller–Breslau principle

The construction of the influence lines of an action can be carried out using the Muller–Breslau principle: "the influence line of an action relating to a section is given by the deformed structure below the moving force obtained by ignoring the action itself, which corresponds to the construction of the influence line of the basic structure by associating a unit displacement (case of a reaction) or a unit rotation (case of a moment) according to the direction of the considered action". This principle does not apply to the construction of influence lines for deflections [10.8].

To show how to apply the Muller-Breslau method, we consider a statically determinate beam stressed by a unitary load moving along the beam (Figure 10.34). The unit force first applies to the section (AC), then to the section (CB). The influence lines of support reaction V_A, the bending moment and the shear force relating to point C can be constructed by sequentially applying the steps of the method.

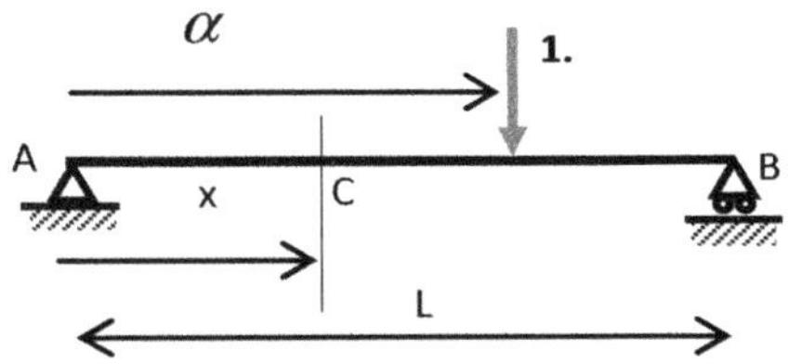

Figure 10.34. *Given beam*

10.7.1. *Influence lines of a support reaction*

To construct the influence lines of the support reaction V_A, we ignore the reaction itself. The hinge at A thus becomes a roller with only 1 degree of freedom. The beam can move in the direction of the force V_A (Figure 10.35).

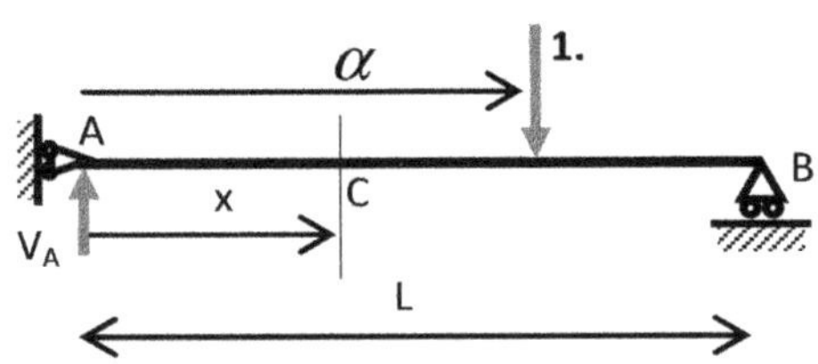

Figure 10.35. *Given beam*

In this case, the actions stressing the beam are the support reaction V_A, the moving force and the support reaction V_B. We associate following the direction of reaction V_A a virtual unit displacement ($\Delta_A = 1.$) (Figure 10.36).

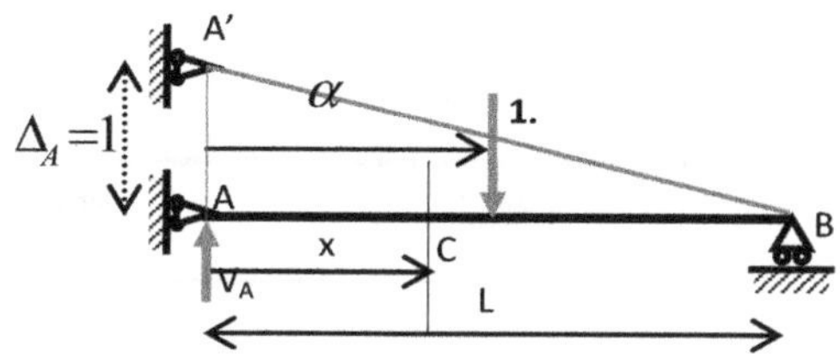

Figure 10.36. *Virtual unit displacement*

The beam is in equilibrium according to the principle of virtual displacements of rigid body movement. Knowing that support B remains stationary alongside support A prevents the horizontal displacement of the beam (Figure 10.36).

We calculate the virtual work of the external forces.

$$W_{v,ext} = V_A(\alpha).\Delta_A - 1.v(\alpha), \tag{10.15}$$

where $v(\alpha)$ is the vertical displacement of the moving force's application point. The displacement depends on α and $\Delta_A = 1$. Equation [10.15] leads to:

$$V_A(\alpha) = v(\alpha) \tag{10.16}$$

The displacement below the moving force is given as:

$$v(\alpha) = 1 - \frac{\alpha}{L} \tag{10.17}$$

Taking into account relationships [10.16] and [10.17], the vertical reaction at A is written as:

$$V_A(\alpha) = 1 - \frac{\alpha}{L} \tag{10.18}$$

The relationship [10.18] is identical to that obtained by the equilibrium method in Figure 10.3. Similarly, the variation of reaction V_A (α) is a function of the position of the moving force.

EXAMPLE 10.6.–

Using the Muller–Breslau principle, construct the influence line of support reaction V_A of the beam (Figure 10.37).

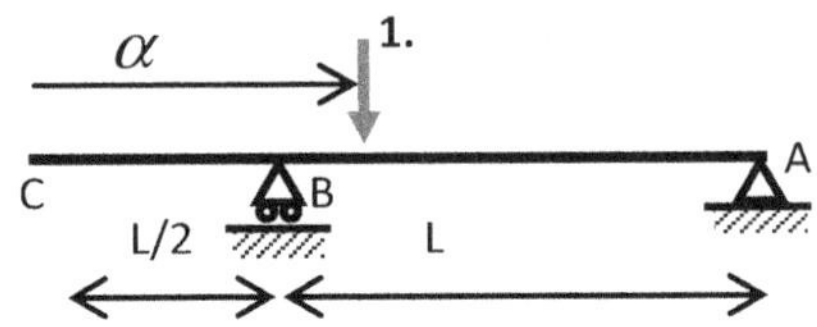

Figure 10.37. *Given beam*

We substitute hinge A with a roller and a vertical reaction V_A (Figure 10.38).

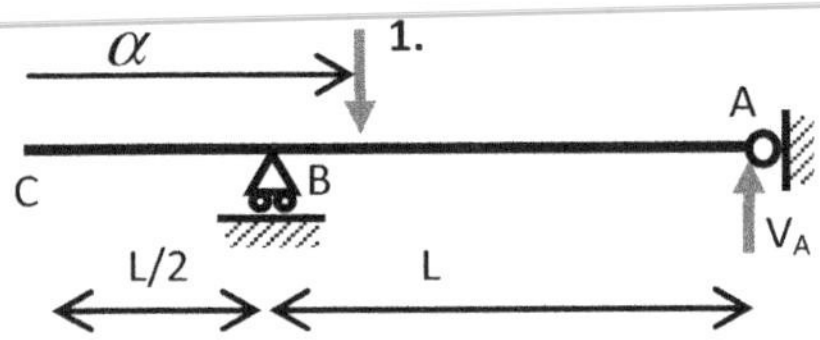

Figure 10.38. *Effect of reaction V_A*

Bar (AC) undergoes a rigid body movement (Figure 10.39) whose displacement $\Delta_A = 1$.

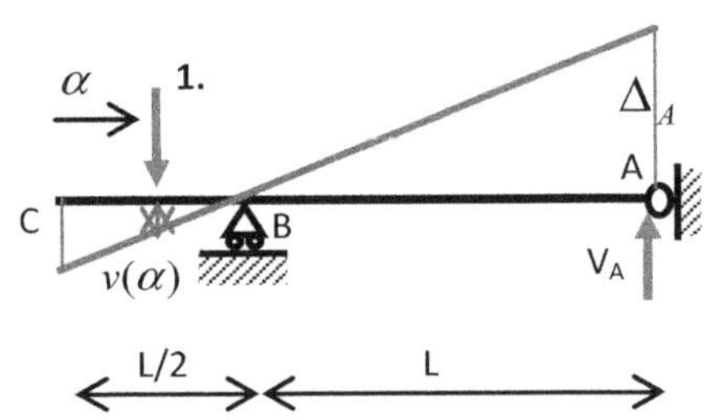

Figure 10.39. *Unit force $0 \leq \alpha \leq L/2$*

$0 \leq \alpha \leq \frac{L}{2}$ (Figure 10.39)

The virtual work of the external forces is zero, so we can write:

$$W_{v,ext} = V_A.\Delta_A - 1.v(\alpha) = 0$$

$$V_A = v(\alpha)$$

The similarity of the displacements (Figure 10.39) makes it possible to write:

$$\frac{-v(\alpha)}{\Delta_A} = \frac{\frac{L}{2} - \alpha}{L}$$

$$V_A(\alpha) = v(\alpha) = -(\frac{1}{2} - \frac{\alpha}{L})$$

$$V_A(0) = -\frac{1}{2}$$

$$V_A(\frac{L}{2}) = 0.$$

$\frac{L}{2} \leq \alpha \leq \frac{3L}{2}$ (Figure 10.40)

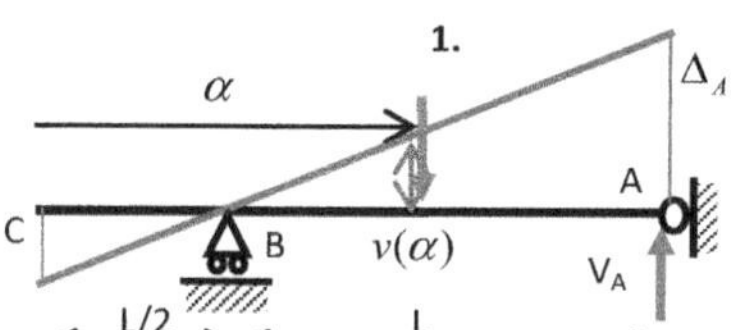

Figure 10.40. *Unit force* $\frac{L}{2} \leq \alpha \leq \frac{3L}{2}$

$$V_A(\alpha) = v(\alpha) = -(\frac{1}{2} - \frac{\alpha}{L})$$

$$V_A(\frac{L}{2}) = 0.$$

$$V_A(\frac{3L}{2}) = 1.$$

The influence line of reaction V_A using the Muller–Breslau principle is given in Figure 10.41.

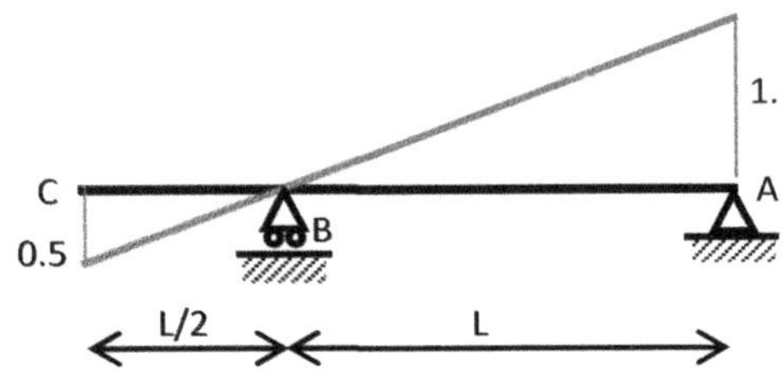

Figure 10.41. *I.L.V$_A$*

10.7.2. *Influence line of a shear force*

To construct the influence line of the shear force at point C using the Muller–Breslau method, we follow the procedure described in section 10.6.1. The beam is cut by a section at point C (Figure 10.42).

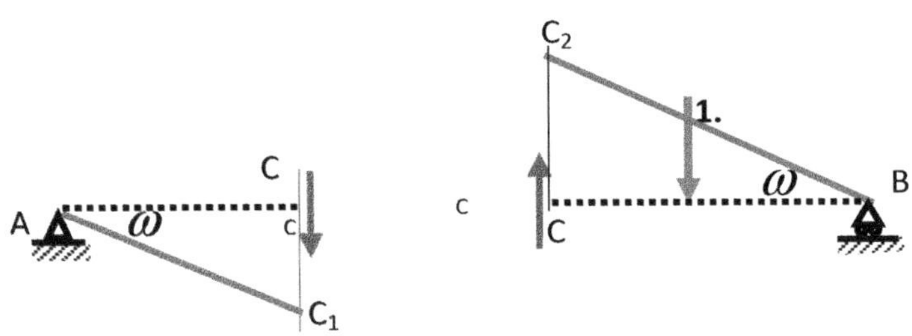

Figure 10.42. *Shear force at point C*

Sections (AC) and (CB) undergo displacements (CC_1) and (CC_2) in the direction of shear force T(x) (Figure 10.42).

A unit displacement is given at point C. The displacements of branches (AC) and (BC) are, respectively, Δ_1 and Δ_2.

$$\Delta = \Delta_1 + \Delta_2 = 1. \qquad [10.19]$$

Displacements Δ_1 and Δ_2 depend on the slope of joints A and B. This consideration makes it possible to conclude that the lines (AC) and (BC) are parallel.

Applying the principle of virtual displacements leads to:

$$W_{v,ext} = T(\mathrm{x}).\Delta_1 + T(\mathrm{x}).\Delta_2 + M(x).\omega - M(x).\omega - 1.v(\alpha) \qquad [10.20]$$

or:

$$W_{v,ext} = T(\mathrm{x}).(\Delta_1 + \Delta_2) - 1.v(\alpha) \qquad [10.21]$$

or:

$$W_{v,ext} = T(\mathrm{x}).\Delta - v(\alpha) \qquad [10.22]$$

Knowing that the virtual work of external actions is zero, we get:

$$T(\mathrm{x}) = v(\alpha) \qquad [10.23]$$

From Figure 10.37, we can write:

$$\Delta_2 = \frac{L-x}{x}\Delta_1 \qquad [10.24]$$

Taking into account the relationships [10.19] and [10.24], we can write:

$$\Delta_1 = \frac{x}{L} \qquad [10.25]$$

$$\Delta_2 = 1 - \frac{x}{L} \qquad [10.26]$$

$0 \le \alpha \le x$ (Figure 10.43)

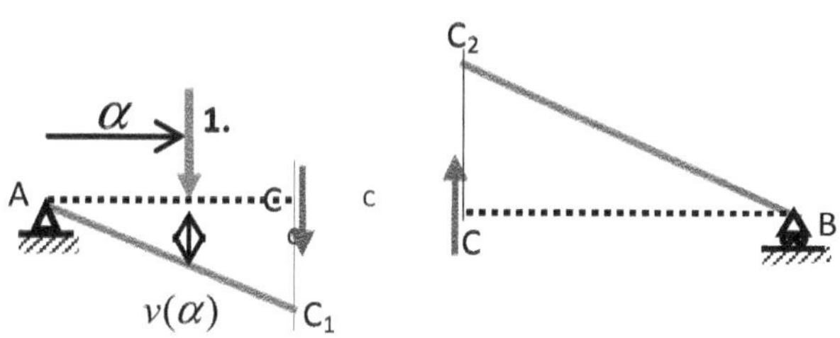

Figure 10.43. *Tc for* $\alpha \le x$

The displacement of the point of application of the unit force is given as:

$$v(\alpha)=\frac{\alpha}{L} \qquad [10.27]$$

Using the relationship [10.23], the expression of the shear force is written as:

$$T(x)=\frac{\alpha}{L} \qquad [10.28]$$

$$T(0)=0.$$

$$T(x)=\frac{x}{L}$$

$x \leq \alpha \leq L$ (Figure 10.44)

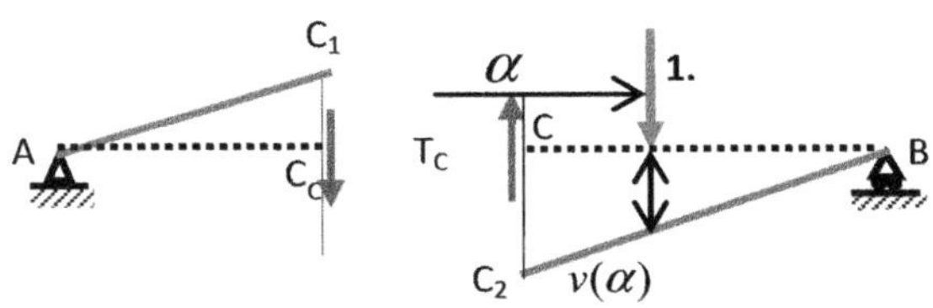

Figure 10.44. *Tc for* $x \leq \alpha \leq L$

In this case, the displacement of the point of application of the moving force is given as:

$$v(x)=1-\frac{\alpha}{L} \qquad [10.29]$$

The shear force expression is:

$$T(x)=v(\alpha)=1-\frac{\alpha}{L} \qquad [10.30]$$

$$T(x)=1-\frac{x}{L}$$

$$T(L)=0.$$

The influence line of the shear force at point C is shown in Figure 10.45.

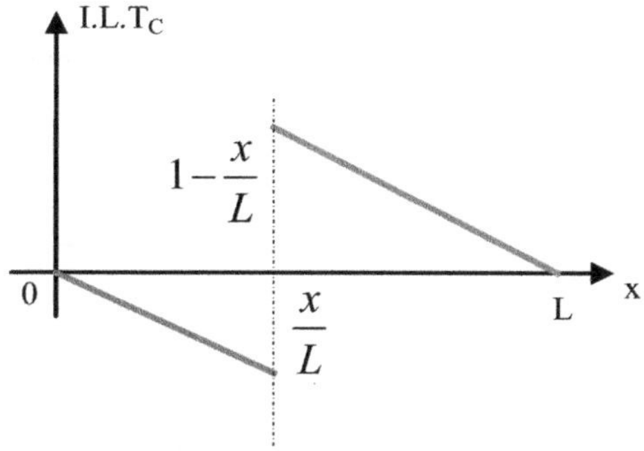

Figure 10.45. *I.L.Tc*

The result obtained using the Muller–Breslau principle is identical to that obtained by the equilibrium method (Figure 10.5).

EXAMPLE 10.7.–

Construct the influence line of the shear force relating to point D located at a distance of x from support B (Figure 10.37).

$0 \le \alpha \le \frac{L}{2}$ (Figure 10.45)

The displacement of the force's point of application is given as:

$$v(\alpha) = \beta(1 - \frac{2\alpha}{L})$$

Substituting the value of β in the previous expression, we obtain:

$$\beta = -\frac{L}{2x}\Delta_1$$

$$v(\alpha) = -\frac{L}{2x}\Delta_1(1 - \frac{2\alpha}{L})$$

$$\Delta_1 = \frac{x}{L}$$

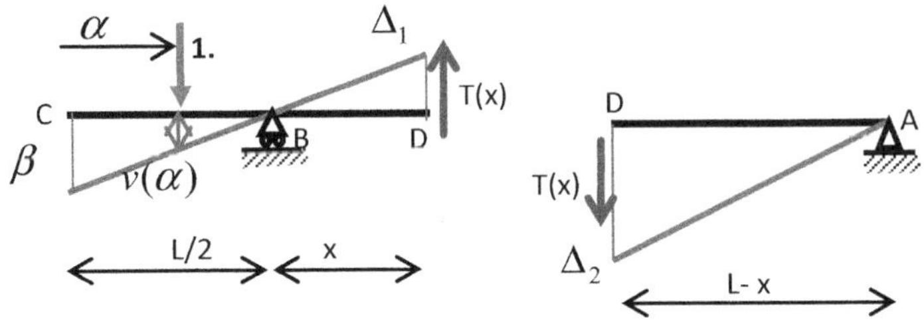

Figure10.46. *Shear force* $0 \le \alpha \le \frac{L}{2}$

Finally, the shear force expression at point D is written as:

$$T(\alpha) = v(\alpha) = -\frac{1}{2}(1 - \frac{2\alpha}{L})$$

$$T(0) = -\frac{1}{2}$$

$$T(\frac{L}{2}) = 0.$$

$\frac{L}{2} \le \alpha \le \frac{L}{2} + x$ (Figure 10.47)

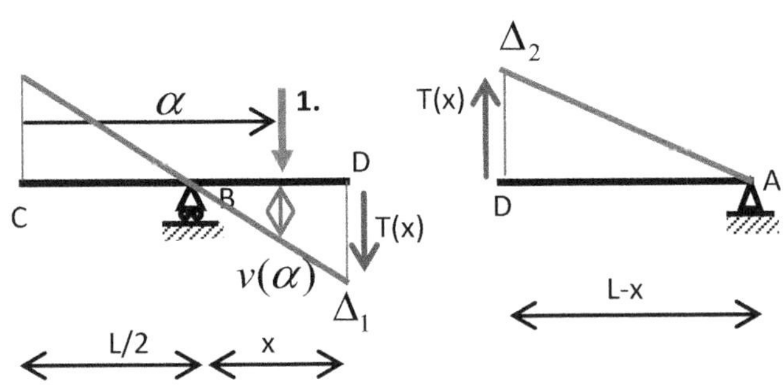

Figure 10.47. *Shear force* $\frac{L}{2} \le \alpha \le \frac{L}{2} + x$

The displacement below the point of application of the moving force is given as:

$$v(\alpha) = \Delta_1 . \frac{\alpha - \frac{L}{2}}{x}$$

Knowing that $\frac{\Delta_1}{x} = \frac{\Delta_2}{L-x}$ and $\Delta_1 + \Delta_2 = 1$.

$$\Delta_1 = \frac{x}{L}$$

$$v(\alpha) = \frac{\alpha}{L} - \frac{1}{2}$$

The shear force expression at point D is written as:

$$T(\alpha) = v(\alpha) = \frac{\alpha}{L} - \frac{1}{2}$$

$$T(\frac{L}{2}) = 0.$$

$$T(\frac{L}{2} + x) = \frac{x}{L}$$

$x \leq \alpha \leq \frac{3L}{2}$ (Figure 10.48)

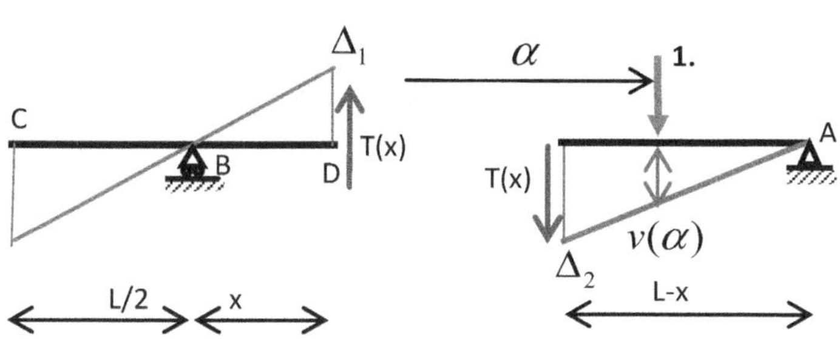

Figure 10.48. *Shear force* $x \leq \alpha \leq \frac{3L}{2}$

$$v(\alpha) = \Delta_2 \frac{\frac{3L}{2} - \alpha}{L - x}$$

and $\Delta_2 = 1 - \frac{x}{L}$

$$v(\alpha) = \frac{3}{2} - \frac{\alpha}{L}$$

The shear force expression at point D is written as:

$$T(\alpha) = v(\alpha) = \frac{3}{2} - \frac{\alpha}{L}$$

$$T(x) = \frac{3}{2} - \frac{x}{L}$$

$$T(\frac{3L}{2}) = 0.$$

Figure 10.49 shows the influence line of a shear force relating to point D.

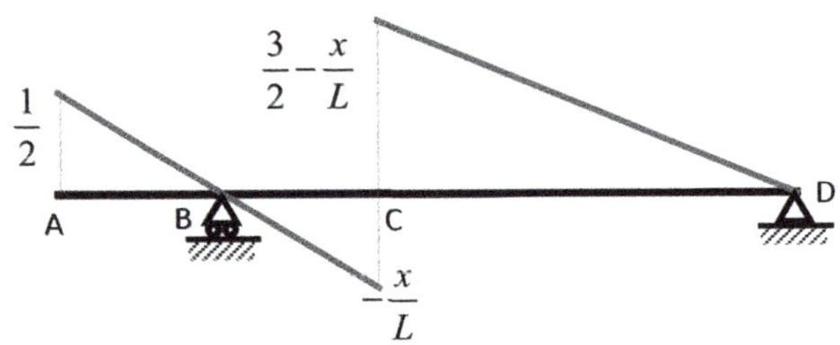

Figure 10.49. *Influence line of a shear force relating to point D.*

10.7.3. Influence line of a bending moment

The influence line of the bending moment is obtained by applying the influence line construction procedure described in sections 10.7.1 and 10.7.2.

The bending moment M_C is ignored by introducing a hinge, which allows a rotation at point C (Figure 10.50).

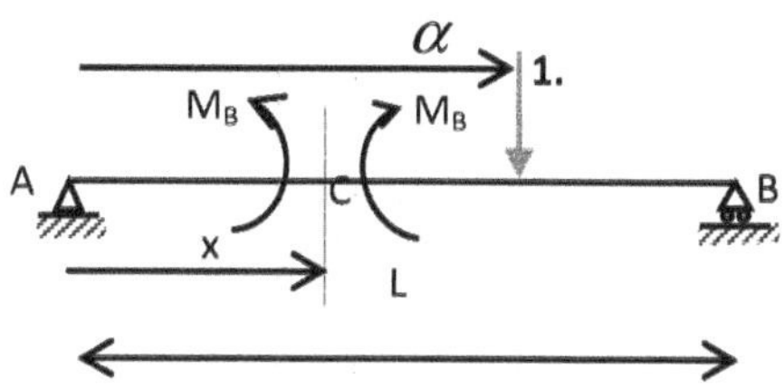

Figure 10.50. *Given beam*

The introduction of a hinge at point C allows the branches (AC) and (CB) to rotate, respectively, at points A and B (Figure 10.51).

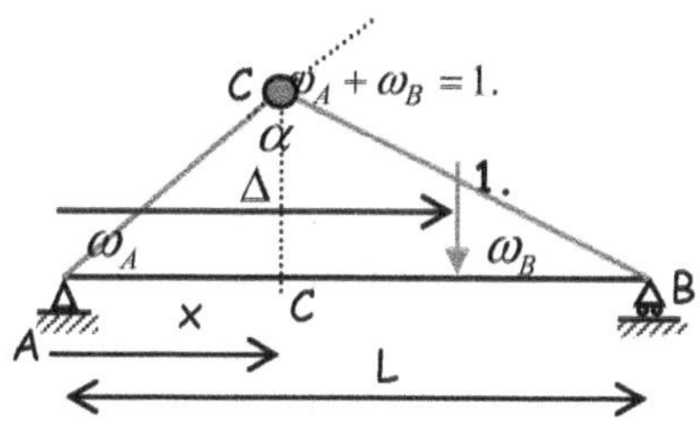

Figure 10.51. *Beam deflection*

The Muller–Breslau principle considers $\omega_C = 1$.

$$\omega_A + \omega_B = 1. \qquad [10.31]$$

We apply the principle of virtual displacements of rigid body movement. The virtual work of external forces is written as:

$$W_{v,ext} = M(\mathrm{x}).\omega_A + M(\mathrm{x}).\omega_B - 1.v(\alpha) \qquad [10.32]$$

The relationship [10.32] is written as:

$$W_{v,ext} = M(\mathrm{x}).(\omega_A + \omega_B) - v(\alpha) \qquad [10.33]$$

We substitute the relationship [10.31] with the relationship [10.33] and knowing that $W_{v,ext} = 0.$, we write:

$$M(\mathrm{x}) = v(\alpha) \qquad [10.34]$$

This relationship shows that the vertical displacement of the point of application of the moving force is equal to the expression of the bending moment at the point where the influence line is required.

The displacement at point C (Figure 10.51) is given as:

$$\Delta = x.\omega_A = (L-x).\omega_B \qquad [10.35]$$

The slopes at points A and B are written as:

$$\omega_A = \frac{L-x}{x}.\omega_B \qquad [10.36]$$

$$\omega_A = 1-\frac{x}{L} \qquad [10.37]$$

$$\omega_B = \frac{x}{L} \qquad [10.38]$$

Substituting the relationship [10.36] with the relationship [10.35], we obtain:

$$\Delta = x.(1-\frac{x}{L}) \qquad [10.39]$$

The expression of the bending moment at point C is:

$$0 \leq \alpha \leq x$$

$$\frac{v(\alpha)}{\Delta} = \frac{\alpha}{x} \qquad [10.40]$$

Substituting the relationship [10.39] with [10.40], we obtain:

$$v(\alpha) = (1-\frac{x}{L}).\alpha \qquad [10.41]$$

$$M(\alpha) = (1-\frac{x}{L}).\alpha \qquad [10.42]$$

$$M(0) = 0.$$

$$M(x) = (1 - \frac{x}{L}).x$$

$$x \leq \alpha \leq L$$

$$\frac{v(\alpha)}{\Delta} = \frac{L - \alpha}{L - x} \quad [10.43]$$

Substituting the relationship [10.39] with [10.43], we obtain:

$$v(\alpha) = x.(1 - \frac{\alpha}{L}) \quad [10.44]$$

$$M(\alpha) = x.(1 - \frac{\alpha}{L}) \quad [10.45]$$

$$M(x) = x.(1 - \frac{x}{L})$$

$$M(L) = 0.$$

The influence line of the bending moment at point C is shown in Figure 10.52.

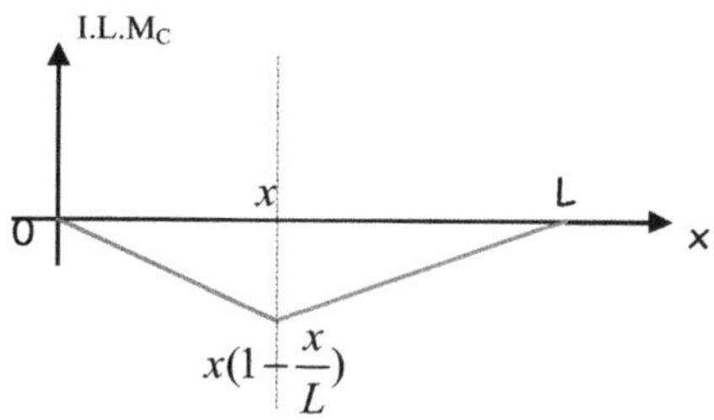

Figure 10.52. *I.L.M$_C$*

EXAMPLE 10.8.–

Using the Muller–Breslau principle, construct the influence line of the bending moment at point D (Figure 10.53).

$$0 \leq \alpha \leq \frac{L}{2}$$

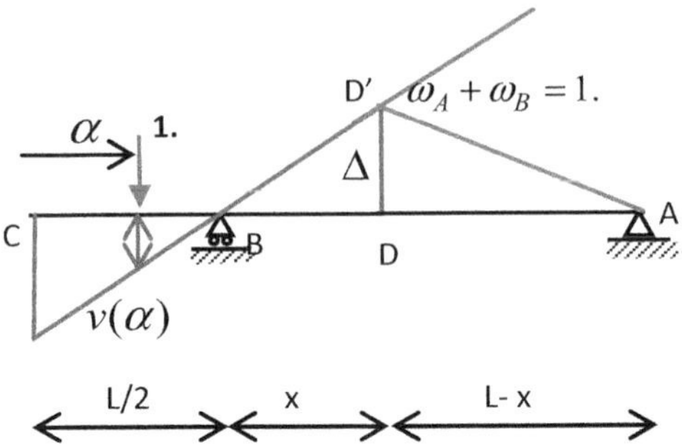

Figure 10.53. *Force* $0 \le \alpha \le \frac{L}{2}$

Given that $\dfrac{-v(\alpha)}{\frac{L}{2} - \alpha} = \dfrac{\Delta}{x}$.

The displacement of the point of application of the moving force is given as:

$$v(\alpha) = -(\frac{L}{2} - \alpha)\frac{\Delta}{x}$$

or

$$v(\alpha) = -(\frac{L}{2} - \alpha).(1 - \frac{x}{L})$$

Hence

$$M(\alpha) = -(\frac{L}{2} - \alpha).(1 - \frac{x}{L})$$

$$M(0) = -\frac{L}{2}(1 - \frac{x}{L})$$

$$M(\frac{L}{2}) = 0.$$

$\frac{L}{2} \le \alpha \le \frac{L}{2} + x$ (Figure 10.54)

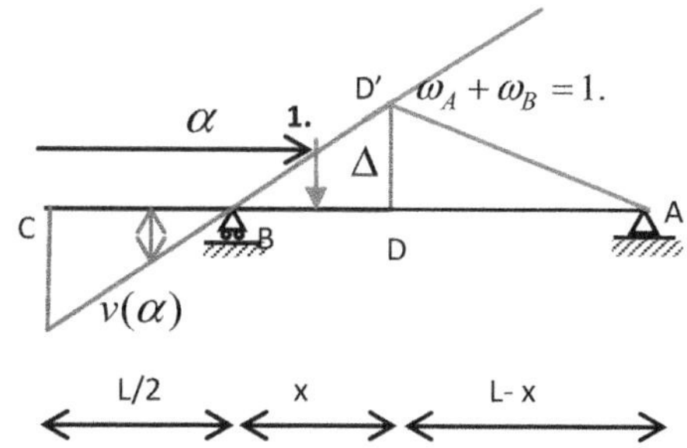

Figure 10.54. *Force* $\frac{L}{2} \leq \alpha \leq \frac{L}{2} + x$

In this case,

$$v(\alpha) = (1 - \frac{x}{L}).(\alpha - \frac{L}{2})$$

Hence, the bending moment at point D is given as:

$$M(\alpha) = (1 - \frac{x}{L}).(\alpha - \frac{L}{2})$$

$$M(\frac{L}{2}) = 0.$$

$$M(\frac{L}{2} + x) = x(1 - \frac{x}{L})$$

$x \leq \alpha \leq \frac{3L}{2}$ (Figure 10.55)

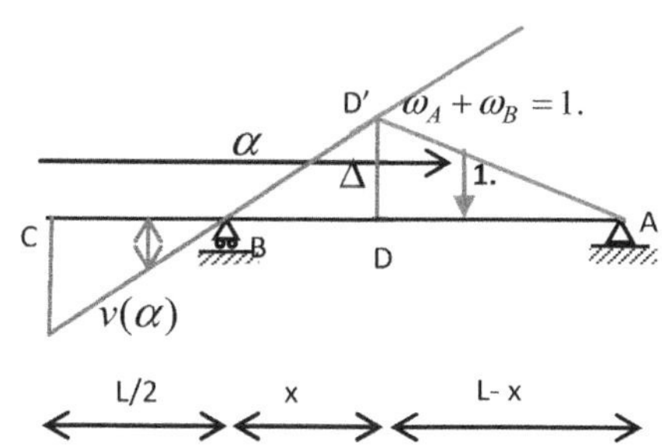

Figure 10.55. *Force* $x \leq \alpha \leq \frac{3L}{2}$

The expression of the bending moment below the moving force is written as

$$M(\alpha) = x.(\frac{3}{2} - \frac{\alpha}{L})$$

$$M(\frac{L}{2} + x) = x.(1 - \frac{x}{L})$$

$$M(\frac{3L}{2}) = 0.$$

Finally, Figure 10.56 shows the influence line of a bending moment relating to point D.

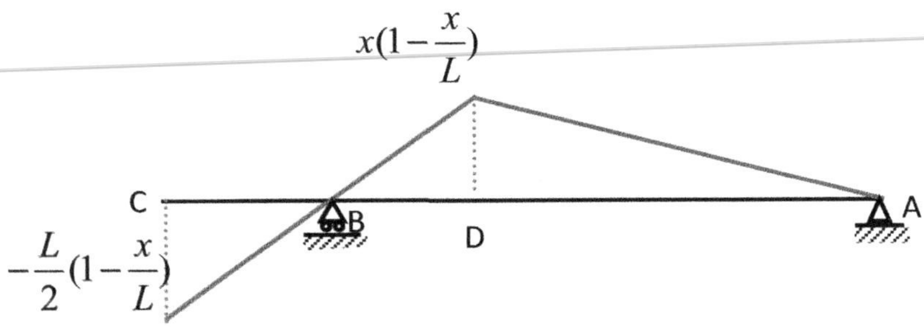

Figure 10.56. *I.L.M*$_D$

10.8. Influence lines of deflections

The influence line of a deflection at a point of a structure is its variation according to the mobility of the unit force. Finding the influence lines of deflections relating to a point is carried out in the same way as that of a shear force or a bending moment.

The construction of the influence line of the vertical displacement at point C (Figure 10.57) is obtained by varying the unit force application point from point A to point B by calculating, each time, the vertical displacement at point C. The methods for calculating deflections (Chapters 6 and 7) make it possible to calculate the vertical displacement at point C.

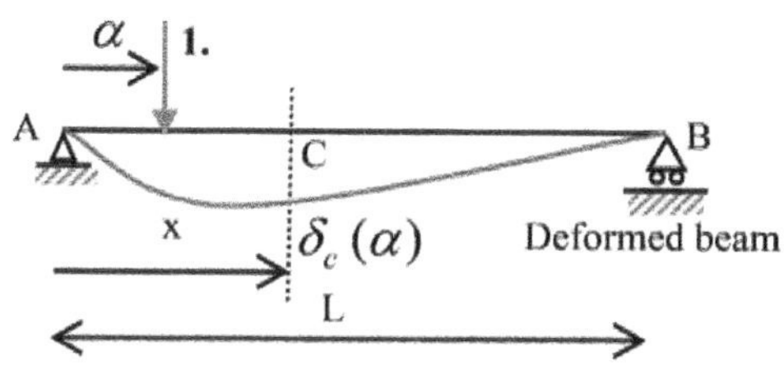

Figure 10.57. *Given beam*

The methods for calculating deflections (Chapters 6 and 7) offer tools for calculating deflections at a point on the beam. Among these methods, the Maxwell–Betti law of reciprocity of deflections is of particular importance when constructing influence lines of deflections in statically determinate beams.

In this case, when a unit force is applied at point D, the vertical displacement at point C is δ_{DC}. Similarly, when a unit force reaches point C, the displacement at point D is δ_{DC} (Figure 10.58). Applying the Maxwell–Betti law allows us to write:

$$\delta_{CD} = \delta_{DC} \qquad [10.46]$$

The relationship [10.46] shows that the value of the influence line at point C when the unit force is applied at point D is equal to the value of the influence line at point D when the unit force is applied at point C.

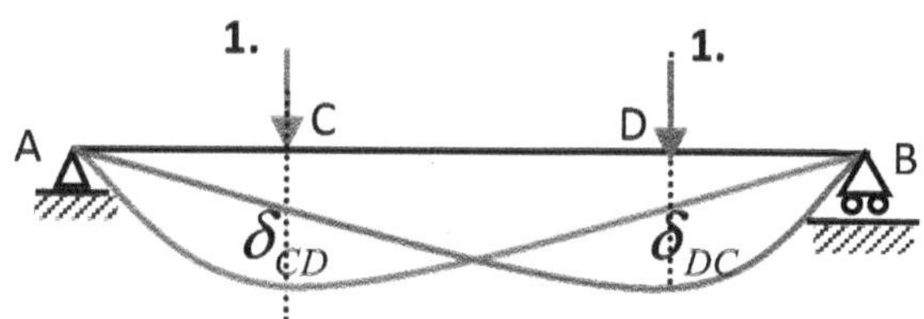

Figure 10.58. *Maxwell–Betti law*

EXAMPLE 10.9.–

Draw the influence line of the vertical displacement and the influence line of the slope at point B of the cantilever beam (Figure 10.59).

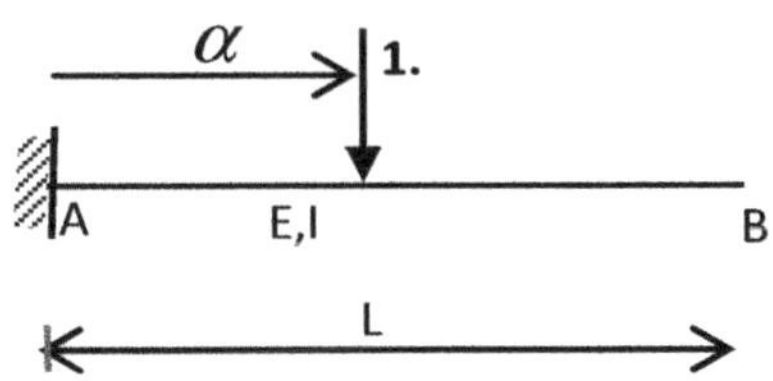

Figure 10.59. *Given cantilever beam*

A unit force is applied at point B. The diagram of the bending moment of the cantilever beam is shown in Figure 10.60.

Figure 10.60. *Bending moment diagram*

The conjugate beam method (Figure 10.61) is used to calculate the displacement at point B.

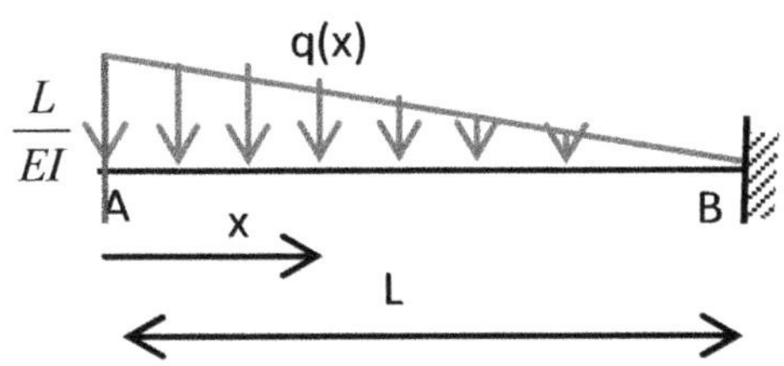

Figure 10.61. *Conjugate beam*

The shear force and the bending moment at any section at a distance x from the support A are written as:

$$q(x)=\frac{L}{EI}(1-\frac{x}{L})\quad M(x)=-\frac{1}{2}\frac{L}{EI}(1-\frac{x}{L}).x^2-\frac{1}{3}\frac{L}{EI}(1-(1-\frac{x}{L})).x^2$$

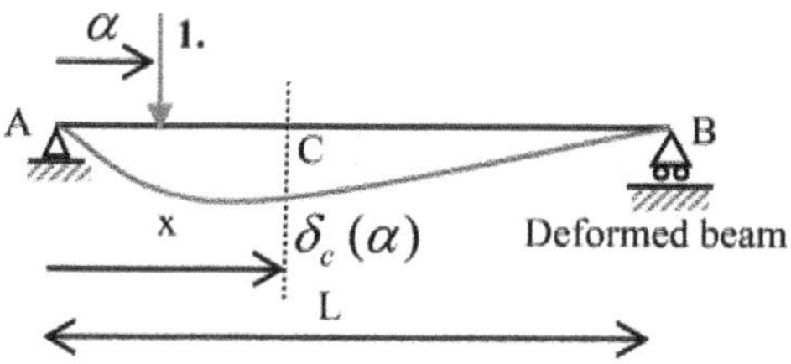

Figure 10.57. *Given beam*

The methods for calculating deflections (Chapters 6 and 7) offer tools for calculating deflections at a point on the beam. Among these methods, the Maxwell–Betti law of reciprocity of deflections is of particular importance when constructing influence lines of deflections in statically determinate beams.

In this case, when a unit force is applied at point D, the vertical displacement at point C is δ_{DC}. Similarly, when a unit force reaches point C, the displacement at point D is δ_{DC} (Figure 10.58). Applying the Maxwell–Betti law allows us to write:

$$\delta_{CD} = \delta_{DC} \qquad [10.46]$$

The relationship [10.46] shows that the value of the influence line at point C when the unit force is applied at point D is equal to the value of the influence line at point D when the unit force is applied at point C.

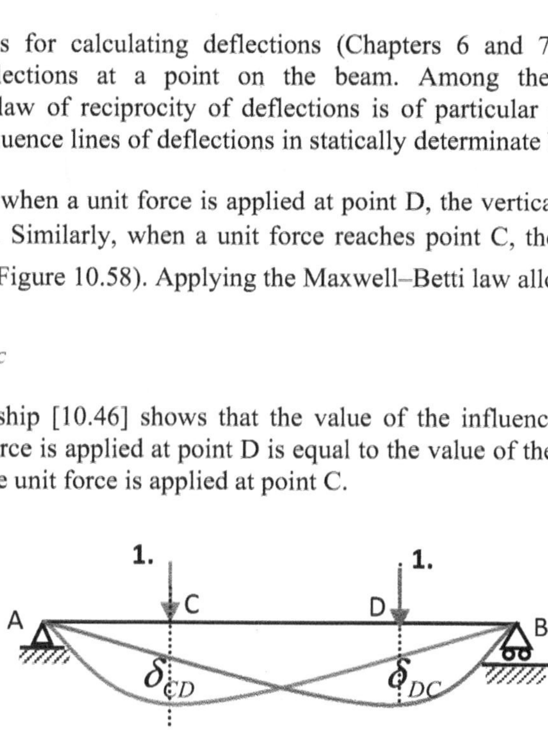

Figure 10.58. *Maxwell–Betti law*

EXAMPLE 10.9.–

Draw the influence line of the vertical displacement and the influence line of the slope at point B of the cantilever beam (Figure 10.59).

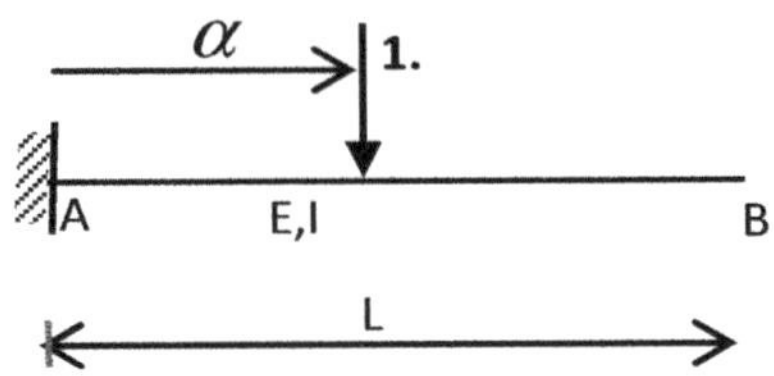

Figure 10.59. *Given cantilever beam*

A unit force is applied at point B. The diagram of the bending moment of the cantilever beam is shown in Figure 10.60.

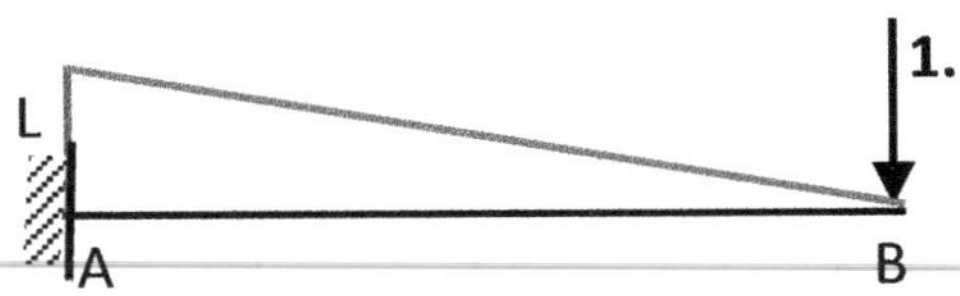

Figure 10.60. *Bending moment diagram*

The conjugate beam method (Figure 10.61) is used to calculate the displacement at point B.

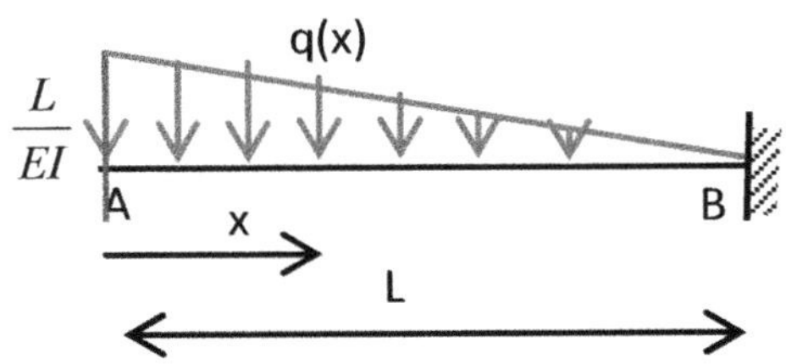

Figure 10.61. *Conjugate beam*

The shear force and the bending moment at any section at a distance x from the support A are written as:

$$q(x) = \frac{L}{EI}(1-\frac{x}{L}) \quad M(x) = -\frac{1}{2}\frac{L}{EI}(1-\frac{x}{L}).x^2 - \frac{1}{3}\frac{L}{EI}(1-(1-\frac{x}{L})).x^2$$

Hence:

$$M(x) = -\frac{1}{6}\frac{L}{EI}(3.x^2 - \frac{x^3}{L})$$

$$\mathrm{T}(x) = \frac{L}{EI}(1-\frac{x}{L}).x + \frac{1}{2}\frac{L}{EI}(1-(1-\frac{x}{L})).x = \frac{L}{EI}(x - \frac{x^2}{2L})$$

So, the displacement at point B according to the point of application of the moving force is given as:

$$\delta(x) = -\frac{1}{6}\frac{L}{EI}(3.x^2 - \frac{x^3}{L})$$

and the corresponding slope is:

$$\omega(x) = \frac{L}{EI}(x - \frac{x^2}{2L})$$

The influence lines of displacement and the slope are shown in Figure 10.62.

Influence line of displacement at point B

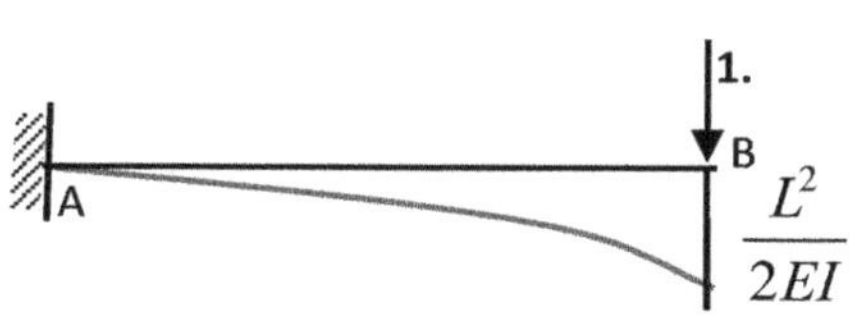

Influence line of the slope at point B

Figure 10.62. *Influence lines of δ_B and ω_B.*

10.9. Conclusion

In this chapter, we have presented the construction of influence lines of an internal or external action of a beam, a frame and a statically determinate truss. The primary objective of this chapter is to draw a curve that links the variation of the internal action according to the position of a mobile unit force along the structure. The conclusions that can be drawn from this chapter are as follows:

– influence lines of an external action and/or internal action can be deduced using the equilibrium method or the Muller–Breslau method;

– influence lines of an external action and/or internal action are found by varying the unit force along the structure, showing the variation of this action according to the position of the application of the unit and moving force;

– the influence line of the shear force relating to a section is the vertical displacement of the point of application of the moving force. The structure is cut by a section at the requested point considering that the shear force at this section generates a unit displacement. The displacement of the point of application of the moving force constitutes the shear force relating to the chosen section. The variation of the shear force is the influence line;

– in the same way, the influence line of the bending moment in relation to any section is the vertical displacement of the point of application of the moving force. The bending moment generates a unit rotation at the chosen section. The vertical displacement of the point of application of the moving force is the bending moment in relation to the considered section. The expression of the bending moment in relation to a section constitutes the influence line of the bending moment;

– the purpose of studying the influence lines of statically determinate structures is to allow the designer to analyze the studied structure under the effect of a moving force.

10.10. Problems

Exercise 1

Determine the influence lines of support reactions V_A and V_B of the following beams:

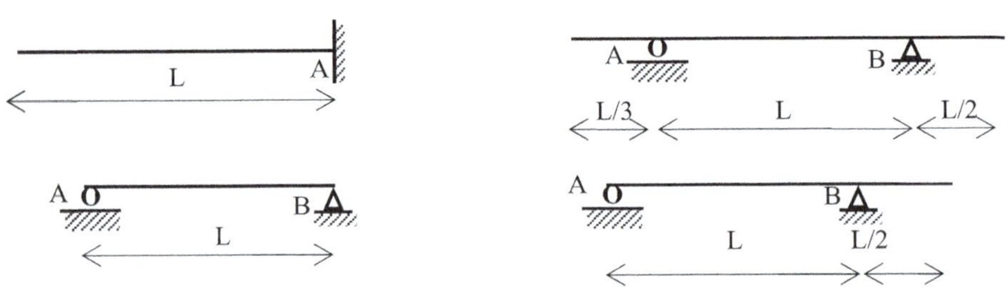

Exercise 2

Determine the influence lines of support reaction V_A of the following beams:

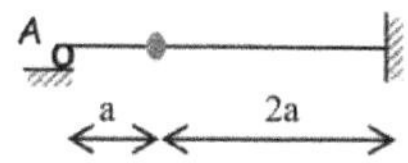

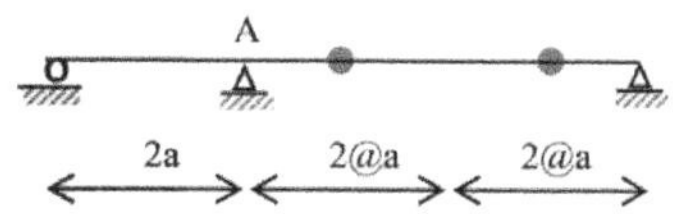

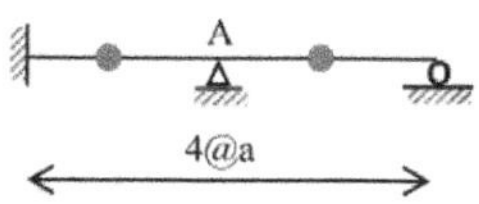

Exercise 3

Construct the influence lines of the shear force and the bending moment in relation to point A of the following beams:

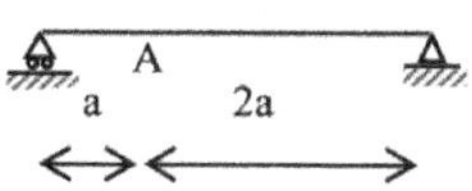

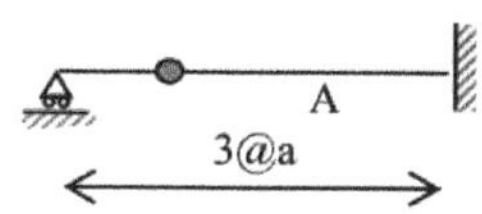

EI is constant for all structures.

Exercise 4

Determine the influence lines of the support reaction R_A and internal force N_1 of bar 1 for all the following trusses. We assume that the unit load is moving on the joints of the upper flange and $E\,\Omega$ is constant.

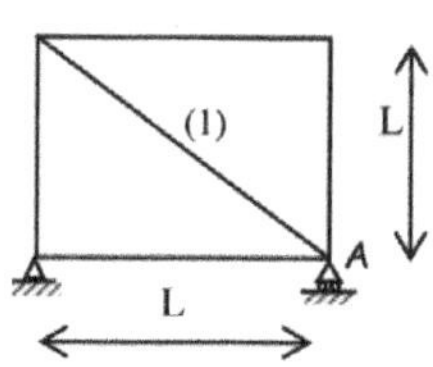

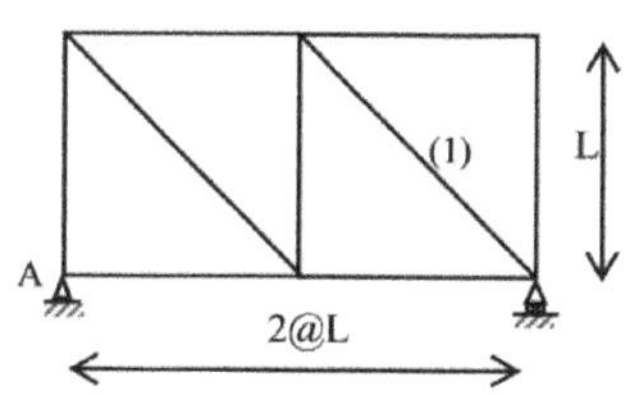

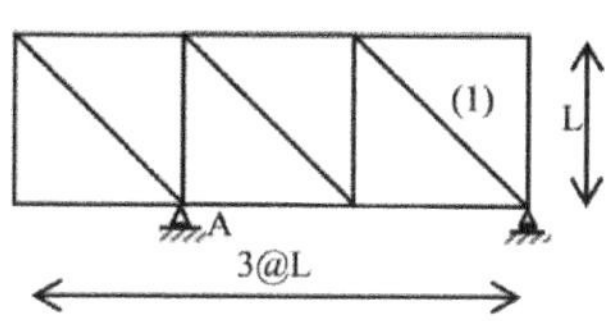

Exercise 5

Draw the influence lines of support reaction R_A, the shear force and the bending moment relating to point B of the following frames:

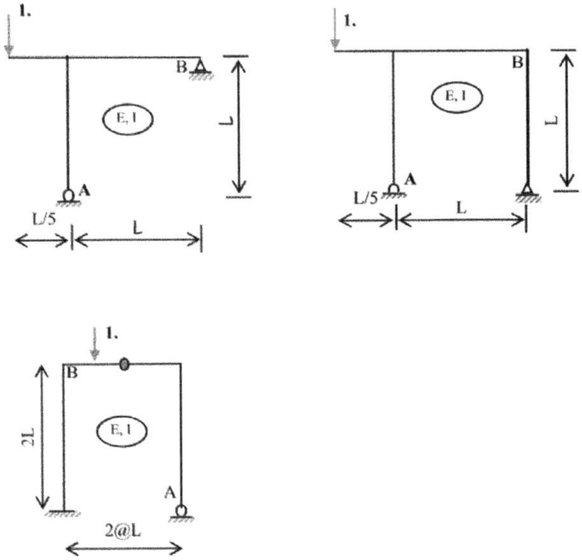

Exercise 6

Repeat Exercises 1–5 using the Muller–Breslau method.

Exercise 7

Draw the influence lines of the displacement at point A and the slope in relation to point B of the following structures:

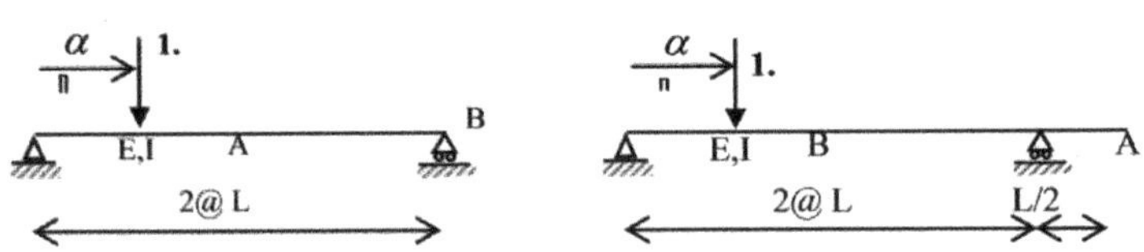

For each structure, use the following method:

1) deformation of the elastic line;

2) principle of virtual work;

3) conjugate beam.

Exercise 8

Draw the influence lines of the force in the spring of the following structures:

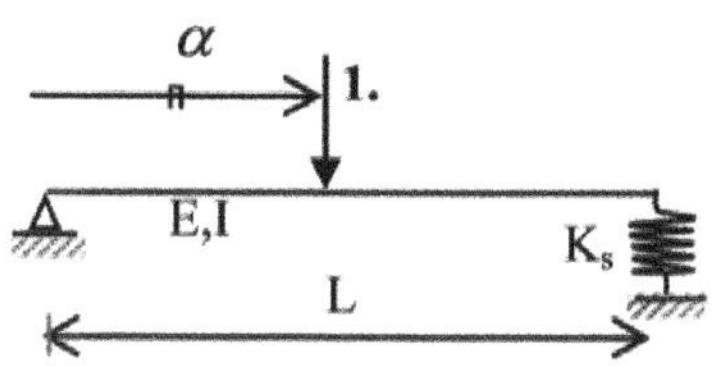

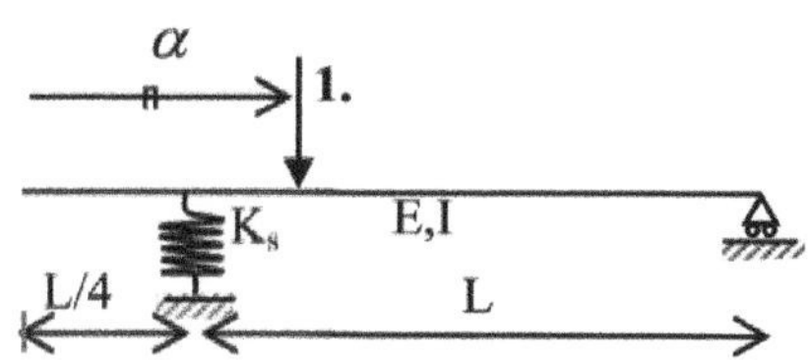

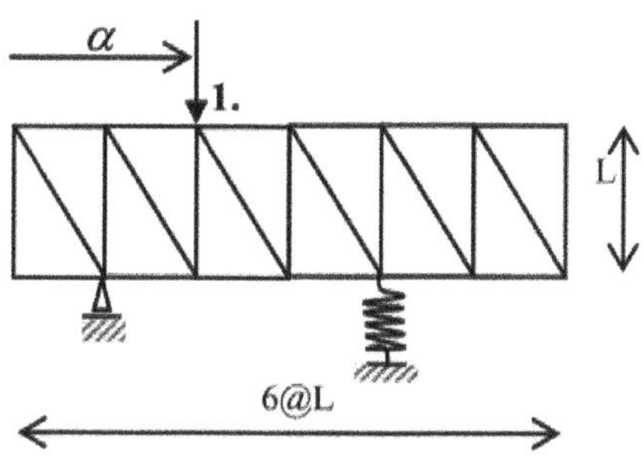

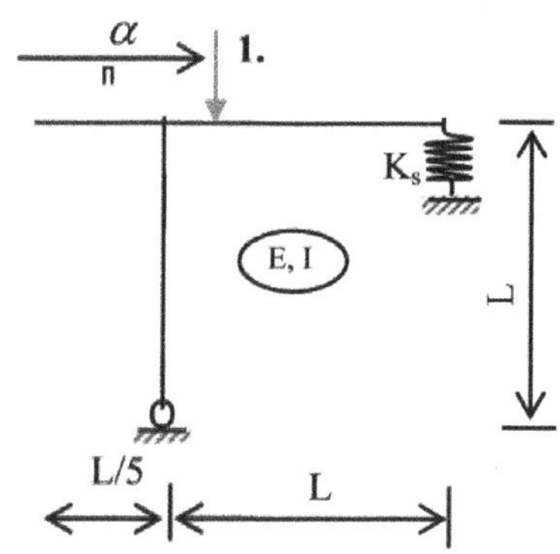

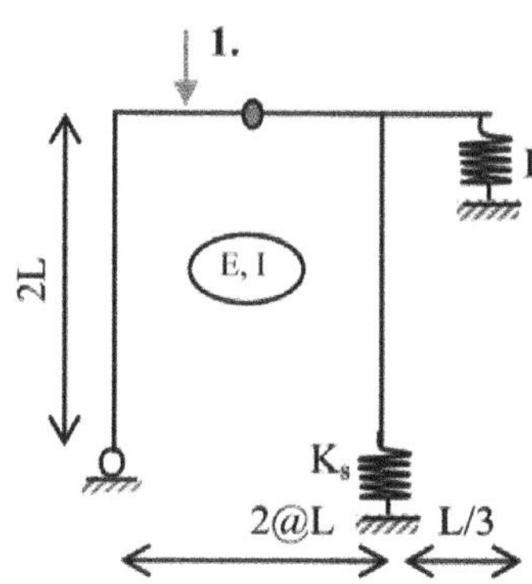

Appendices

A.1. Standard structural deflections

The beams have a constant flexural rigidity EI and a length L.

Structure	Deformation
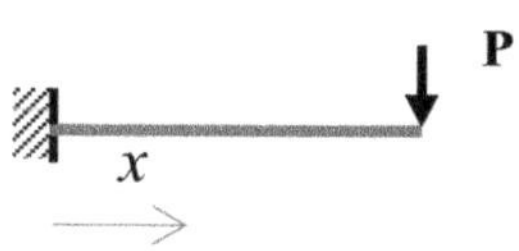	$v(x) = -\frac{PL^3}{6EI}(x^3 - 3Lx)$
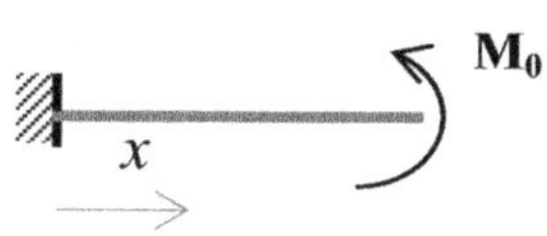	$v(x) = \frac{M_0}{2EI}x^2$
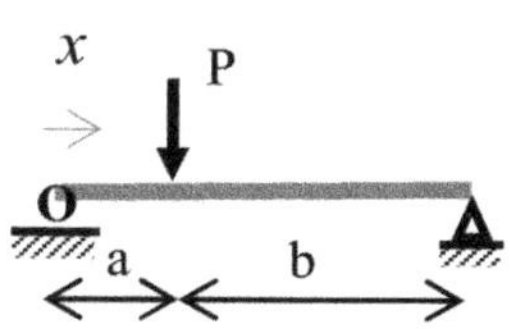	$v(x) = -\frac{Pbx}{6EI}(L^2 - b^2 - x^3)$ $0 \le x \le a$
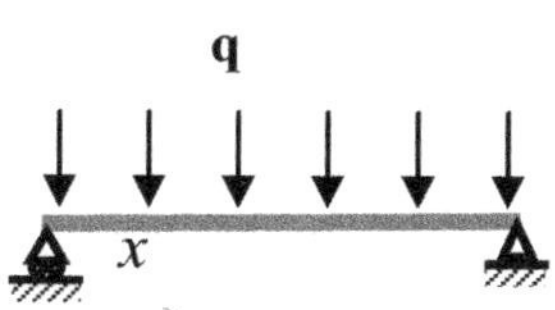	$v(x) = -\frac{qx}{24EI}(x^3 - 2Lx^2 + L^3)$

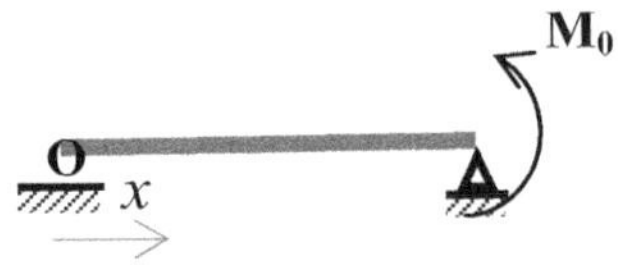	$v(x) = -\frac{M_0 x}{6EIL}(L^2 - x^2)$
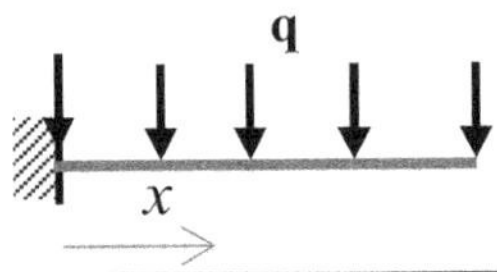	$v(x) = -\frac{q}{24EI}(x^4 - 4Lx^3 + 6L^2x^2)$

A.2. Evaluation of the integral $\int_0^L M(x).m(x).dx$

$\int_0^L M(x).\mu(x).dx$	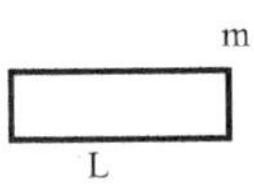	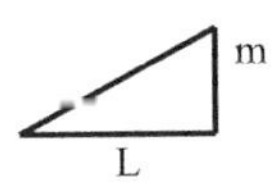	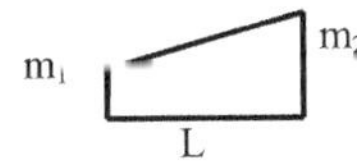	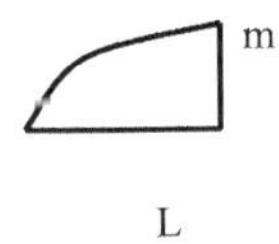
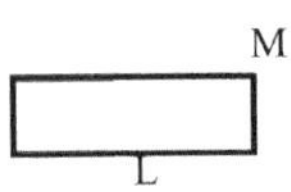	$M.m.L$	$\frac{1}{2}M.m.L$	$\frac{1}{2}M.(m_1 + m_2).L$	$\frac{2}{3}M.m.L$
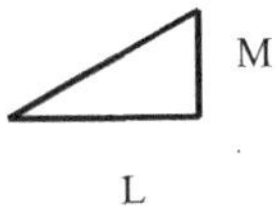	$\frac{1}{2}M.m.L$	$\frac{1}{3}M.m.L$	$\frac{1}{6}M.(m_1 + 2m_2).L$	$\frac{5}{12}M.m.L$
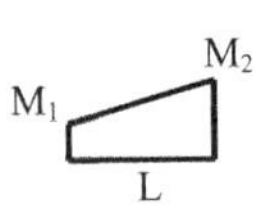	$\frac{1}{2}m.(M_1 + M_2).L$	$\frac{1}{6}\text{m}.(\text{M}_1 + 2M_2).L$	$\frac{1}{6}\begin{bmatrix} m_1(2M_1 + M_2) \\ +m_2(M_1 + 2M_2) \end{bmatrix}L$	$\frac{1}{6}\left[m'(3M_1 + 5M_2)\right]L$
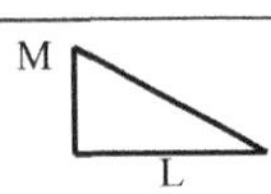	$\frac{1}{2}M.m.L$	$\frac{1}{6}M.m.L$	$\frac{1}{6}M.(2m_1 + m_2).L$	$\frac{1}{4}M.m.L$

Bibliography

[ANT 16] ANTONELLI A., DESODT C., MOLINARD H.H., Conception et constructions des arcs, ENS de Paris, 2016.

[BAS 79] BASSIN M.G., BRODSKY S.M., WOLKOFF H.W., *Statics and Strength of Materials*, McGraw Hill, New York, 1979.

[BLA 89] BLAKE L.S., *Civil Engineer's: Reference book*, Butterworths, Oxford, 1989.

[BOU 17] BOUVET C., *Mechanics of Aeronautical Solids, Materials and Structures*, ISTE Ltd, London and John Wiley & Sons, New York, 2017.

[BRO 94] BROUGHTON P., NDUMBARO P., *The Analysis of Cable and Centenary Structures*, Thomas Telford, London, 1994.

[BRO 16] BROCATO M., *Statique : De la géométrie à la conception des structures*, Presses des Ponts, Paris, 2016.

[CAL 13] CALGARO J.A., GULVANESSIAN H., HOLICKY M., *Bases de calcul des structures selon l'Eurocode 0 NF EN 1990*, Le Moniteur, Paris, 2013.

[CHE 10] CHENAF M., BELLOIR E., FLORENCE C., *Actions du vent sur les bâtiments d'après l'Eurocode 1 – Calcul des pressions et forces de vent sur l'enveloppe des bâtiments*, CSTB, Marne-La-Vallée, 2010.

[DEL 15] DELAPLACE A., FABRICE GATUINGT F., FREDERIC RAGUENEAU F., *Aide-mémoire Mécanique des structures*, Dunod, Paris, 2015.

[DEL 12] DEL PEDRO M., GMÜR T., BOTSIS J., *Introduction à la mécanique des solides et des structures*, Presses Polytechniques et Universitaires Romandes, Lausanne, 2012.

[DOU 13] DOUBERE J.C., *Résistance des matériaux: Cours – Exercices corrigés*, Eyrolles, Paris, 2013.

[EIS 11] EISLEY J.G., WAAS A.M., *Analysis of Structures: An Introduction Including Numerical Methods*, John Wiley & Sons, New York, 2011.

[ENG 07] ENGEL H., *Les systèmes structurels*, Dergham, Beirut, 2007.

[ERI 47] ERICKSEN B., *Influence Lines for Thrust and Bending Moments in Fixes Arcs*, Concrete Publication Limited, London, 1947.

[FRE 05] FREY F., *Analyse des structures et milieux continus: Tome 1, Statique appliquée*, Presses Polytechniques et Universitaires Romandes, Lausanne, 2005.

[FRE 11] FREY F., *Analyse des structures et milieux continus – Statique appliquée*, Presses Polytechniques et Universitaires Romandes, Lausanne, 2011.

[GHA 09] GHALI A., NEVILLE M., BROWN T.G., *Structural Analysis: A Unified Classical and Matrix Approach*, Spon Press, London, 2009.

[GHA 15] GHANEM S., NAJM H., "Structural performance of two-hinged Islamic arches", *Journal of Engineering and Architecture*, vol. 3, no. 1, pp. 38–52, 2015.

[GOU 11] GOURINAT Y., *Exercices et problèmes de mécanique des solides et des structures,* Dunod, Paris, 2011.

[HIB 12] HIBBELER R.C., *Structural Analysis*, Prentice Hall, Upper Saddle River, 2012.

[IRV 92] IRVINE H.M., *Cables Structures*, MIT Press, Cambridge, 1992.

[KAS 11] KASSIMALI A., *Structural Analysis*, Cengage Learning, Boston, 2011.

[KHA 15] KHALFALLAH S., *Méthodes d'analyse des structures hyperstatiques*, Edilivre, Paris, 2015.

[KHA 16] KHALFALLAH S., *Analyse des structures hyperstatiques : recueil d'exercices résolus*, Edilivre, Paris, 2016.

[LEE 05] LEET K.M., UANG C.M., *Fundamentals of Structural Analysis*, McGraw Hill, New York, 2005.

[LES 14] LESTUZZI P., PLUG L., *Analyse des structures et milieu continu : Structures en barres et poutres*, Presses Polytechniques Universitaires Romandes, Lausanne, 2014.

[MAI 17] MAITRE P., *Calcul pratique des actions sur les structures*, Le Moniteur, Paris, 2017.

[MEG 05] MEGSON T.H.G., *Structural and Stress Analysis*, Elsevier, Oxford, 2005.

[NAY 00] NAYFEH A.H., PAI F.R., *Linear and Nonlinear Structural Mechanics*, Wiley Interscience, New York, 2004.

[NIC 10] NICOT F., LIMAM A., *Neige, paravalanches et constructions*, Hermès Science-Lavoisier, Paris, 2010.

[NOR 76] NORRIS C.H., WILBUR J.B., UTKU S., *Elementary Structural Analysis*, McGraw-Hill, New York, 1976.

[OLL 17] OLLIVIER P.Y., *La conception des structures : Matériaux, dimensionnement et aspects constructifs*, Dunod, Paris, 2017.

[PAI 16] PAILLE J.M., *Calcul des structures en béton : Guide d'application de l'Eurocode 2*, Eyrolles, Paris, 2016.

[PAU 11] PAULTRE P., *Structures en béton armé: Analyse et dimensionnement*, Presses Internationales Polytechniques, Montreal, 2011.

[PEV 79] PEVROT A.H., GOULOIS A.M., "Analysis of cable structures", *Computers & Structures*, vol. 10, no. 5, pp. 805–813, 1979.

[PIC 92] PICARD A., *Analyse des structures*, Elsevier-Masson, Paris, 1992.

[POP 68] POPOV E.P., *Introduction to Mechanics of Solids*, Prentice Hall, Upper Saddle River, 1968.

[PRA 97] PRAKASH RAO D.S., *Graphical Methods in Structural Analysis*, Universities Press, Hyderabad, 1997.

[REG 00] REGLES NV 65, Regles nv 65 modifiées 99 et N 84 modifiées 2000, Eyrolles, Paris, 2000.

[REG 09] REGLES NV 65, Règles définissant les effets de la neige et du vent sur les constructions et annexes, DTU P06-002, AFNOR, 2009.

[REG 96] REGLES PS-96, Règles de construction parasismiques, Eyrolles, Paris, 1996.

[SAT 98] SATHYAMOORTHEY M., *Nonlinear Analysis of Structures*, CRC Press, Boca Raton, 1998.

[SMI 88] SMITH J.C., *Structural Analysis*, Harper & Row, New York, 1988.

[SON 14] SONAVANE T., Analysis of Arches, Master's thesis, University of Virginia, 2014.

[STU 97] STUDER M.A., FREY F., *Introduction à l'analyse des structures*, Presses Polytechniques et Universitaires Romandes, Lausanne, 1997.

[THE 14] THEORIN A., FORET G., *Calcul des structures*, Ecole Polytechnique Universitaire, Paris, 2014.

[VOL 08] VOLDOIRE F., BAMBERGER Y., *Mécanique des structures Initiation – Approfondissements – Applications*, Presses des Ponts, Paris, 2008.

[WES 89] WEST H.H., *Analysis of Structures: An Integration of Classical and Modern Methods,* John Wiley & Sons, New York, 1989.

[WIL 09] WILLIAMS A., *Structural Analysis: In Theory and Practice*, Elsevier, Oxford, 2009.

[YAN 94] YANG Y.B., KUO S.R., *Nonlinear Framed Structures*, Prentice Hall, Upper Saddle River, 1994.

Index

L, M, P

R, S

T, V, W

Other titles from

iSTE

in

Civil Engineering and Geomechanics

2018

VERBRUGGE Jean-Claude, SCHROEDER Christian
Geotechnical Correlations for Solis and Rocks

2017

LAUZIN Xavier
Civil Engineering Structures According to the Eurocodes

PUECH Alain, GARNIER Jacques
Design of Piles Under Cyclic Loading: SOLCYP Recommendations

SELLIER Alain, GRIMAL Étienne, MULTON Stéphane, BOURDAROT Éric
Swelling Concrete in Dams and Hydraulic Structures: DSC 2017

2016

BARRE Francis, BISCH Philippe, CHAUVEL Danièle *et al.*
Control of Cracking in Reinforced Concrete Structures: Research Project CEOS.fr

PIJAUDIER-CABOT Gilles, LA BORDERIE Christian, REESS Thierry, CHEN Wen, MAUREL Olivier, REY-BETHBEDER Franck, DE FERRON Antoine
Electrohydraulic Fracturing of Rocks

TORRENTI Jean-Michel, LA TORRE Francesca
Materials and Infrastructures 1
Materials and Infrastructures 2

2015

AÏT-MOKHTAR Abdelkarim, MILLET Olivier
Structure Design and Degradation Mechanisms in Coastal Environments

MONNET Jacques
In Situ Tests in Geotechnical Engineering

2014

DAÏAN Jean-François
Equilibrium and Transfer in Porous Media – 3 volume series
Equilibrium States – Volume 1
Transfer Laws – Volume 2
Applications, Isothermal Transport, Coupled Transfers – Volume 3

2013

AMZIANE Sofiane, ARNAUD Laurent
Bio-aggregate-based Building Materials: Applications to Hemp Concretes

BONELLI Stéphane
Erosion in Geomechanics Applied to Dams and Levees

CASANDJIAN Charles, CHALLAMEL Noël, LANOS Christophe, HELLESLAND Jostein
Reinforced Concrete Beams, Columns and Frames: Mechanics and Design

GUÉGUEN Philippe
Seismic Vulnerability of Structures

HELLESLAND Jostein, CHALLAMEL Noël, CASANDJIAN Charles, LANOS Christophe
Reinforced Concrete Beams, Columns and Frames: Section and Slender Member Analysis

LALOUI Lyesse, DI DONNA Alice
Energy Geostructures: Innovation in Underground Engineering

LEGCHENKO Anatoly
Magnetic Resonance Imaging for Groundwater

2012

BONELLI Stéphane
Erosion of Geomaterials

JACOB Bernard *et al.*
ICWIM6 – Proceedings of the International Conference on Weigh-In-Motion

OLLIVIER Jean-Pierre, TORRENTI Jean-Marc, CARCASSES Myriam
Physical Properties of Concrete and Concrete Constituents

PIJAUDIER-CABOT Gilles, PEREIRA Jean-Michel
Geomechanics in CO_2 Storage Facilities

2011

BAROTH Julien, BREYSSE Denys, SCHOEFS Franck
Construction Reliability: Safety, Variability and Sustainability

CREMONA Christian
Structural Performance: Probability-based Assessment

HICHER Pierre-Yves
Multiscales Geomechanics: From Soil to Engineering Projects

IONESCU Ioan R. *et al.*
Plasticity of Crystalline Materials: from Dislocations to Continuum

LOUKILI Ahmed
Self Compacting Concrete

MOUTON Yves
Organic Materials for Sustainable Construction

NICOT François, LAMBERT Stéphane
Rockfall Engineering

PENSÉ-LHÉRITIER Anne-Marie
Formulation

PIJAUDIER-CABOT Gilles, DUFOUR Frédéric
Damage Mechanics of Cementitious Materials and Structures

RADJAI Farhang, DUBOIS Frédéric
Discrete-element Modeling of Granular Materials

RESPLENDINO Jacques, TOUTLEMONDE François
Designing and Building with UHPFRC

2010

ALSHIBLI A. Khalid
Advances in Computed Tomography for Geomechanics

BUZAUD Eric, IONESCU Ioan R., VOYIADJIS Georges
Materials under Extreme Loadings / Application to Penetration and Impact

LALOUI Lyesse
Mechanics of Unsaturated Geomechanics

NOVA Roberto
Soil Mechanics

SCHREFLER Bernard, DELAGE Pierre
Environmental Geomechanics

TORRENTI Jean-Michel, REYNOUARD Jean-Marie, PIJAUDIER-CABOT Gilles
Mechanical Behavior of Concrete

2009

AURIAULT Jean-Louis, BOUTIN Claude, GEINDREAU Christian
Homogenization of Coupled Phenomena in Heterogenous Media

CAMBOU Bernard, JEAN Michel, RADJAI Fahrang
Micromechanics of Granular Materials

MAZARS Jacky, MILLARD Alain
Dynamic Behavior of Concrete and Seismic Engineering

NICOT François, WAN Richard
Micromechanics of Failure in Granular Geomechanics

2008

BETBEDER-MATIBET Jacques
Seismic Engineering

CAZACU Oana
Multiscale Modeling of Heterogenous Materials

HICHER Pierre-Yves, SHAO Jian-Fu
Soil and Rock Elastoplasticity

JACOB Bernard *et al.*
HVTT 10

JACOB Bernard *et al.*
ICWIM 5

SHAO Jian-Fu, BURLION Nicolas
GeoProc2008

2006

BALAGEAS Daniel, FRITZEN Claus-Peter, GÜEMES Alfredo
Structural Health Monitoring

DESRUES Jacques *et al.*
Advances in X-ray Tomography for Geomaterials

FSTT
Microtunneling and Horizontal Drilling

MOUTON Yves
Organic Materials in Civil Engineering

2005

PIJAUDIER-CABOT Gilles, GÉRARD Bruno, ACKER Paul
Creep Shrinkage and Durability of Concrete and Concrete Structures CONCREEP – 7

Printed and bound by CPI Group (UK) Ltd, Croydon, CR0 4YY

15/09/2026

14962349-0005